Civilian Specialist

Britain's Transport Experts
and the First World War

 New Historical
PERSPECTIVES

Civilian Specialists at War

Britain's Transport Experts and the First World War

Christopher Phillips

LONDON
ROYAL HISTORICAL SOCIETY
INSTITUTE OF HISTORICAL RESEARCH
UNIVERSITY OF LONDON PRESS

Published in 2020 by

UNIVERSITY OF LONDON PRESS
SCHOOL OF ADVANCED STUDY
INSTITUTE OF HISTORICAL RESEARCH
Senate House, Malet Street, London WC1E 7HU

Available to download free or to purchase the hard copy edition at https://www.sas.ac.uk/publications/.

ISBNs
978-1-909646-90-2 (hardback edition)
978-1-909646-97-1 (paperback edition)
978-1-909646-92-6 (PDF edition)
978-1-909646-91-9 (ePub edition)
978-1-912702-45-9 (.mobi edition)

DOI 10.14296/420.9781909646926

New Historical
PERSPECTIVES

Cover image: A Royal Engineers working party coming down from the line by light railway, Ypres, 7 January 1918. © Imperial War Museum (Q 8404).

Contents

List of illustrations

Tables

Acknowledgements

This book is in many ways a story about a collaborative project. Its existence is also in many ways the result of a collaborative project. My name may be the only one to appear on the front cover, but many people have contributed to its conception, development and completion – both professionally and personally – in the nearly fifteen years that have passed since I submitted a timid proposal for an undergraduate dissertation on 'something to do with logistics in the First World War' at the University of Birmingham. Throughout that period I have been lucky enough to have been able to rely upon the moral and financial support of my family. My mother Diane and brother Andrew have been constant sources of warmth and patience, whilst without my wife Claire's sacrifices, encouragement and valuable advice this book would never have been completed. I am principally indebted to them and dedicate it to all three of them with my gratitude and love. My brother's partner Helen, my niece Jessica and my friend James Hyde have each provided welcome distractions from the First World War in their own ways – and for that I am also profoundly grateful.

That the book following these acknowledgements is barely recognizable from the proposal from which it sprang is thanks to a host of inspirational scholars. During my time at the University of Birmingham I was lucky enough to be a part of the Centre for First World War Studies under the stewardship of John Bourne. He guided me into the care of Rob Thompson, whose influence upon the subject of this book is testament to his boundless enthusiasm and expertise in the field. I would like to thank them both for their support, along with the assistance provided by Peter Simkins and the stimulating conversations undertaken with my fellow postgraduate students in the British First World War Studies class of 2009. All were pivotal in my decision to continue my studies at a higher level.

My doctoral research would not have been possible without the financial assistance provided to me by a PhD studentship in First World War Studies, awarded by the Faculty of Arts at the University of Leeds. From Leeds I would like to thank: my PhD supervisors, Holger Afflerbach and Alison Fell for giving me the benefit of their experience and knowledge; David Stevenson and Edward Spiers for their advice and guidance during my viva voce examination; my fellow students, including Dominic Berry, Laura Boyd, Hannah Coates, Henry Irving, Claire Martin, Simone Pelizza, Philippa Read, Danielle Sprecher and Mark Walmsley, whose work across a variety of subjects created a vibrant and supportive atmosphere; the numerous colleagues and community partners involved in the Legacies of War project between 2011 and 2015, which opened my horizons to the breadth and scale of interest in the First World War as a subject of historical interest; and all those who contributed to the organisation and delivery of the Legacies of War seminar series.

Beyond the University of Leeds, the ideas contained within this book have been shaped by opportunities to present on aspects of my research in a variety of settings. I have been privileged enough to discuss elements of this work at conferences organised by the British Commission for Military History, the Association of Business Historians, the National Railway Museum and the International Railway History Association among others, and am grateful to the many conferences, workshops and symposia at which my thoughts on the war have generated constructive, thought-provoking comments and feedback. I am also grateful to David Turner, Kristian Coates Ulrichsen, Sandra Gittins, William Philpott, Jim Greaves and the late Elizabeth Greenhalgh for either reading excerpts of the manuscript or providing me with valuable information on some of the individuals and events included within it.

Since completing my doctoral thesis, like many others at the outset of what they hope will be an academic career, I have worked at multiple institutions on precarious contracts. The pressures facing early career researchers in the current higher education environment are severe and multi-directional, and impossible to survive without the understanding and assistance of senior colleagues. I am hugely grateful to the former academics of the History department and academic administration staff at Leeds Trinity University – particularly the indefatigable Rosemary Mitchell – for their support and encouragement during and after my time in Horsforth, to Rebecca Gill and Ashley Firth at the University of Huddersfield, and to Warren Dockter, Farrah Hawana, Gillian McFadyen and everyone else within the Department of International Politics at Aberystwyth University who have welcomed me into their world with open arms since January 2019.

I have also been privileged to have found support from the First World War Network, an international hub dedicated to support and connect postgraduate students and early career researchers with research interests in any aspect of the First World War. I am grateful to my colleagues on the steering committee, to Sarah Lloyd and Nick Mansfield, to the participants in our various workshops and conferences, and our financial supporters at the Arts and Humanities Research Council – both for their sustained interest in providing opportunities for those at the start of their careers and for helping me to maintain a connection to academia during periods of unemployment.

Acknowledgements

The perils of precarity have been a great source of delay to the completion of this book. Consequently, I am thoroughly appreciative of the Royal Historical Society and the Institute for Historical Research for their continued and unstinting support in this direction. Penny Summerfield and Jane Winters have been exceptionally patient series convenors. Jonathan Newbury, Philip Carter, Emily Morrell and Kerry Whitston proved instrumental in sustaining my passion for the book during the writing phase, and Cath D'Alton created the excellent maps. This book is immeasurably more polished following discussion with D'Maris Coffman, Simon Newman, Anthony Heywood and Jonathan Boff under the auspices of an 'author workshop' in May 2018. The written comments provided by the latter two scholars have been instrumental in my avoidance of a number of errors and omissions. All those that remain are mine alone.

Finally, I would like to acknowledge the patient, unstinting, professional work of the many archivists, librarians and administrators who have wittingly or unwittingly contributed to this book by storing, preserving, cataloguing and making available the material I have consulted over the course of this project. Thanks to the vital work of staff at the National Archives, the Imperial War Museum, the Greenlands Academic Resource Centre, the Modern Records Centre at the University of Warwick, the National Library of Scotland, the Liddell Hart Centre for Military Archives, the Australian War Memorial, the Houses of Parliament and the special collections of both Keele University and the University of Leeds, my arguments are based upon having had access to a diverse range of original documents. The quality of those arguments I leave to the reader to judge.

Christopher Phillips
Llangawsai
December 2019

Abbreviations

AEF	American Expeditionary Force
ASC	Army Service Corps
BEF	British Expeditionary Force
BLSC	Brotherton Library Special Collections
BSF	British Salonika Force
CID	Committee of Imperial Defence
CIGS	Chief of the imperial general staff
DGMR	Director-general of military railways
DGT	Director-general of transportation
DMO	Director of military operations
DOM	Director of movements
DRT	Director of railway transport
EEF	Egyptian Expeditionary Force
ERSC	Engineer and Railway Staff Corps
FSR	*Field Service Regulations*
GARC	Greenlands Academic Resource Centre
GHQ	General headquarters
GQG	*Grand Quartier Général*
IATC	Inter-allied Transportation Council
ICE	Institution of Civil Engineers
IGC	Inspector-general of communications
IWM	Imperial War Museum
IWT	Inland water transport
KUSCA	Keele University Special Collections and Archives
LBSCR	London, Brighton and South Coast Railway
LHCMA	Liddell Hart Centre for Military Archives
LNWR	London and North-Western Railway
LSE	London School of Economics
LSWR	London and South-Western Railway

NCL	Nuffield College Library
NLS	National Library of Scotland
PA	Parliamentary Archives
QMG	Quartermaster-general
RAMC	Royal Army Medical Corps
REC	Railway Executive Committee
ROD	Railway Operating Division
SECR	South-Eastern and Chatham Railway
SWC	Supreme War Council
TNA	The National Archives of the United Kingdom
UWMRC	University of Warwick Modern Records Centre
WF	With France

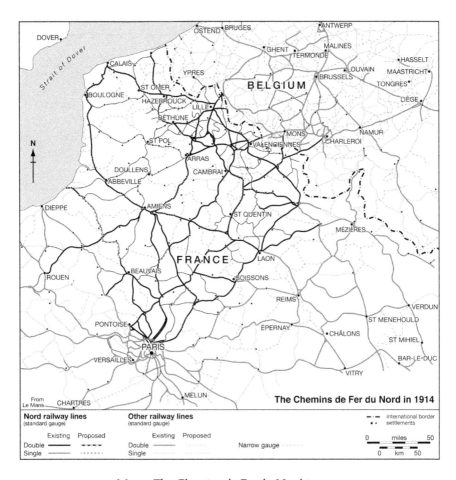

Map 1. The Chemins de Fer du Nord in 1914.

Source: A. M. Henniker, *History of the Great War: Transportation on the Western Front, 1914–1918* (London, 1937). Map drawn by Cath D'Alton.

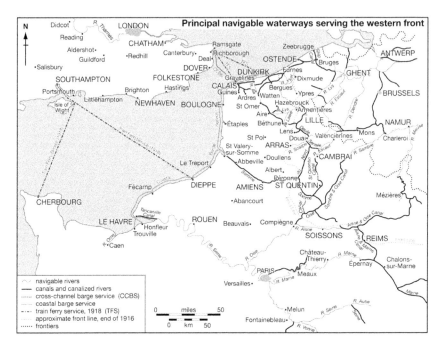

Map 2. Principal navigable waterways serving the western front.

Source: A. M. Henniker, *History of the Great War: Transportation on the Western Front, 1914–1918* (London, 1937). Map drawn by Cath D'Alton.

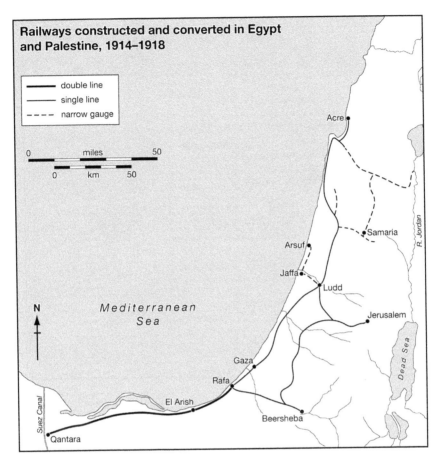

Map 3. Railways constructed and converted in Egypt and Palestine, 1914–18.

Source: 'The Palestine campaign', *Railway Gazette: Special War Transportation Number*, 21 Sept. 1920, pp. 119–28, at p. 120. Map drawn by Cath D'Alton.

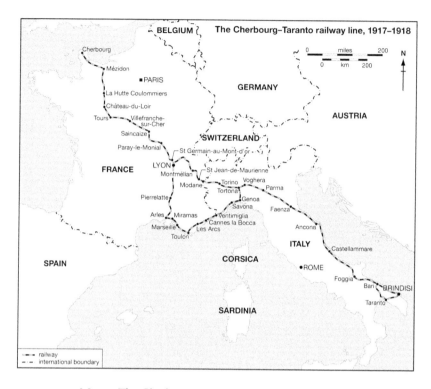

Map 4. The Cherbourg–Taranto railway line, 1917–18.

Source: 'The Mediterranean line of communication', *Railway Gazette: Special War Transportation Number*, 21 Sept. 1920, pp. 119–28, at p. 120. Map drawn by Cath D'Alton.

Introduction

Looking back upon the operations of 1916, and in anticipation of the battles to come, the French prime minister, Aristide Briand, remarked that military offensives had become 'really like a great industrial undertaking. There were so many miles of front, so many troops, and so many guns required; all had to be calculated to a nicety, and all kinds of preparations made'.[1] In the wake of 1917's inconclusive campaigning season, *The Times* provided a more concise observation: 'Modern war is modern industry, organised for a single definite purpose'.[2] The years between 1914 and 1918 witnessed the 'advent of a totalising war strategy that pitted industrial nations and their citizenries against each other'.[3] The conflict's dimensions made it impracticable for all of the belligerents to rely exclusively upon their cadre of professionally trained soldiers both for its conduct and its coordination. All turned to the manpower of civil society to enhance the size and strengths of their martial forces. In Britain, an influx of volunteers and conscripts provided the world's foremost naval power with an army capable of matching the vast forces raised in continental Europe. Their presence imbued almost every aspect of the British war effort. Examples of their bravery, sacrifices and eventual mastery of modern warfare on the industrial battlefield have inspired a prodigious literature in the century since the war was fought. Yet civilian brain power, as well as muscle power, played a vital role in the prosecution of the First World War.

When did the potential utility of civilian expertise find acceptance within the higher political and military administration of the British war effort? How were the skills and aptitudes possessed by the members of a highly industrialized society like pre-war Britain applied to the conduct of an industrial war? Was the relevance of non-military experience recognized and valued within the War Office and the army's various theatres of operations? This book addresses these questions. It comprises a detailed investigation of the roles performed by Britain's transport experts in support of the British army's military operations during the war. Two things are

---

[1] TNA, CAB 28/2, papers I.C. 13–32, Secretary's notes of allied conferences held at the Consulta, Rome, 5–7 Jan. 1917, p. 5.

[2] 'An army of labour. Behind the lines in France', *The Times*, 26 Dec. 1917, p. 8.

[3] T. M. Proctor, *Civilians in a World at War, 1914–1918* (New York, 2010), p. 3.

of particular importance to this study: the contributions made by senior employees of the British empire's transport concerns to the character and conduct of the war; and what these individuals' experiences can tell us both about how the First World War was conceptualized as it unfolded and about the relationship between soldiers and civilians at the summit of an international coalition. Through an examination of civilian specialists at war, this book illustrates how the British army leveraged modern business methods – developed for the administration of a global trading empire and with the aim of improving profitability – and discusses the ways in which the business of killing and the intensification of military power were shaped by the input of non-military figures during the twentieth century's first great conflagration.

The First World War did not create the phenomenon of civilian expertise augmenting the work of the state's military apparatus in the prosecution of war. Following the so-called military revolution of the mid sixteenth century, merchants known as sutlers followed the armies of early modern Europe and provided soldiers with the opportunity to purchase non-military items such as sugar, tobacco and coffee – a practice that continued until as recently as the American Civil War.[4] Britain's efforts in the French Revolutionary and Napoleonic wars were supported by a vast network of contractors who produced tents, kettles, knapsacks, uniforms, boots and sundry other items for the British troops, which provided the material foundations for the national war effort against France.[5] Following the advent of rail travel, the civilian contractors Samuel Morton Peto, Edward Betts and Thomas Brassey provided staff and materials for the construction of a fourteen-mile-long railway between the port of Balaklava and the British lines when the existing transport infrastructure's inadequacies jeopardized the siege of Sevastopol during the Crimean War.[6] Such activities, undertaken either to sustain the morale or increase the fighting abilities of the armies they served, were carried out on a far larger scale and with a higher degree of integration during the industrial wars of the twentieth century.

After 1914, civilians were drawn into the war in a way that was profoundly more immersive than in previous conflicts. The British government found work for the employees of railway companies in the construction of railway

---

[4] M. van Creveld, *Supplying War: Logistics from Wallenstein to Patton* (Cambridge, 1977), pp. 5–8.

[5] R. Knight, *Britain against Napoleon: the Organization of Victory, 1793–1815* (London, 2014); J. Uglow, *In These Times: Living in Britain Through Napoleon's Wars, 1793–1815* (London, 2015), pp. 46–8.

[6] T. Coleman, *The Railway Navvies* (London, 1981), pp. 212–20.

lines in various theatres, the maintenance and operation of locomotives, and the prosecution of duties in the 'War Office, the Admiralty, the Home Office, the Colonial Office, the Board of Agriculture and Fisheries, the Ministry of Munitions … the Ministry of Shipping, the Ministry of Food, the Wool Transport Office, the Trench Warfare Department, Army Canteens, [and] the Petrol Committee' among others.[7] Whereas in the past civilians in the proximity of the battlefield remained a distinct entity, the First World War cultivated military efforts in which the 'soldier-civilian relationship' was 'intermeshed' to an unprecedented degree.[8] In some of the cases within this book, Britain's transport experts assumed military rank and acquired recognizable positions within the military chain of command. In others, particularly on the home front, they operated on the fringes of the army and fulfilled quasi-military, quasi-civilian functions. Throughout the war effort, this book argues, their exertions helped to sustain and ultimately bring about a successful end to the bloodshed.

The various ways in which civilians navigated the First World War have attracted increasing scholarly interest in recent years. Historians have addressed the question of why non-military participants in the war became involved in the conflict and considered wider questions about 'the variety of meanings of "civilian" in wartime'.[9] Laura Ugolini has explored the motivations and experiences of middle-class men on the English home front, a category applicable to many of the individuals discussed within this book. For Ugolini, the war forced middle-class men, particularly those who were middle-aged or older, to confront popular understandings of manliness and patriotic duty. Those who attempted to emulate the military commitments of younger men were 'often not an object of admiration, but of ridicule and contempt'.[10] Yet through the direction of their skills and expertise into activities deemed valuable to the war effort, men unable to fight made contributions that – although not as dangerous as those of their sons – provided them with a sense that they had borne their 'fair share' of the war's burden. Sir Sam Fay, general manager of the Great Central Railway and the director of movements (DOM) at the War Office from 1917 onwards, provides evidence that reinforces Ugolini's argument. The dedication page of Fay's memoir, *The War Office at War*, confirmed the

[7] E. A. Pratt, *British Railways and the Great War: Organisation, Efforts, Difficulties and Achievements* (2 vols., London, 1921), i, p. xiii.

[8] H. Jones, 'The Great War: how 1914–18 changed the relationship between war and civilians', *RUSI Journal*, clix (2014), 84–91, at p. 90.

[9] Proctor, *Civilians in a World at War*, p. 8.

[10] L. Ugolini, *Civvies: Middle-Class Men on the English Home Front, 1914–18* (Manchester, 2017), p. 109.

book's purpose as a permanent record of what 'Grandad' had done during the war.[11]

Tammy Proctor's work has gone beyond middle-class men to discuss civilians' contributions on a broader scale. However, while recognizing the varied roles taken on by non-soldiers between 1914 and 1918, her analysis concentrated upon those in care-giving or humanitarian capacities rather than those drawn into the conflict for more overtly aggressive purposes. Consequently, although Proctor acknowledged that the First World War 'spawned the modern phenomenon of "expert" assistance in the management and maintenance of war',[12] her examination of civilian influence over the conduct of operations was largely restricted to the manufacture of chemical weapons. In doing so, Proctor's work joins a corpus of literature that has considered the technical and moral implications of the mobilization of scientific knowledge during the conflict.[13] This book seeks to draw business knowledge and managerial acumen into the same discussions. In an era before the establishment of private military contractors able to provide states with what Marc Lindemann has referred to as 'expertise at a price',[14] during the First World War civilians were deeply embedded in the war machine and performed tasks both directly and indirectly related to their peacetime specializations. Businessmen were less inhibited by the values of internationalism that had served as guiding principles for leading scientists in the century prior to 1914,[15] but their contributions to the technical and organizational aspects of the war had a similarly profound effect on the scale and duration of the violence that the opposing armies inflicted upon one another.

The place of civilian expertise within the British war effort merits further study, as the subject has been embroiled within one of the conflict's

[11] S. Fay, *The War Office at War* (London, 1937). Fay's eldest son, Samuel Ernest, served on the western front with the 111th Railway Company.

[12] Proctor, *Civilians in a World at War*, p. 177.

[13] R. MacLeod, 'The scientists go to war: revisiting precept and practice, 1914–1919', *Journal of War and Culture Studies*, ii (2009), 37–51; R. M. MacLeod, 'The "arsenal" in the Strand: Australian chemists and the British munitions effort 1916–1919', *Annals of Science*, xlvi (1989), 45–67; R. MacLeod, 'The chemists go to war: the mobilization of civilian chemists and the British war effort, 1914–1918', *Annals of Science*, l (1993), 455–81; D. C. Richter, *Chemical Soldiers: British Gas Warfare in World War I* (London, 1994); P. Doyle, *Disputed Earth: Geology and Trench Warfare on the Western Front 1914–18* (London, 2017); G. Hartcup, *The War of Invention: Scientific Developments, 1914–1918* (London, 1988).

[14] M. Lindemann, 'Civilian contractors under military law', *Parameters: US Army War College Quarterly*, xxxvii (2007), 83–94.

[15] E. Crawford, 'Internationalism in science as a casualty of the First World War: relations between German and allied scientists as reflected in nominations for the Nobel prizes in physics and chemistry', *Social Science Information*, xxvii (1988), 163–201.

most noxious historiographical legacies. In his *War Memoirs*, the wartime prime minister David Lloyd George asserted that it was only through his 'forcing' of 'unwanted civilians' upon the army in the summer of 1916 that the military reluctantly agreed to engage with the myriad talents and abilities prevalent in Britain's sophisticated industrial economy.[16] In Lloyd George's version of events the British army was handicapped in its operations by the predominance of insular, incompetent 'inexperts' within its senior ranks. The army's high command was incapable of understanding the organizational and conceptual implications of modern warfare, unable to offer effective solutions to the battlefield challenges it faced and unwilling to accept the advice of those who possessed skills and experience with demonstrable applicability to the prosecution of a multi-dimensional and complex war effort.[17] In contrast to the image of upper-class senior officers wedded to an outdated Victorian model of warfare that has become a standard trope in the popular memory of the war, Lloyd George's premiership comprised – in the words of his most prominent biographer –'vitality, urgency, [and] improvisation', and was dominated by an 'astonishing disregard for convention'.[18] As prime minister, Lloyd George oversaw the creation of a small war cabinet to oversee the higher direction of the war, established new government ministries to manage crucial areas such as shipping, food, information and reconstruction, and maintained his commitment – initially demonstrated at the Ministry of Munitions in 1915 – to the use of men with 'first class business experience' in positions of great responsibility.[19] He campaigned and won the khaki election of December 1918 on the platform of having been the man who won the war. Many on the other side of the civil–military divide took great issue with that slogan between the two world wars.

The so-called battle of the memoirs, fought during the 1920s and 1930s between those who had led Britain's political and military war efforts, established the terms of a debate that largely dominated the British military history of the war for generations.[20] Lloyd George's stringent criticisms of

---

[16] D. Lloyd George, *War Memoirs of David Lloyd George* (2 vols., London, 1938), i. 474.

[17] Lloyd George, *War Memoirs*, i, pp. v–vi.

[18] J. Grigg, *Lloyd George: War Leader, 1916–1918* (London, 2003), p. 11.

[19] C. Wrigley, 'The Ministry of Munitions: an innovatory department', in *War and the State: the Transformation of British Government, 1914–1919*, ed. K. Burk (London, 1982), pp. 32–56; K. Grieves, 'Improvising the British war effort: Eric Geddes and Lloyd George, 1915–18', *War & Society*, vii (1989), 40–55; D. Crow, *A Man of Push and Go: the Life of George Macaulay Booth* (London, 1965).

[20] I. Beckett, 'Frocks and brasshats', in *The First World War and British Military History*, ed. B. Bond (Oxford, 1991), pp. 89–112. Key works within the 'battle of the memoirs' include W. S. Churchill, *The World Crisis* (6 vols., London, 1923); E. Grey, *Twenty-Five Years,*

the army's conduct of the war – built upon foundations provided by the output of his political colleague Winston Churchill and that other prolific wordsmith of the age, Basil Liddell Hart – fixed the terms of discussion upon a polarized examination of two men and the decisions they made during the war: Field Marshal Sir Douglas Haig, commander-in-chief of the British Expeditionary Force (BEF) through the Somme, Passchendaele and the final year of the war; and Lloyd George, director of Britain's global war effort from December 1916 and a prominent member of the cabinet throughout the conflict.[21] Whereas Lloyd George embodied drive, energy and innovation, and espoused the industrial approach to war, Haig came to personify an army command that was reactive, technophobic, unimaginative and isolationist. The work of Tim Travers has exemplified a historical focus on the latter of these deficiencies. Across a series of books and articles, Travers argued that the conservatism that pervaded the BEF's general headquarters (GHQ) produced a concept of warfare that had been rendered obsolete by technological developments.[22] In addition, Haig's

---

1892–1916 (2 vols., London, 1925); H. H. Asquith, *Memories and Reflections* (2 vols., London, 1928); M. Aitken, *Politicians and the War, 1914–1916* (2 vols., 1928); R. B. Haldane, *Richard Burdon Haldane: an Autobiography* (London, 1929); D. Lloyd George, *War Memoirs of David Lloyd George* (6 vols., London, 1933); D. Haig, *Sir Douglas Haig's Despatches (December 1915– April 1919)*, ed. J. H. Boraston (London, 1919); J. D. P. French, *1914* (London, 1919); W. R. Robertson, *Soldiers and Statesmen, 1914–1918* (2 vols., London, 1926); H. Gough, *The Fifth Army* (London, 1931).

[21] For a range of judgments on Haig, see B. H. Liddell Hart, *Reputations, Ten Years After* (1928); A. Duff Cooper, *Haig* (2 vols., London, 1935); J. Terraine, *Douglas Haig: the Educated Soldier* (1963); J. Laffin, *British Butchers and Bunglers of World War One* (Gloucester, 1988); G. J. De Groot, *Douglas Haig, 1861–1928* (London, 1988); D. Winter, *Haig's Command: a Reassessment* (1991); W. Reid, *Douglas Haig: Architect of Victory* (Edinburgh, 2006); J. P. Harris, *Douglas Haig and the First World War* (Cambridge, 2008); *Haig: a Reappraisal 80 Years On*, ed. B. Bond and N. Cave (Barnsley, 2009); G. Sheffield, *The Chief: Douglas Haig and the British Army* (London, 2011). On Lloyd George's wartime career, see K. O. Morgan, 'Lloyd George's premiership: a study in "prime ministerial government"', *Hist. Jour.*, xiii (1970), 130–57; R. J. Q. Adams, *Arms and the Wizard: Lloyd George and the Ministry of Munitions, 1915–1916* (London, 1978); J. Grigg, *Lloyd George: From Peace to War, 1912–1916* (London, 1985); Grigg, *War Leader*; D. French, *The Strategy of the Lloyd George Coalition, 1916–1918* (Oxford, 1995); G. H. Cassar, *Lloyd George at War, 1916–1918* (London, 2011).

[22] T. Travers, 'The hidden army: structural problems in the British officer corps, 1900–1918', *Jour. Contemp. Hist.*, xvii (1982), 523–44; T. Travers, 'A particular style of command: Haig and GHQ, 1916–18', *Journal of Strategic Studies*, x (1987), 363–76; T. Travers, *The Killing Ground: the British Army, the Western Front and the Emergence of Modern Warfare, 1900–1918* (London, 1990); T. Travers, 'The evolution of British strategy and tactics on the western front in 1918: GHQ, manpower, and technology', *Jour. Military Hist.*, liv (1990), 173–200; T. Travers, 'Could the tanks of 1918 have been war-winners for the British Expeditionary Force?', *Jour. Contemp. Hist.*, xxvii (1992), 389–406; T. Travers, *How the War was Won: Command and Technology in the British Army on the Western Front, 1917–1918* (London, 1992).

personality 'prevented him from easily accepting innovation and change ... [which] led to his own isolation [and] the isolation of GHQ as a whole from the rest of the BEF'.[23]

Lloyd George's 'forcing' of an 'unwanted civilian' into this cloistered environment could be expected to have met a furious backlash from such an inflexible and 'inner-directed' character.[24] However, the unwanted civilian in question, Sir Eric Geddes, formed an instant working bond and a lasting personal friendship with the taciturn professional soldier. Haig appointed Geddes as the BEF's director-general of transportation (DGT) in October 1916, sought to retain his services as a transport advisor following Geddes's appointment as controller of the Royal Navy in May 1917, singled him out for praise within his wartime and post-war despatches[25] and even asked Geddes to act as godfather to his son in 1918.[26] The depth and warmth of the personal relationship between Haig and Geddes was exceptional among the soldiers and civilians who were thrust into close proximity between 1914 and 1918, but this book contends that the portrayal of a hidebound, aloof, narrow-minded and deeply jealous officer class cannot be supported by an examination of the interactions between the military and Britain's transport experts. While *individual* examples of animosity on both sides of the civil–military divide will be discussed below, no evidence can be found to support the concept of *institutional* insularity that has formed a sturdy bedrock for criticisms of Britain's military approach to the conflict's organizational challenges.

Britain's military leaders from the period surrounding the First World War are not alone in having been traduced in the historical record. In 1986 Donald Coleman and Christine MacLeod unearthed 'a mountain of apparently damning evidence' as to the 'incompetence ... ignorance, indifference, hostility [to new technology], prejudice and complacency' of British businessmen in the century after 1850.[27] Both at the time and in the historical analysis of the period that followed, the industrial elite in turn-of-the-century Britain have been largely denigrated in comparison to their counterparts in advanced nations such as the United States and Germany. American methods and processes in industries as diverse as railways, shoe

---

[23] Travers, *The Killing Ground*, p. 104.

[24] For a critical assessment of Haig's character, see Travers, *The Killing Ground*, pp. 85–118. For a more balanced account, see Sheffield, *The Chief*.

[25] Haig, *Despatches*, pp. 77, 351.

[26] TNA, ADM 116/1807, Sir Eric Geddes – private correspondence, letters from Lady Haig to Geddes, 1918.

[27] D. C. Coleman and C. Macleod, 'Attitudes to new techniques: British businessmen, 1800–1950', *Economic History Review*, xxxix (1986), 588–611, at p. 588.

making, printing and machine tool manufacturing were acknowledged as superior to the procedures followed by British companies, while in pre-war debates over Britain's economic competitiveness 'Germany assumed the dual role of model and enemy'.[28] In his position as commandant of the British army's staff college at Camberley between 1906 and 1910, Brigadier-General Henry Wilson delivered a series of lectures that implored his students to learn from close studies of the German language, people and army.[29] The outbreak of war did nothing to shift perceptions that Germany was 'the best organised community in the world, the best organised whether for war or peace', and that Britain had 'been employing too much the haphazard, leisurely, go-as-you-please methods, which, believe me, would not have enabled us to maintain our place as a nation even in peace very much longer'.[30] According to Lord Esher, Britain's principal enemy possessed 'the concentrated, unified and organised capacity, both scientific, military, philosophical, etc., of the highest developed nation the world has ever known'.[31]

Examples of a 'civil servant being ignorant of technology, a businessman not investing in a modern machine, or a soldier doubting the efficacy of new weapons', have been used to create an impression of Edwardian Britain as 'congenitally short-sighted' and incapable of responding effectively to the diffusion of new techniques, equipment and working methods that took place in the years before the war.[32] Admirers of foreign dynamism and critics of perceived domestic deficiencies have been held up as beacons of unheeded prescience, commentators who foresaw the predictable decline of Britain's status as a great power. Correlli Barnett's 1986 study, *The Audit of War*, typified such material.[33] Barnett's pre-war Britain comprised a workforce of unskilled 'coolies', a managerial cadre hostile towards professional education

---

[28] G. Paish, *The British Railway Position* (London, 1902); S. B. Saul, 'The American impact on British industry 1895–1914', *Business History*, ii (1960), 19–38; G. R. Searle, *The Quest for National Efficiency: a Study in British Politics and Political Thought, 1899–1914* (Oxford, 1971), pp. 54–57.

[29] Imperial War Museum (IWM), private papers of Field Marshal Sir Henry Wilson, HHW 3/3/7, 'Intelligence in peace and war: knowledge in power', 13 Nov. 1907, p. 13.

[30] D. Lloyd George, *Through Terror to Triumph: Speeches and Pronouncements of the Right Hon. David Lloyd George, M.P., since the Beginning of the War*, ed. F. L. Stevenson (London, 1915), p. 104.

[31] Quoted in E. Greenhalgh, *Victory through Coalition: Britain and France during the First World War* (Cambridge, 2005), p. 29.

[32] D. Edgerton, 'The prophet militant and industrial: the peculiarities of Correlli Barnett', *Twentieth Century British History*, ii (1991), 360–79, at p. 366.

[33] C. Barnett, *The Audit of War: the Illusion and Reality of Britain as a Great Nation* (London, 1986).

and a decadent and irresponsible governing class of 'romantics'.[34] The mass rejection of urban volunteers for service in the South African War due to their lack of physical fitness embodied the decline of British manpower. The low output of graduate scientists and engineers created a 'crisis' of British industry exemplified by an ongoing dependence on 'rule-of-thumb' and a rejection of systematic management principles. In comparison to the rigorous sponsorship and application of scientific knowledge in Germany, and the emergence of standardization and mechanization in the United States, British industry appeared to be backward, stagnant and primitive.

Such a pessimistic outlook raises a series of tough questions. If Britain was in such a relatively weak position in 1914 – populated by an unfit, uneducated, unskilled workforce and directed by an elite more interested in cricket and classics than the latest technological advances – how then was the country able to organize the largest, most wide-ranging, most total war effort in its military history? How were the complexities and scales of industrial warfare recognized, comprehended and coordinated with such success against Germany, the apparent model of national and military efficiency? How was all of this achieved despite the absence of a mass army drawn from the entire cross-section of British society in 1914?

That the ethos, workforce and managerial ability capable of meeting this challenge existed in Britain has been central to arguments advanced by David Edgerton. In the opening decades of the twentieth century, Edgerton identified in Britain 'a military-industrial-scientific complex which was … second to none'.[35] The absence of a large British army to match those put into the field by the other great powers in 1914 was not evidence of a liberal aversion to defence. Instead, it was a manifestation of Britain's desire to wage war using its own chosen means: a naval force capable of dominating world trade and depriving its enemies of the means to live. Britain's success in both world wars, Edgerton argued, depended on the 'pre-war international, modern and capital-intensive orientations of British armed force and the British ability to harness the resources not only of the nation but of much of the world … The British empire was victorious … because it was rich and could and did use its unique position in the world to fight wars of steel and gold'.[36] Before 1914, no nation on earth either imported or exported more than Britain. From 1914 onwards, the trading

---

[34] The idea of a spread of 'rural romanticism' among the British upper classes is developed further in M. J. Wiener, *English Culture and the Decline of the Industrial Spirit, 1850–1980* (2nd edn., Cambridge, 2004).

[35] D. Edgerton, *Warfare State: Britain, 1920–1970* (Cambridge, 2006), p. 1.

[36] D. Edgerton, *The Rise and Fall of the British Nation: a Twentieth-Century History* (London, 2018), p. 47.

network that connected the wheat fields and cattle ranches of the Americas to Britain, and the British cotton factories and coal fields to the world, was redirected to the application of steel and gold to combat.

The existence and maintenance of a global trading empire centred on London demonstrates that Britain possessed the human and material resources required for the mass transportation that industrial war demanded. Over 620,000 employees operated an integrated railway network that linked the major urban centres to each other, and joined the great export- and import-dependent industries to ports able to handle colossal tonnages of goods. At the outbreak of the First World War the London and North-Western Railway (LNWR) employed around 110,000 people, distributed across a network that stretched from London to Carlisle and from Swansea to Leeds. The next largest railways – the Great Western, Midland and North-Eastern – each employed in excess of 50,000 men across a range of specialist occupations. In fields as diverse as locomotive engineering, marketing, ticket sales and hospitality, the growth of the railways in the nineteenth century had necessitated the establishment of new methods for the mobilization, control, supervision and direction of large and intricate operating systems.[37] By 1911, the national transport system (comprising the rail and canal networks) possessed the ability to handle some 1,500 million passenger journeys and 560 million tons of goods per year.[38] This book analyses the manner in which the men, materials and methods that maintained this colossal peacetime traffic were redeployed and repurposed to the task of transporting Britain's military power during four years of conflict.

Transportation provides an ideal lens through which to observe Britain's application of civilian expertise to the demands of the First World War for three reasons. First, the popular image of the war in Britain may revolve around the static, rigid line of trenches that stretched across Belgium and France from the English Channel to the Swiss border, but the fighting relied upon movement and transportation to an unprecedented degree. The war drew in participants from all continents and consumed raw materials sourced from all over the globe. As the BEF expanded in size between 1914

---

[37] A. D. Chandler, 'The railroads: pioneers in modern corporate management', *Business History Review*, xxxix (1965), 16–40; M. Campbell-Kelly, 'The Railway Clearing House and Victorian data processing', in *Information Acumen: the Understanding and Use of Knowledge in Modern Business*, ed. L. Bud-Frierman (London, 1994), pp. 51–74.

[38] L. Shaw-Taylor and X. You, 'The development of the railway network in Britain 1825–1911', *The Online Historical Atlas of Transport, Urbanization and Economic Development in England and Wales c.1680–1911* (2018), p. 26 <https://www.campop.geog.cam.ac.uk/research/projects/transport/onlineatlas/railways.pdf> [accessed 19 July 2018].

and 1917 – and as British troops went into action at Gallipoli, in Italy, Macedonia, Palestine and elsewhere – they were entirely dependent upon complex production and distribution networks that connected the factories of Britain and the world to the front line. The colossal volumes of material required to sustain offensive operations – during periods of heavy fighting a total of 1,934 tons of supplies (including ammunition, engineering stores and food) were required each day for every mile of front held by the BEF – could not be sourced locally.[39] Therefore, the rail, road and waterway networks surrounding the areas where fighting took place were essential both to the soldiers' sustenance and to the evolution of material-intensive combat methodologies.

Old methods of transportation based upon the carrying capacities of horse and cart – and designed to supply armies unburdened by vast quantities of barbed wire, sandbags, duckboards, spare parts for aircraft and tanks and myriad other *impedimenta* – were inadequate to handle the demands of modern war.[40] In previous eras, except in the case of sieges, fighting had been conducted between opposing bodies of men armed with weapons they could carry to and from the battlefield. During the First World War, the principal combat role was taken by machines that 'require[d] a constant supply of material in the form of shells and cartridges to render them of any use'.[41] The bald statistics of the war effort can be converted into detailed tables and graphs, which chart the belligerents' capacities to devour resources during the conflict. However, numbers and images alone cannot convey crucial information about the development of the material war. As successive problems of battlefield supply were identified and solved between 1914 and 1918, fresh challenges emerged. Viewing the war from the perspective of Britain's transport experts provides a platform from which those challenges can be examined, both in terms of their impact on the British army's ability to wage war successfully and in relation to the applicability of industrial methods to the industrial battlefield.

Second, by focusing upon a factor of the war that was ubiquitous – that is to say that the limitations of the available transport infrastructure were an inescapable reality to the military and political leaders of 1914–18 regardless of their nationality – this book is able to move beyond the vituperative, personalized debating chamber occupied by the 'frocks and

---

[39] A. M. Henniker, *History of the Great War: Transportation on the Western Front, 1914–1918* (London, 1937), p. 157.

[40] J. Thompson, *The Lifeblood of War: Logistics in Armed Conflict* (Oxford, 1991), pp. 40–41; Van Creveld, *Supplying War*, p. 141.

[41] R. Bonham-Smith, 'Railway transport arrangements in France', *Royal United Services Institution. Journal*, lxi (1916), 47–62, at p. 47.

brass hats'. Instead, it contributes to an ongoing scholarly campaign to shift attention away from the great or not-so-great individuals in military history and towards a more complete understanding of the vast organizations and complex, integrated bureaucratic systems that were constructed for the task of winning the war.[42] In doing so, it complements studies that have analysed Britain's mobilization of economic and human resources, and have begun to identify *how* civilian expertise and technologies were applied to the challenges of the industrial battlefield. This process is particularly advanced in terms of the wartime exploitation of advances in communications and myriad branches of science.[43] According to Martin van Creveld, the allies' victory was significantly influenced by their more efficient management and administration of the gigantic organizational systems that underpinned the war. During what van Creveld described as 'the age of systems', technological advances between 1830 and 1945 'very largely turned war itself into a question of managing complex systems'. This process forced senior politicians, military commanders, scientists and business leaders to consider warfare through a series of interconnected, delicately balanced organizations that were highly responsive to change.[44] Throughout the First World War, allied leaders were faced with the identification and resolution of a host of conceptual, technical and administrative conundrums that could not be tackled in isolation. As will be seen, comprehending this truth proved a slow process. Until the nations (not just one nation, nor solely the military forces created by those nations) engaged in the struggle accurately identified the likely scale of effort and organization required to bring about victory, all too frequently individuals and institutions attempted to solve Rubik's cube while only able to view one face of the puzzle.

This book catalogues some of the difficulties grappled with by Britain's transport experts over the course of an evolving four-year conflict. It augments recent work that has analysed the British army's ability to learn, adapt, implement and innovate between 1914 and 1918. Building upon

---

[42] An example of this approach, applied to the Second World War, is P. Kennedy, *Engineers of Victory: the Problem Solvers Who Turned the Tide in the Second World War* (London, 2013).

[43] See the texts listed above in n. 14, and B. N. Hall, 'The "life-blood" of command? The British army, communications and the telephone, 1877–1914', *War & Society*, xxvii (2008), 43–65; B. N. Hall, 'Technological adaptation in a global conflict: the British army and communications beyond the western front, 1914–1918', *Jour. Military Hist.*, lxxviii (2014), 37–71; B. N. Hall, *Communications and British Operations on the Western Front, 1914–1918* (Cambridge, 2017); E. Bruton and G. Gooday, 'Listening in combat – surveillance technologies beyond the visual in the First World War', *History and Technology*, xxxii (2016), 213–26.

[44] M. van Creveld, *Technology and War: from 2000 B.C. to the Present* (New York, 1991), p. 161.

broadly sympathetic analyses – certainly in comparison to earlier works of the 'lions led by donkeys' persuasion – of the BEF's evolution in combat tactics and battlefield command published from the 1980s onwards,[45] both Jonathan Boff and Aimée Fox have demonstrated that learning in the army was a multifaceted process. New methods were developed but applied neither universally nor consistently across the British military campaigns, while knowledge itself was disseminated throughout the organization by a combination of formal structures and informal networks.[46] Their work has added a much-needed layer of sophistication to the hitherto over-simplified concept of a learning curve or learning process within the wartime army, one that is supplemented by the conclusions drawn in this book.[47]

Alongside supporting the continued examination of the British army and state's approach to industrial war, this book hopes to provide a framework for further studies on the development of the transportation systems in the war efforts of Britain's allies and enemies. Britain fought throughout the war in a coalition and against another coalition. The individual elements of both coalitions acted independently and in concert; they responded to their enemy's activities and sought to outwit them with carefully guarded innovations, such as poison gas and the tank. At no point in the war did one national effort exist in isolation from those of other nations: 'there was a continuous dynamic of push and pull, measure and counter-measure, between [and within] the two sides'.[48] On the allied side alone, French and British forces – and at various points, and with varying degrees of influence, those of other allied or associated nations – coexisted on the western front and required access to sufficient warehouses, trains, railheads, road space and vehicles to maintain the fighting efficiency and health of their troops. Further afield, the gap between Russian demands for material and financial assistance, and the ability (and willingness) of its partners to fulfil them,

---

[45] See, e.g., S. Bidwell and D. Graham, *Fire-Power: British Army Weapons and Theories of War, 1904–1945* (Barnsley, 2004); P. Griffith, *Battle Tactics on the Western Front: the British Army's Art of Attack, 1916–18* (New Haven, Conn., 1994); J. B. A. Bailey, *The First World War and the Birth of the Modern Style of Warfare* (Camberley, 1996); G. Sheffield, *Forgotten Victory. The First World War: Myths and Realities* (London, 2001); A. Simpson, *Directing Operations: British Corps Command on the Western Front, 1914–18* (Stroud, 2006).

[46] J. Boff, *Winning and Losing on the Western Front: the British Third Army and the Defeat of Germany in 1918* (Cambridge, 2012); A. Fox, *Learning to Fight: Military Innovation and Change in the British Army, 1914–1918* (Cambridge, 2017).

[47] For a synthesis of the existing literature and a deconstruction of the learning curve thesis's limitations, see Boff, *Winning and Losing*, pp. 11–12, 247–9; Hall, *Communications and British Operations*, pp. 3–5.

[48] J. Boff, *Haig's Enemy: Crown Prince Rupprecht and Germany's War on the Western Front* (Oxford, 2018), p. 5.

shrank and grew as the war unfolded and assessments of Russian capabilities changed.[49] The war engaged the allies in a constant process of re-evaluation and reconfiguration of the human, material and financial commitments required to maintain and improve the coalition's martial qualities. At the same time, national interests and the coalition's priorities were sometimes incompatible – and between 1914 and 1918 inter-allied relations were further complicated by a plethora of linguistic and cultural factors.[50]

Understanding how these crucial issues influenced the relationship between coalition partners has been central to relatively few studies of the First World War. In recent years William Philpott, Elizabeth Greenhalgh and Chris Kempshall have addressed various dimensions of the political, military, social and administrative mechanics of the Franco-British (and, in Kempshall's case, the Franco-British-American) coalition.[51] Each have illustrated how victory was ultimately achieved *through* coalition but also charted the 'muddled perceptions, stifled communications, disappointed expectations, [and] paranoid reactions' that underscored inter-allied relations.[52] Britain's transport experts – men used to operating within the competitive atmosphere of a capitalist economy and occupying executive positions of great influence – were not always temperamentally suited to an environment that demanded conciliation, negotiation and a sympathetic approach to harmonize inter-allied disputes. The transport systems that developed on the western front and in other theatres populated by coalition forces were never the result of British ingenuity and resources alone. Therefore, the influence of, in particular, French attitudes towards the demands and desires of Britain's transport experts feature in much of the narrative that follows.

Third, comparatively few full-length studies of transportation's influence over the conduct of the war have been published. Transportation, as contemporary observers understood, was 'so interwoven with modern commerce and industry' that it could not be separated from the history

---

[49] K. Neilson, *Strategy and Supply: the Anglo-Russian Alliance, 1914–17* (London, 1984).

[50] G. Sheffield, '"Not the same as friendship": the British empire and coalition warfare in the era of the First World War', in *Entangling Alliances: Coalition Warfare in the Twentieth Century*, ed. P. Dennis and J. Grey (Canberra, 2005), pp. 38–52.

[51] W. Philpott, *Anglo-French Relations and Strategy on the Western Front, 1914–18* (Basingstoke, 1996); Greenhalgh, *Victory through Coalition*; C. Kempshall, *British, French and American Relations on the Western Front, 1914–1918* (London, 2018). On the complexities of coalition warfare more broadly, see *Coalition Warfare: an Uneasy Accord*, ed. K. Neilson and R. A. Prete (Waterloo, ON, 1983); S. Bidwell and D. Graham, *Coalitions, Politicians and Generals: Some Aspects of Command in Two World Wars* (London, 1993); *Britain and France in Two World Wars: Truth, Myth and Memory*, ed. R. Tombs and E. Chabal (London, 2013).

[52] Richard E. Neustadt, quoted in Greenhalgh, *Victory through Coalition*, pp. 1–2.

of such matters.[53] Yet as the experience of the First World War receded, the intricacies and minutiae of transport details were eclipsed by more glamorous and controversial debates over the British army's strategic and tactical evolution. Two of the most enduring histories of the war, both repackaged for new audiences during the centenary, paid scant attention to the supply and movement challenges that taxed the military authorities between 1914 and 1918.[54] Recent scholarship has improved in this direction, but analyses of supply issues remain for the most part superficial and subordinated to narratives that focus upon combat operations. Paul Harris's account of the western front's climactic battles – during which time transport factors exerted the defining influence over the allies' ability to pursue the retreating German army – exemplifies the trend. Harris acknowledged the 'essential' importance of logistics and military engineering to the BEF's achievements between August and November 1918, but devoted a mere handful of pages to discussion of these topics.[55] As the railway journalist Edwin Pratt lamented while the war was in progress, military historians 'have too often disregarded such matters of detail as to how the armies got [to the battlefield] and the possible effects of good or defective transport conditions, including the maintenance of supplies and communications, on the whole course of a campaign'.[56]

Martin van Creveld's pioneering work has underlined the importance of logistical support as a precursor to successful military operations,[57] but scholars of the First World War have yet to adequately address Pratt's challenge. Van Creveld argued that logistical factors fixed the parameters of what an army could or could not achieve on the battlefield, and highlighted the dangers to modern armies of both scarcity and superabundance. At the most basic level, an army's failure to provide a steady supply of 3,000 calories per day meant that men would 'very soon cease to be of any use as soldiers'.[58] Conversely, the *impedimenta* of the modern army – guns, tanks, aeroplanes, tractors, road stone, barbed wire, sandbags, duckboards, petrol, spare parts, tools and sundry other items – created a situation in which gargantuan volumes of material occupied the capacity of road, rail and

---

[53] C. Travis, 'The science of railroading: a further plea for the establishment of a transport institute', *Great Central Railway Journal*, xiii (1917), 40–2, at p. 40.

[54] B. H. Liddell Hart, *The Real War: 1914–1918* (London, 1930); A. J. P. Taylor, *The First World War: an Illustrated History* (London, 1963).

[55] J. P. Harris, *Amiens to the Armistice: the BEF in the Hundred Days' Campaign, 8 August–11 November 1918* (London, 1998), pp. 54–5, 218.

[56] E. A. Pratt, *The Rise of Rail Power in War and Conquest, 1833–1914* (London, 1916), p. vii.

[57] Van Creveld, *Supplying War*.

[58] Van Creveld, *Supplying War*, p. 1.

Table 0.1. Resource and development ratios, allies:central powers.

	Population	Territory	Territory per head	Gross domestic product
November 1914	5.2:1	11.5:1	2.2:1	2.9:1
November 1916	5.5:1	12.1:1	2.2:1	3.2:1
November 1918	8.1:1	13.5:1	1.7:1	4.6:1

Note: Figures display the capacities of the allies' and central powers' coalitions at various points of the war, according to the populations, sizes and incomes of each power within the coalitions for the year 1913.

Source: S. Broadberry and M. Harrison, 'The economics of World War I: an overview', in *The Economics of World War I*, ed. S. Broadberry and M. Harrison (Cambridge, 2005), pp. 3–40, at p. 11.

waterway networks in the vicinity of the troops. Consequently, the armies' abilities to reduce dependency upon their lines of communications were severely restricted. In this sense, the global reach and financial superiority of the allies over the central powers were less pronounced than the figures imply. The allies could, and did, considerably out-spend and out-produce their opponents throughout the war. Yet their ability to bring their human, territorial and economic advantages to bear upon the central powers (see Table 0.1) was constrained by the efficiency with which the transport networks in France and Belgium, Russia, Macedonia and elsewhere were operated.

As yet, attempts to remedy the historical ignorance of how transportation functioned during the First World War have largely focused upon the infrastructure behind the western front.[59] The most prominent single-volume treatment of the British experience, by Colonel A. M. Henniker, appeared in 1937 as part of the official history of the conflict.[60] It remains the most thorough overview of the BEF's evolving challenge and a vital source of organizational details. However, according to one reviewer, *Transportation on the Western Front* was 'not for the casual reader in search

---

[59] For exceptions to this trend, see L. J. Hall, *The Inland Water Transport in Mesopotamia* (London, 1921); K. Roy, 'From defeat to victory: logistics of the campaign in Mesopotamia, 1914–1918', *First World War Studies*, i (2010), 35–55; K. C. Ulrichsen, *The Logistics and Politics of the British Campaigns in the Middle East, 1914–22* (Basingstoke, 2011). Further material, albeit of uneven coverage, can be obtained from the relevant sections in R. H. Beadon, *The Royal Army Service Corps: a History of Transport and Supply in the British Army* (2 vols., Cambridge, 1931), ii; *History of the Corps of Royal Engineers*, ed. H. L. Pritchard (11 vols., Chatham, 1952), vi; W. J. K. Davies, *Light Railways of the First World War: a History of Tactical Rail Communications on the British Fronts, 1914–18* (Newton Abbot, 1967).

[60] Henniker, *History of the Great War*.

of easy entertainment. Its subject is wanting in popular appeal, and its lack of the human touch will make it unattractive even to the serious-minded'.[61] Published towards the end of the 'battle of the memoirs', Henniker's text was unashamedly coloured by its author's background and made little impact on a public debate framed by Churchill's and Lloyd George's livelier accounts. Henniker, a career soldier who served in a variety of transportation roles between 1914 and 1918, argued that many of the difficulties experienced by the BEF were the result of insufficient foresight on the government's part coupled with a lack of faith in the soldiers' abilities to discharge their duties. As summarized by Sir James Edmonds's acerbic comment in the volume's introduction – a clear rejoinder to Lloyd George's attacks upon the military's supposed incompetence – the official history's view was that 'what soldiers had been denied was freely accorded to a civilian. Similarly, all his ideas for expansion were accepted'.[62] Whereas Henniker and his colleagues had been forced to make the best of inadequate resources early in the war, Lloyd George's attempts to circumvent the military leadership by appointing civilians to senior positions within the British war effort in 1916 were accompanied by a new willingness to commit manpower and materials to the BEF's rearward organization.

The subject of Edmonds's statement was Sir Eric Geddes, who occupied a central role in the first scholarly assessment of the BEF's transportation services. In *British Logistics on the Western Front*, Ian M. Brown argued that the BEF's evolution in combat tactics and battlefield command was predicated on superb leadership in the fields of administration and logistics.[63] Administrative excellence from mid 1917 onwards, built upon foundations established by Geddes, freed the BEF's senior commanders from having to concern themselves with questions of supply. Following Geddes's reorganization of the transport services in France, the material requirements of the BEF's 'teeth' were satisfied by an increasingly efficient 'tail'. Reliable logistics were the conduit that permitted the allies to make effective use of their resource advantage over the central powers, fostering what Hew Strachan has described as 'prodigality in munitions expenditure'.[64]

[61] W. J. Wilgus, 'Review of transportation on the western front, 1914–1918. Compiled by Colonel A. M. Henniker', *American Historical Review*, xliv (1939), 386–8, at p. 386. Wilgus was far from an uninterested observer of wartime transportation. He had served on the headquarters staff of the American Expeditionary Force's transportation service in France, and authored a volume on America's transportation effort that 'bristle[d] with dry and dusty figures and statistics'. See W. J. Wilgus, *Transporting the A.E.F. in Western Europe, 1917–1919* (New York, 1931), p. xxv.

[62] Edmonds' introduction in Henniker, *History of the Great War*, p. xxii.

[63] I. M. Brown, *British Logistics on the Western Front, 1914–1919* (London, 1998).

[64] H. Strachan, *The First World War: To Arms* (Oxford, 2001), p. 999.

Brown's work was the first to thoroughly acknowledge the wide-ranging influence that an individual civilian, aside from political leaders, could have over the conduct and character of the fighting on the western front. However, Brown's study focused predominantly upon the army's response to the challenge of supplying the industrial battlefield rather than the interaction between civilian and military expertise. Consequently, his text – which echoed Lloyd George in its accusation that the BEF displayed 'anti-civilian' phobia prior to 1916 – created an impression that Geddes emerged that summer from a hitherto undervalued and largely untapped pool of talent.[65] The narrow terms of Keith Grieves's 1989 biography of Geddes provide little material to alter that conclusion.[66] While Grieves emphasized the applicability of his subject's business experience to the many wartime roles that Geddes occupied, the constraints of the biography's structure conspired to limit opportunities for Geddes's wartime endeavours to be placed in their proper context. A key aim of this book is to redress this shortcoming in the existing literature by looking beyond Geddes and considering the contributions of Britain's transport experts more broadly. It does not attempt to minimize Geddes's impact or suggest that he was not a pivotal figure in the history of the war, but rather argues that Geddes was far from unique as a manifestation of civil–military cooperation in the British war effort. He was not the first senior transport expert to work closely with the British army to improve the force's supply system, nor was he the last to occupy a prominent position in the upper echelons of the military hierarchy. Instead, Geddes represents a crucial link within a chain of civil–military connections between the army, government and principal transport enterprises, which was forged long before the outbreak of hostilities in August 1914.

Charting the existence of the longstanding professional relationship between some of Britain's largest private companies and the state – in both its political and military forms – in the decades before the First World War is the central focus of this book's first part. In chapter one the image of the pre-war British army as an insular, conservative institution, unreceptive to outside influences, is challenged. Professional soldiers and civilian experts participated in a mutually beneficial exchange of knowledge and experience, which produced officers proficient in the skills necessary for the operation of a modern army's lines of communications and a cadre of transport professionals cognizant of an industrial army's logistics requirements.

[65] Brown, *British Logistics*, p. 89.
[66] K. Grieves, *Sir Eric Geddes: Business and Government in War and Peace* (Manchester, 1989).

Chapter two considers the implications of this working relationship for the army's mobility in wartime. The 'with France' (WF) scheme, which placed the BEF on what became the western front in 1914, incorporated within its preparations the most thorough example of civil–military cooperation in peacetime British military history. Working in conjunction with the army, navy and each other, Britain's privately owned, competitive railway companies propelled the BEF to the continental mainland in time to help stem the tide of the German invasion.

Part two of the book covers the war before October 1916, and charts the remarkable and unplanned expansion of the British war effort. The creation and maintenance of a mass army – to supplement and rival the conscripted forces of France and Germany respectively – coupled with the spread of hostilities far beyond the fields of France and Flanders, presented the British authorities with a series of colossal organizational challenges. Chapter three examines the multifarious contributions made to the prosecution of an increasingly global war by Britain's transport experts. It emphasizes the depth and breadth of the civilian talents available to an imperial power like Britain, and demonstrates the restrictions placed upon British freedom of action by the priorities and requirements of its coalition partners. The latter theme recurs in chapters four and five, which comprise two case studies that demonstrate both the extent to which the army recognized and appreciated civilian expertise – prior to Lloyd George's 'forcing' of Eric Geddes upon them – and the limited extent to which the complexity and interconnectedness of wartime transportation was understood within the British war effort ahead of the battle of the Somme. Chapter six outlines both the BEF's underdeveloped conceptual awareness of transport's function in military operations and the existence of professional suspicion towards 'outsiders' within GHQ during the Somme. It argues that Geddes's transportation mission played a pivotal role in 'show[ing] what *transportation* meant, and how each variation in one process of movement must inevitably have its effects on the others'.[67] The Geddes mission, and Sir Douglas Haig's response to it, established the platform upon which the material-intensive warfare of 1917 and 1918 was fought.

The third and final part of this book discusses the implications of Haig's decision to appoint Geddes as DGT and investigates the manner in which Britain's transport experts directly and indirectly contributed to the prosecution of a war effort of unprecedented ferocity. In successive chapters it analyses: the material and methodological implications of the

---

[67] M. G. Taylor, 'Land transportation in the late war', *Royal United Services Institution. Journal*, lxvi (1921), 699–722, at p. 705. Emphasis in original.

'civilianization' of Britain's war effort; the global diffusion of British expertise and the pursuit of a more harmonic, inter-allied approach to the conflict; and the impact of civilian specialists upon the industrial battlefield. These chapters argue that civilian intervention was not a panacea to the complex organizational problems posed by the First World War. Disagreements and disputes over the application of Britain's human and material resources continued to permeate the civilian and military leadership of the war effort, but by the late summer of 1918 the transportation services behind Britain's armed forces were 'good enough' to ensure the allies' advantages could be brought to bear upon the exhausted Germans. Overall, this book contends that Britain's ultimately successful war effort was the result of a synthesis between civilian and military expertise, which had been gathered and applied to the challenges involved in the acquisition and maintenance of a global empire during the previous century.

The civilians who operated the transport network that connected Britain to the world in peacetime became a central pillar of the British response to the conundrum of an industrial war. They joined their colleagues from the political, military, industrial and scientific communities in the task of seeking a pathway to victory on the western front and beyond. In documenting their contributions, this book addresses just a few of the myriad ways in which the 'existing structures, organizations and modes of thinking' in pre-war Britain influenced the direction and administration of warfare in the early twentieth century.[68] It is those existing structures, assembled collaboratively by Britain's transport experts, the government and the military, which will be considered first.

[68] A. S. Fell and J. Meyer, 'Introduction: untold legacies of the First World War in Britain', *War & Society*, xxxiv (2015), 85–9, at p. 87.

I. Preparation

1. Forging a relationship: the army, the government and Britain's transport experts, 1825–1914

On 18 February 1915 the Liberal member of parliament for Glasgow College, Henry Watt, asked the Board of Agriculture if farmers in England's southern counties had raised concerns about the availability of Scottish seed potatoes for the coming crop. Four days later his colleague, James Hogge of Edinburgh East, questioned the president of the Board of Trade, Walter Runciman, over the availability of the 'usual railway facilities' for the upcoming flat-racing season on account of the ongoing war. On 4 May the Unionist member for Armagh North, William Moore, directed Runciman's attention to complaints that bleaching, finishing and dyeing firms in Ulster were on the verge of closure thanks to restrictions on the amount of cotton allowed to leave Manchester. Moore implored Runciman to do whatever was required to avoid 'unemployment and distress among the workers' of his constituency.[1] In each of the above cases, and many more besides – despite the seemingly disparate nature of the enquiries – the government either referred to information already received from the Railway Executive Committee (REC) or assured the questioner that the matter would shortly receive the REC's attention.

The REC was formally established less than two years before the outbreak of the First World War, but it embodied a longstanding professional link between the British state, its army and the principal means of transport available for the movement of the army on both home and foreign soil. Composed ultimately of twelve general managers, drawn from the nation's most prominent railway companies, its existence helped ensure that political, military, economic and social questions with the potential to affect transportation in Britain received the consideration of technically proficient industrial specialists. By the summer of 1915 the REC had become an integral component of the nation's ongoing organizational response to the emerging conflict, a position it retained for the duration of the war.

This chapter examines the interactions between the British army and the emerging railway companies at the organizational and operational levels

---

[1] Hansard, *Parliamentary Debates*, 5th ser., lxix (18 Feb. 1915), col. 1313; lxx (22 Feb. 1915), cols. 7–8; lxxi (4 May 1915), cols. 976–7.

'Forging a relationship: the army, the government and Britain's transport experts, 1825–14', in C. Phillips, *Civilian Specialists at War: Britain's Transport Experts and the First World War* (London, 2020), pp. 23–61. License: CC-BY-NC-ND 4.0.

before the First World War, contacts that emphasize the army's willingness to seek out and exploit expertise from civilian sources. The partnership that emerged between the army and the railways was a component of a wider practice of knowledge exchange – particularly prevalent across the engineering profession – that was of mutual benefit to the civilian and military spheres. Members of the Institution of Civil Engineers (ICE) were frequent speakers at the school of military engineering in Chatham and the authors of articles in the *Royal Engineers Journal*, while their military counterparts were regular lecturers at ICE's headquarters in London.[2] Both parties were beneficiaries of the close links established between the railways and the army, although recent research has raised questions about the prevalence of former soldiers within senior roles in the early railway companies.[3] As Di Drummond has observed, 'the military element' exemplified early railway management, while the idea of a 'railway servant' inspired by the concept of service in the armed forces remained popular across the industry before the war.[4] Even as the railways matured, the bonds between the army and the railway companies did not break. British and imperial railways provided opportunities for soldiers to obtain practical experience in the repair, maintenance and operation of lines, and across the empire British railwaymen worked alongside the military to protect and project London's power. British soldiers actively sought out and engaged with the experience and expertise that railway managers and their workforces possessed. Together, they contributed to the creation of an army proficient in the skills required to work 'a vast business organization' under the most intense pressure.[5]

A cooperative endeavour: the government, the army and the railways

A connection between the government, the army and the railways of Britain was established as early as 15 September 1830 at the opening of the Liverpool and Manchester Railway. The duke of Wellington, Sir Robert

---

[2] A. Fox, *Learning to Fight: Military Innovation and Change in the British Army, 1914–1918* (Cambridge, 2017), p. 165.

[3] D. Turner, 'Unlocking the early railway manager – a project to follow', *Turnip Rail*, 2011 <http://turniprail.blogspot.com/2011/04/unlocking-early-railway-manager-project.html> [accessed 11 Sept. 2018].

[4] D. K. Drummond, *Crewe: Railway Town, Company, and People, 1840–1914* (Aldershot, 1995), pp. 59–60; F. McKenna, *The Railway Workers, 1840–1970* (London, 1980), pp. 30–4; TNA, ZLIB 29/691, 'Education and advancement of the railway clerk' by E. C. Geddes, 1910, p. 1.

[5] H. A. Young, 'Practical economy in the army', *Royal United Services Institution. Journal*, l (1906), 1281–5, at p. 1282.

Peel and the Liverpool MP William Huskisson were among the dignitaries in attendance. The new method of conveyance got off to an inauspicious start. Huskisson was fatally injured by a passing train, while Wellington was subjected to 'hooting, cat-calling and shouting' from a hostile Manchester crowd.[6] The day's events 'prejudiced the Duke for ever against railways', but their 'practical utility' for the carriage of troops was clear to the hero of Waterloo. Shortly after its infamous opening day the Liverpool and Manchester became the first railway line in the world to carry soldiers on active service, when newly arrived soldiers from Ireland were sent inland by rail on 10 July 1832. The 91st (Argyllshire) Regiment were saved a two-day march on the reverse journey before their embarkation to Dublin on two steamers.[7] From these small beginnings, over the next eighty years 'the potential of the railways … captured the imagination of many soldiers, not merely Royal Engineers but also senior commanders both at home and overseas'.[8] In the decades prior to 1914, soldiers of the British army and servants of the country's largest railway companies participated in a range of civil–military ventures designed both to improve the nation's security and increase the army's understanding of a novel and complex industrial tool.

The influence of soldiers over the development of railways in Britain and beyond was considerable. Across Europe, the presence of uncertain land frontiers between emerging nation states and the collective memory of the Napoleonic Wars fuelled military and political interest in the strategic potential of railways. From the 1830s until the eve of war railway construction on either side of the French-German frontier was eyed with deep suspicion, as the laying of more lines and the expansion of railway capacities near the border provided the foundations for the deployment of ever larger forces at the outset of a major war.[9] Pamphlets and treatises written in France and Germany prophesied the sudden appearance of hostile forces, or championed the railways' potential as a defensive resource even before widespread construction of lines had taken place.[10] Shortly before his death, General Lamarque declared to the French chamber of deputies that the strategic use of railways would lead to a 'revolution in military science as great as that which had been brought about by the

[6] McKenna, *The Railway Workers*, pp. 22–3.

[7] E. M. Spiers, *Engines for Empire: the Victorian Army and its use of Railways* (Manchester, 2015), p. 6; J. N. Westwood, *Railways at War* (London, 1980), p. 6.

[8] Spiers, *Engines for Empire*, p. 149.

[9] D. Stevenson, *Armaments and the Coming of War: Europe, 1904–1914* (Oxford, 1996), p. 15.

[10] For a brief discussion of the literature produced during this period, see E. A. Pratt, *The Rise of Rail Power in War and Conquest, 1833–1914* (London, 1916), pp. 2–8.

use of gunpowder'.[11] In *Die Eisenbahnen als militärische Operationslinien betrachtet und durch Beispiele erläutert*, published under the sobriquet of 'Pz.' in 1842 by Captain Karl Eduard Pönitz – like Lamarque a veteran of the 1813–15 campaigns – the author argued in favour of the employment of railways for military purposes before he elaborated upon a scheme for the construction of a network of strategic lines to serve the whole of Germany. The protection of Germany's frontiers from French or Russian attack was the central motivation behind Pönitz's proposal, and an examination of traffic and equipment within Germany and its neighbours was undertaken by the general staff in Berlin before 1850. However, as Denis Showalter observed, the advocacy of authors such as Pönitz and the relatively puny capabilities of early railways stymied the wholesale embrace of the medium within Germany until Prussian troops demonstrated the practical utility of railways during the counter-revolution of 1848–9. The Prussian forces 'crisscrossed Germany by rail despite the network's shortcomings' and the railways were integrated into the state's mobilization plans shortly after.[12]

Britain's situation as an island power afforded its railway promoters a degree of insulation from the strategic considerations that exercised French and German leaders, which reduced the efficiency of the British railway network from the point of view of national defence. Whereas the 1871 German constitution gave the military a standing right to supervise railway construction through the Imperial Railway Office, British soldiers lacked similar influence over – and their naval colleagues sufficient interest in – the provision of lines for principally strategic rather than economic roles. Concerns over a potential French invasion in the 1840s inspired a range of proposals for new railways to improve Britain's defences on the south coast. However, although the inspector-general of fortifications, Sir John Fox Burgoyne, affirmed in evidence to a royal commission that 'the whole safety of the kingdom' depended upon the nation's ability to concentrate men swiftly – and his stance was bolstered by Wellington's endorsement – the government declined to act.[13] At the opposite end of the country, Lord Fisher's decision to move the Grand Fleet to Scapa Flow in the 1910s was taken without consideration of the railway requirements of a modern navy. While Moltke the elder implored his countrymen to build railways rather than fortresses, '[n]othing whatever was done by the State to improve the land approaches to Scapa Flow' in the decade before the First World War.[14]

---

[11] Quoted in Pratt, *Rise of Rail Power*, pp. 6–7.
[12] D. E. Showalter, 'Soldiers and steam: railways and the military in Prussia, 1832 to 1848', *Historian*, xxxiv (1972), 242–59.
[13] Spiers, *Engines for Empire*, pp. 24–7.
[14] E. A. Pratt, *British Railways and the Great War: Organisation, Efforts, Difficulties and*

The Highland Railway, an unavoidable link between the Grand Fleet and London, remained single-tracked for more than three-quarters of its length; it also lacked sufficient siding space and employees to discharge the extra burdens thrown upon it by the flows of wartime traffic.[15]

Yet despite their failure to influence the composition of the British railway network, military experts played a considerable role in the industry's governance prior to the First World War. The unexpected popularity of the Liverpool and Manchester line among passengers prompted the spread of railways across the country. Between 1835 and 1837, parliamentary acts for the creation of fifty new lines – comprising 1,600 miles of track – were passed in the nineteenth century's first railway mania. By 1848 Britain possessed more than 13,000 miles of track, over which passed some £10 million worth of traffic.[16] As the network expanded the government began to take a more active role in its administration. The railways had, according to the vice president of the Board of Trade, Henry Labouchere, 'bound the country in chains of iron'.[17] However, the subject of government supervision of the railways divided opinion among the talents of the emerging industry. On the one hand, Isambard Kingdom Brunel warned that 'Government inspectors would receive no cooperation from railwaymen [who understood very well] … how to look after the public safety'. On the other hand, George Stephenson advised Labouchere to establish a railway department for which 'supervision without interference' could provide a guiding principle.[18] The 1840 Railway Regulation Act followed Stephenson's advice and authorized the Board of Trade to 'appoint any proper person or persons to inspect a railway'. To guard against potential abuse from those with vested interests the act disqualified anyone who had been a director of, or 'held any office of trust or profit' in, a railway company from being appointed as an inspector of railways.[19] Such stringent requirements rendered the vast majority of the small number of railway engineers in Britain ineligible for service with the railway inspectorate. Consequently, the Board of Trade turned to the

---

Achievements (2 vols., London, 1921), ii. 958–60; D. Stevenson, 'War by timetable? The railway race before 1914', *Past & Present*, clxii (1999), 163–94, at pp. 171–2.

[15] H. A. Vallance, *The Highland Railway: the History of the Railways of the Scottish Highlands* (5 vols., Newton Abbot, 1969), ii. 102–6.

[16] P. S. Bagwell and P. J. Lyth, *Transport in Britain: from Canal Lock to Gridlock* (London, 2006), p. 54; W. C. Lubenow, *The Politics of Government Growth: Early Victorian Attitudes toward State Intervention, 1833–1848* (Newton Abbot, 1971), pp. 108–9.

[17] Quoted in H. Parris, *Government and the Railways in Nineteenth-Century Britain* (London, 1965), p. 26.

[18] L. T. C. Rolt, *Red for Danger: a History of Railway Accidents and Railway Safety* (4th edn., Newton Abbot, 1982), pp. 17–18.

[19] *An Act for Regulating Railways* (Parl. Papers 1840 [97], xcvii).

Royal Engineers to supervise the expansion of Britain's railway network across the country and to investigate the causes of accidents on parliament's behalf. Beginning with Lieutenant-Colonel Sir Frederick Smith in 1840, from the 1840s until the 1960s every chief inspector of railways possessed a background in the Royal Engineers. Their relationship with the nascent industry was, in the words of one historian, 'characterised less by hostility than by signs of partnership'.[20]

Alongside the government, the railway companies themselves provided opportunities for ex-servicemen in the industry's first twenty years. Several of the early railways in Britain benefited from former military and naval officers' experiences in accounts, bookkeeping and the coordination of large bodies of men. Captain Mark Huish, who became the LNWR's general manager in 1846, was not an isolated case. The Manchester and Birmingham, London and Birmingham, Great North of England, Caledonian and Manchester and Leeds all claimed former captains and lieutenants among their senior staff prior to the era of the 'railway professional'. Huish's experience typified the transition. His tendency towards micro-management provided him with a remarkable knowledge of the growing industry, but produced friction that isolated him from his subordinate officers and the LNWR's directors.[21] George Neele, the LNWR's former superintendent, recalled in 1904 that managers with military backgrounds demanded 'quarter deck discipline … which compared unfavourably in public estimation with the business capabilities of the more practical and less pretentious Carrier'.[22] After twelve years at the company's helm Huish was replaced by William Cawkwell, a man who had begun his career as a clerk at Brighouse station in 1840 and risen through the Lancashire and Yorkshire Railway before moving to the LNWR in 1858.[23]

As the railway companies gradually erased their connections to individual soldiers, significant developments across the English Channel provided a catalyst that drew the railway industry and the army into closer contact. In 1858, a new rail link between Paris and Cherbourg opened, accompanied by a significant enlargement of the latter's capacity as a port. The provision of 'cannons, cannons, cannons, wherever you turned' at Cherbourg – in conjunction with the increased capacity of the line from Paris to the coast

[20] Lubenow, *The Politics of Government Growth*, p. 108.
[21] T. R. Gourvish, *Mark Huish and the London & North Western Railway: a Study of Management* (Leicester, 1972).
[22] G. P. Neele, *Railway Reminiscences* (London, 1904), p. 8.
[23] Gourvish, *Mark Huish*, pp. 54–5, 177–8, 259–60; 'Obituary: William Cawkwell, 1807–1897', *Minutes of the Proceedings of the Institution of Civil Engineers*, cxxix (1897), 398–400.

– inspired a reassessment of the national defences across the Channel.[24] In August 1859, a royal commission under General Sir Harry Jones was appointed by Palmerstone's new Liberal government to consider Britain's defences, in view both of recent improvements in naval technology and the global distribution of the Royal Navy's assets.[25] The commission's report opened with the stark admission that the navy's resources were 'insufficient' and 'could not … be entirely relied upon' to maintain naval superiority and eliminate the threat of foreign invasion. The commission considered the coastline from the Humber to Penzance to be at risk, and recommended that fortifications with an estimated cost of £11,850,000 be constructed across southern England and Wales to protect Britain's most important dockyards and arsenals.[26]

The challenges of national defence – coupled with the government's authorization of the formation of volunteer rifle corps for the first time since the threat of invasion had passed in 1814 – caught the attentions of the ICE. Recruitment for the Volunteer Force had been swift, despite the requirement that all members had to pay a subscription fee to enlist. Whereas the Volunteers of the Napoleonic Wars had no option but to march across country, the British railway network of the 1860s comprised thousands of miles of track upon which men, horses and equipment could be moved to any point of danger.[27] However, that track was owned and operated by a byzantine configuration of large and small companies, many with their own locomotives and rolling stock, rather than a single corporate entity. In anticipation of the complications likely to be encountered in the event that a large-scale movement of the Volunteers was required, the ICE's honorary secretary, Sir Charles Manby, wrote to the institution's council on 2 July 1860 to suggest the formation of a 'Volunteer Engineering Staff Corps for the Arrangement of the Transport of Troops and Stores, the Construction of defensive works and the destruction of other works, in case of Invasion'. A month later Manby proposed a new 'Volunteer Corps of Engineers', which could organize the railways when war threatened. The corps' officers, Manby suggested, could be drawn from the railways and would serve

---

[24] G. B. Sinclair, *The Staff Corps: the History of the Engineer and Logistic Staff Corps RE* (Chatham, 2001), p. 11.

[25] *Report of the Commissioners Appointed to Consider the Defences of the United Kingdom,* (Parl. Papers 1860 [2682], xxiii), p. x. On the Anglo-French rivalry in naval innovation during the 19th century, see B. Wilson, *Empire of the Deep: the Rise and Fall of the British Navy* (2014), pp. 483–93.

[26] *Defences of the United Kingdom,* pp. ix–x, xviii.

[27] C. E. C. Townsend, *All Rank and No File: a History of the Engineer and Railway Staff Corps RE, 1865–1965* (London, 1969), p. 2.

under the command of a member of the ICE's council. Sidney Herbert, the secretary of state for war, responded to Manby's proposal warmly and 'asked that a small committee of members of the Institution's Council and officers from the War Office be formed to consider the proposal fully'.[28]

Surprisingly, given the prevailing image of the Victorian army under the duke of Cambridge as a stagnant and conservative organization, the most difficult obstacles to the establishment of the new corps came from civilian rather than military figures. Some opposition to Manby's proposal came from members of the ICE's council while, despite the War Office's encouragement, 'the Railway Companies could not be brought to understand the necessity for, or the advantages of, the proposed system'.[29] A full four years elapsed before Manby acquired the names of twenty men who were willing to join him in the new corps, and a list of twelve civil engineers and nine general managers of railway companies was submitted to the War Office in November 1864. The Engineer and Railway Staff Corps (ERSC) was eventually formed the following January, and all twenty-one men were commissioned as lieutenant-colonels in the new corps. The ERSC's role was to:

> Secure unity and action throughout the Railway system of the United Kingdom in time of invasion to the end that troops and material may be transported in any required direction with certainty and the utmost rapidity ... [and to ensure] that works of construction and destruction in connection with railway communications which the exigencies of war may render necessary should be carried out with equal certainty and rapidity.[30]

The ERSC was thus conceived in 1865 in purely defensive terms, as a bulwark against invasion – a position it retained for the rest of the nineteenth century.

The absence of an invasion did not equate to a paucity of work for the new corps. In the twenty years that followed its formation the ERSC was presented with a range of scenarios by the War Office, and directed to prepare timetables for the transportation of varying numbers of troops to different locations around Britain. The specificity and scale of the exercises emphasized the recognized importance of the railway network both to the mass transportation of large bodies of troops and the sustenance of the national economy. For the ERSC's first exercise, presented to the corps on 3 April 1865, the scenario called for a swift movement of 280,000 men – drawn from across the country – to forestall an invasion between the Thames and the Wash. Alongside ensuring the movement of the troops

[28] Sinclair, *The Staff Corps*, p. 14.
[29] Townsend, *All Rank and No File*, p. 4.
[30] Sinclair, *The Staff Corps*, p. 15.

'with the utmost rapidity and certainty', the general managers of the railway companies were asked to 'give special consideration' to the maintenance of food supplies to London and other large towns 'which were wholly dependent on the railways for their daily supply'.[31] The complexity of the movements required to concentrate a force in East Anglia took over a year to disentangle, and when printed by the London, Chatham and Dover Railway the timetable linked to the exercise ran to 311 pages. Over a period of just eighty hours following the companies' receipt of an order to begin the move, the timetable called for the movement of 962 trains – more than double the number required for the BEF's embarkation in August 1914.

The ERSC's response to the War Office's hypothetical scenarios was time-consuming and expensive. The surviving documents make it impossible to accurately identify the number of man-hours employed in the creation of the timetables, but the costly nature of the activities undertaken by the ERSC was noted by Sir John Burgoyne shortly after the first exercise was complete. 'It is evident', Burgoyne wrote, 'that every Memoir and Paper … must have required much labour not only from the officer himself but his assistants – draughtsmen, clerks, etc., and that many were also the occasion for considerable other expense, such as for Travelling, Books, Maps and other matters'.[32] Burgoyne's comments demonstrate that even when movements were confined to internal distribution – and, therefore, did not involve the coordination of land and sea transport or consideration of many of the accoutrements that accompanied the BEF in 1914 – the military machine's requirements for transportation demanded an industrial response that could not be restricted to a tiny cadre of trusted individuals.

Changes in the international situation greatly affected both the nature of the scenarios that the ERSC was asked to consider and the urgency with which their responses were developed. The threat of a French invasion, which inspired the first three exercises issued to the corps between 1865 and 1870, receded in the wake of the Franco-Prussian War. Consequently, according to one author, the inspiration for the fourth exercise was the invasion of Britain by a German-speaking enemy as depicted in George Chesney's contemporaneous novella *The Battle of Dorking*. The scenario called for the concentration of over 100,000 fewer troops than had been the case in 1865, but discussions on the fourth exercise meandered on for four years before a timetable was produced.[33] A fifth and final exercise appeared in June 1882, after which the reduced threat of invasion eroded the principal

[31] Townsend, *All Rank and No File*, p. 7.
[32] Townsend, *All Rank and No File*, p. 8.
[33] Sinclair, *The Staff Corps*, pp. 21–3.

reason for the ERSC's existence. The corps was reduced to an establishment of sixty officers in August 1907 – the initial complement of twenty-one having ballooned after 1865 to over one hundred as entry to the corps was extended to new professions and more railway companies – and by 1910 the corps existed largely on paper rather than as a vibrant civil–military exchange.[34] According to Sir Sam Fay, an officer in the corps from June 1902 onwards, by the time war was declared in August 1914 the only function for which the ERSC met was an annual dinner at the War Office.[35]

One of its members may have perceived the corps to have been moribund by 1914, but the continuation of the annual dinner at the War Office emphasizes both the army's desire to maintain links with highly qualified civilian specialists and the social status ascribed to voluntary military service by the senior executives of the private firms. (The modern-day Engineer and Logistics Staff Corps retains the same voluntary ethos.)[36] Furthermore, the reduced workload assigned to the ERSC did not mean that the professional link between the War Office and Britain's transport experts was severed after the final timetable was submitted in March 1885. Instead, the working relationship between the military and the railways migrated to a specialized forum that – like the ERSC before it – owed its birth to the formation of war clouds over Europe.

The REC, eventually formed in the wake of the Agadir crisis of 1911, underwent a long gestation. The ERSC's first call for a 'central and responsible authority' to be created for the purpose of coordinating the military's railway needs was made in 1870, and repeated in response to the fourth mobilization exercise in 1876.[37] However, the reduced frequency with which the corps met after the Franco-Prussian War meant that the question lapsed for more than a decade until it was rekindled by the War Office in July 1888. An initial memorandum on the subject of railway coordination in the event of a national emergency was produced the following year by Sir George Findlay, the LNWR's general manager and a lieutenant-colonel in the ERSC. Findlay's paper opened with a clear statement in support of the corps' involvement in railway operations in the event of a war, and emphasized that a portion of the ERSC's council should sit *en permanence* at the War Office.[38] The pace of change was slow. Eight years passed before a civil–military committee composed of Frederick Harrison, Sir Henry

---

[34] TNA, WO 114/114, territorial force: establishment and strengths, 1908–1914.

[35] S. Fay, *The War Office at War* (London, 1937), p. 39.

[36] M. Stancombe, 'The Staff Corps: a civilian resource for the military', *ICE Proceedings*, clvii (2004), 22–6.

[37] Townsend, *All Rank and No File*, p. 17.

[38] Sinclair, *The Staff Corps*, p. 28.

Oakley, George Henry Turner and Sir Charles Scooter – general managers of the LNWR (Findlay died in March 1893), Great Northern, Midland and London and South-Western (LSWR) railways respectively – was established to advise the secretary of state for war on railway matters in the event of a national emergency.[39] All four were lieutenant-colonels in the ERSC and advocated that their role, alongside providing a permanent council to examine railway matters on the secretary of state's behalf, should be to draw up 'a detailed scheme for the movement of the different troops on mobilization on [sic] data supplied to them by the War Office'.[40]

The Permanent Railway Council, later known as the Army Railway Council and subsequently the War Railway Council, met in full just four times between 1897 and 1910. However, it established the principles upon which the British army and the railway companies interacted before the First World War and undertook voluminous quantities of statistical work on the War Office's behalf.[41] From the council's inception all communications between it and the army were directed through the Quartermaster-general's (QMG) office to avoid duplication of effort, and until 1903 the council's railway representatives were drawn solely from among the companies with headquarters located in London.[42] In 1903 the comparative insularity of the council was rectified, and membership was widened to acknowledge both the Royal Navy's mobilization requirements (which necessitated the name change from Army Railway Council to War Railway Council) and the presence of important railways based outside the capital. The North-Eastern Railway, whose head office overlooked York station, and the Manchester-based Lancashire and Yorkshire Railway both employed huge workforces and operated large shipping fleets in addition to their railway interests. Yet neither had been represented on the Army Railway Council. The railways of Scotland and Ireland, both crucial to the mobilization procedures of the army and navy, were similarly notable absentees. The War Railway Council's expansion amended these deficiencies: the North-Eastern and Lancashire and Yorkshire railways were added to the English contingent, while Robert Millar and Henry Plews – general managers of the Caledonian and Great Northern Railway

[39] TNA, WO 33/56, War Office: reports, memoranda and papers (O and A series), report of committee assembled to consider working of railways of Great Britain and Ireland in the event of general mobilisation, 22 May 1896.

[40] TNA, WO 33/56, report of committee assembled to consider working of railways of Great Britain and Ireland, p. 2.

[41] Pratt, *British Railways and the Great War*, i. 13.

[42] TNA, WO 32/9184, formation of permanent council to advise on railway matters, Wood to Brodrick, 15 Apr. 1897.

(Ireland) respectively – were appointed to represent the Scottish and Irish railways.[43]

Alongside expanding the council's membership, the 1903 reorganization extended the scope of its deliberations. For the first time the naval, military and railway elements of the mobilization process could be coordinated, although considerations of national security demanded that only the railway companies directly affected by any contemplated manoeuvres were invited to attend council meetings. Each of the companies represented on the council nominated employees to coordinate the technical work involved in the creation of railway timetables and to liaise with the other companies concerned with the movements of individual trains. Through these methods, timetables for home defence that identified the numbers of troops, horses, vehicles and stores to be moved, the departure and arrival locations for each train involved and the date after mobilization upon which the movement was to take place were compiled and submitted to the War Office within two years of the War Railway Council's establishment. Following a process of consultation and amendment between the War Office and the railways – one that was mirrored after 1911 in the production and revision of the BEF's mobilization scheme – by 1909 the railway companies were able to prepare what they regarded as 'mobilization timetables proper' for the national defence.[44]

In addition to providing the environment in which a comprehensive mobilization scheme could be devised, the War Railway Council's expansion presented opportunities for a forensic examination of the possible effects of modern warfare upon Britain's economic prosperity. As the world's largest exporter and importer, pre-war Britain lay at the heart of a vast maritime communications network. Britain's ports played a central role both in the delivery of British products to the globe and the world's foodstuffs to Britain. Any dislocation to the traffic flows in and out of the country had potentially profound implications for the flows of traffic around the British transport network. On 19 January 1909 the ship owner and Lloyd's chairman, Sir Frederic Bolton, wrote to the prime minister about Britain's food security in the event of war with Germany. Bolton warned Herbert Asquith that 'a suspension of imports [from ports on the North Sea or Channel coast] would divert trade to ports farther from the danger zone, which are not used or adapted to handling import trade. A difficulty would therefore arise in supplying the area normally dependent on the port into which trade had temporarily ceased to flow'.[45]

---

[43] TNA, WO 32/9185, reorganisation of council, Lake to Nicholson, 21 Oct. 1902; Clarke to Harrison, 16 Jan. 1903; the ARC, 7 Feb. 1903.

[44] Pratt, *British Railways and the Great War*, i. 14–15.

[45] Nuffield College Library (NCL), papers of John Edward Bernard Seely, Lord Mottistone,

Bolton's letter to the prime minister did not represent a serendipitous alignment between the state and private enterprise. Instead, it emerged from another example of the professional union between the armed forces and Britain's transport experts. As chairman of Lloyd's, Bolton had been part of a special committee of prominent figures in the shipping industry that assisted with the organization of the Admiralty's naval manoeuvres in 1906. The manoeuvres had been designed 'with a view to studying the important question of the Attack and Defence of Commerce' at sea, and called for the participation of privately owned vessels alongside the Royal Navy's forces. Between December 1905 and June 1906 Bolton worked out many of the details required in relation to the insurance of the commercial vessels, presented ship owners with an indemnity package to cover their involvement and helped to 'weed out' unfit vessels within those offered to the Admiralty for use in the manoeuvres. His work for the committee earned him a knighthood and sparked an interest in the potential dislocation of shipping that inspired his 1909 letter to Asquith.[46]

Bolton's observations were sufficiently concerning to compel the Committee of Imperial Defence (CID) to investigate further, and Bolton joined representatives of the Admiralty in an examination of the challenges involved in the distribution of food supplies and raw materials around Britain in the event of war with another maritime power. The conclusions of their interim report, issued by the CID in December 1909, were pessimistic. 'The problems involved' in an investigation of such magnitude and complexity, Bolton began, were 'many and far reaching':

> The range of the influence of the principal ports must be determined, both in general and in particular; the population dependent on the ports for supplies; the interchange of both home and foreign produce between different parts of the Kingdom by land and water, especially the latter; the quantity of imports and exports for each port, in weight and cubic measurement; the possibility of securing the needed supplies from fresh directions; the ability of the railway companies to provide the rolling stock that would be required; the capacity of the lines to take the large increase of traffic that would be thrown upon them; the arrangements that would be necessary as regards labour; the difficulties of distribution within the area itself under entirely new conditions; and many other points, fresh ones constantly presenting themselves as the investigation proceeds.[47]

Mottistone 11/6, sub-committee to consider the desirability of an enquiry into the question of local transportation and distribution of food supplies in time of war, Appendix 1: note submitted by the secretary for the consideration of the sub-committee, 17 Jan. 1910, p. 5.

[46] S. Cobb, *Preparing for Blockade 1885–1914: Naval Contingency for Economic Warfare* (Farnham, 2013), pp. 189, 194–204, 222–3.

[47] NCL, Mottistone papers, Mottistone 11/6, sub-committee, Appendix 2: preliminary

The movement of coal provided just one example of the difficulties that might have to be addressed in the event of wartime interference with Britain's maritime traffic. In 1908 Greater London had received 8.1 million tons of coal by rail. Over the same period just over eight million tons arrived in the capital by ship, the majority of which was despatched from ports on the north-east coast.[48] In the event of a prolonged closure to traffic of the North Sea, Britain's railways would be required to provide a vast quantity of rolling stock to undertake journeys far in excess of the distances typically covered by goods trains on the British railway network. As John Armstrong's analysis of freight traffic in 1910 has demonstrated, the average distance of a goods haul on the railway network was less than fifty miles. South Shields, the port of origin for roughly one-quarter of London's coal in the same year, was almost three hundred miles away from the capital.[49] Bolton remarked in his preliminary report that Britain's strategic planners had hitherto given 'no thought' to the requirements of the civil population, and stressed that the naval, military and civilian interests of the nation had to be dealt with in concert.[50] At the end of his year-long investigation, Bolton expressed doubts as to whether the railway companies could 'cope with the extra strain that would be thrown on them in time of war'.[51] In response, a further CID sub-committee was established to ascertain whether – given the scenario that all ports from Hull in the north-east, past the Thames estuary, and as far along the south coast as Portsmouth were closed to traffic – the railways of Britain could adapt sufficiently to ensure the delivery of adequate supplies of food and raw materials to the nation's urban centres.

Such a detailed and complex investigation demanded expert contributions. The railway companies provided them. In January 1911, the general managers of companies involved in the supply of London and

---

report by Sir Frederic Bolton on an investigation into some of the conditions of supplies of commodities to and from the United Kingdom, and, in particular, as to how these would be affected by any interference with the trade of our ports in time of war, and the measures which might be taken to avert, or deal with, the difficulties which would arise under such circumstances, 7 Dec. 1909, p. 7.

[48] See Table 1 and explanatory notes in J. Armstrong, 'The role of coastal shipping in UK transport: an estimate of comparative traffic movements in 1910', *Journal of Transport History*, viii (1987), 164–78, at pp. 167–8; NCL, Mottistone papers, Mottistone 11/7, interim report of the sub-committee (CID) on the local transportation and distribution of supplies in time of war, 24 Jan. 1911, p. 6.

[49] Armstrong, 'The role of coastal shipping in UK transport', pp. 166–8.

[50] NCL, Mottistone papers, Mottistone 11/6, sub-committee, Appendix 2: preliminary report by Sir Frederic Bolton, p. 12.

[51] NCL, Mottistone papers, Mottistone 11/6, sub-committee, Report of the sub-committee, 22 March 1910, p. 3.

the transport of goods to and from Britain's major commercial ports were instructed to consider how best the railways could address the hypothetical challenge placed before them. The sub-committee presented its findings eight months later, during a summer in which Franco-German disputes over Morocco illustrated the fragility of European peace and a railway strike at home emphasized the industry's importance to British economic life.[52] Fay, the Great Central's general manager and one of the sub-committee's participants, recalled that:

> We had to take into consideration the fact that the closing of the ports on the eastern coast would greatly increase the demands on the Liverpool and Manchester Docks in dealing with foodstuffs normally supplied through Hull and Grimsby to the populous districts of the North-East of England. We calculated that the situation could be met by the terminal facilities of Southampton, Bristol, Liverpool, Birkenhead and Manchester, but pointed out that if large movements of troops and material took place concurrently with the demand for the conveyance of increased provisions to London [as would inevitably be the case were the BEF to be despatched to the continent], congestion would occur.[53]

The sub-committee's report contained a stark conclusion: if no effective arrangements were in place to connect London by rail to the ports on the southern and western coasts, then 'famine prices would soon be reached' in the capital due to lack of supplies. To minimize the risks of dislocation and congestion on the railway network – and to ensure that London's shops remained stocked and its factories fuelled – the sub-committee recommended that 'the General Managers who have already been consulted be formed into a Permanent Committee with power to add to their number'. This committee, it continued, should be authorised to control movement over the railway network as a whole in the event of an emergency – to ensure that government priorities received the immediate attention of the railway companies. Fay and his colleagues stressed the importance of ensuring that the 'controlling body should be in close touch with the military and naval authorities in order that military movements by railway should not clash with special working for provisioning London, and that the same carts and horses should not be requisitioned by both bodies'.[54]

---

[52] C. M. Clark, *The Sleepwalkers: How Europe Went to War in 1914* (London, 2012), pp. 204–13; Pratt, *British Railways and the Great War*, i. 31–2.

[53] Fay, *The War Office at War*, p. 18.

[54] NCL, Mottistone papers, Mottistone 11/175, report from the general managers of the Great Central, Great Northern, Great Western, London and North-Western, London and South-Western and Midland Railway Companies to the Right Hon. Lieutenant-Colonel J. Seely, on the provisioning of London in the event of war, 1 Aug. 1911, pp. 2–3.

The report's final observation revisited a familiar theme, as it diverged little from the request first made some forty years earlier by the ERSC:

> During the course of this enquiry we have been impressed by the desirability of having some central body at which matters from time to time referred to railway companies by various government departments may be considered as a whole. At present it frequently happens that some question is referred to the railway companies by, e.g., the War Office, and when this has been dealt with, some other question is referred by some other department which alters the standpoint from which the first question should be considered.[55]

The contents of the statements made by Britain's transport experts in the 1870s and 1910s may have been similar, but the contexts in which they were made differed profoundly. France's humiliation in 1871 had reduced the imminent threat of invasion in the minds of Britain's military authorities. The entente cordiale of 1904 had removed the French invasion threat entirely, and set in motion a concerted effort between the militaries of France and Britain to understand each other's working practices. The Agadir crisis in 1911 provided the catalyst for a reinforcement of Britain's military commitment to the French army – solidified by the Wilson–Dubail memorandum on 21 July, which stated that six British infantry divisions, one cavalry division and two mounted brigades would be deployed on the left flank of the French army by the fifteenth day of mobilization.[56] The potential for the railways to be called upon to effect the swift movement of 150,000 men and 67,000 horses to the coast, the provision of adequate stocks of coal for the nation's fleet and factories, and the preservation of goods and commuter services across the country simultaneously offered compelling reasons for the establishment of a permanent link between the government, the armed forces and Britain's transport experts.

The REC as eventually formed in November 1912 reflected the government's acknowledgement of the unique skills possessed by Britain's transport experts. Stanley Baldwin, the president of the Board of Trade, was named the committee's nominal chairman, but responsibility for the operation of the railways in the event of war was unambiguously left in the hands of the professionals. At the REC's first meeting Baldwin emphasised that the government 'had no idea of running the railways themselves. All they wanted was to place the State in such a position that it would be able to give binding instructions and to require separate railways to cooperate as part of a single system … The control of the system would be vested in the

---

[55] NCL, Mottistone papers, Mottistone 11/175, report from the general managers, p. 3.
[56] Clark, *The Sleepwalkers*, p. 213.

... Executive Committee'.[57] In addition, the establishment of a permanent body comprising the general managers of the country's most prominent railways afforded the War Office and Admiralty with the opportunity to educate the companies 'in matters relating to naval and military transport before any emergency arose'.[58] To prevent confusion and duplication of effort the War Railway Council was disbanded in February 1913, and a communications board was established to coordinate the activities of each party in the REC.[59]

The composition of the communications board illustrated the importance of the railways to an industrial nation on a war footing. The railwaymen – in the first instance comprising the general managers or nominated representatives of the LNWR, Midland, Great Western, North-Eastern, Lancashire and Yorkshire, Great Central, Great Northern, LSWR and Caledonian railways – were joined on the board by prominent individuals with executive responsibilities in their respective departments of state. The QMG, Sir John Cowans, was chairman, and other members of the board included: Sir Edward Troup, the permanent under-secretary of state at the Home Office; Rear Admiral Alexander Duff, the director of the Admiralty war staff's mobilization division; Brigadier-General D. Henderson, the director of military transport; and William Marwood, assistant secretary of the Board of Trade's railway department.[60] The board's members were officials 'of such high standing that ... they were able to decide on their own authority, most of the matters that the Railway Executive Committee ... brought up for consideration. In this way a large number of questions, including many of the highest importance, were settled in advance'.[61]

The fitting out of the REC's offices provides an example of the small but important details that the board grappled with in the final months of peace. Sir Frank Ree, general manager of the LNWR and the REC's first acting chairman, argued at the board's first meeting that 'ample offices' were required for the REC and its staff to coordinate the programme of movements required on the declaration of a general mobilization. The offices, he suggested, needed to be 'connected by telephone directly to the government departments concerned and with each railway'. The board concurred, and Marwood 'undertook to bring the matter before the

[57] Pratt, *British Railways and the Great War*, i. 41–2.
[58] Pratt, *British Railways and the Great War*, i. 39.
[59] TNA, WO 32/9188, re-constitution of council, proceedings of communications board, Brade to Cooper, 13 Feb. 1913.
[60] TNA, WO 32/9188, re-constitution of council, Communications Board.
[61] Pratt, *British Railways and the Great War*, i. 44.

Board of Trade'.[62] Two months later Marwood reported that the General Post Office was 'preparing an estimate for the installation of the suggested telephone connection' with the LNWR, whose Westminster office had been identified as the most suitable location for the REC's headquarters.[63] By November 1913 the work was 'practically complete', ensuring that the REC entered the mobilization period before the First World War with a direct connection both to the latest requirements of the state and the situation across the British railway network.[64]

By drawing together Britain's transport experts, the armed forces and key government departments, the REC's communications board provided an opportunity for each group to gain valuable insights into the others' concerns and requirements. Close and frequent access to senior naval and military authorities ensured that senior figures in the railway industry were kept informed of the armed forces' evolving transport requirements, while contact with the railway companies' senior executives allowed government and military officials to develop a clearer understanding of the national network's possibilities and limitations. Yet in the decade prior to the First World War these cross-sectoral exchanges of knowledge were restricted neither to those at the pinnacle of their chosen professions, nor solely to the development of plans for national defence. The railway industry played an active role both in the reform of British government and the education of Britain's military transport authorities.

The transmission of transport expertise

In a 1973 article, Terry Gourvish analysed the origins, careers and social statuses of the chief executives of Britain's fifteen leading railway companies between 1850 and 1922. His study revealed that 'the economic position of the chief executives … showed a steady improvement as the industry developed, and this improvement was in time reflected in a higher social status'. By 1900 the senior managerial figures who went on to populate the REC had already broken down the barriers that separated the 'paid official from the big business capitalist', a process that created opportunities for the ambitious railway manager to 'involve himself in a wide range of industrial,

---

[62] TNA, WO 32/9188, re-constitution of council, notes of proceedings of the first meeting held at the Board of Trade at 11:30 a.m., 30 May 1913, p. 3.

[63] TNA, WO 32/9188, re-constitution of council, notes of proceedings of the second meeting held at the Board of Trade at 3:30 p.m., 22 July 1913, p. 2; J. A. B. Hamilton, *Britain's Railways in World War I* (London, 1947), p. 24.

[64] TNA, WO 32/9188, re-constitution of council, notes of proceedings of the third meeting held at the Board of Trade at 3 p.m., 4 Nov. 1913, p. 2.

commercial, and even governmental activities'.[65]

Sir George Gibb exemplified the latter of these endeavours. As general manager of the North-Eastern Railway and then managing director of the Underground Electric Railways Company of London, Gibb received his knighthood in acknowledgement of his contributions to two royal commissions in the first decade of the twentieth century. In the first, Gibb sat alongside prominent figures from the political and military spheres, and examined the organization of the War Office following the army's bleak performance at the beginning of the South African War. In addition to hearing evidence from senior military and political figures with obvious links to the War Office's operations, the committee obtained information from the Admiralty – about the working practices of the Royal Navy – from 'railway companies, from important manufacturing companies, and from large cooperative societies with reference to their business procedures'.[66] In the second, Gibb joined a panel of parliamentarians, civil servants and selected experts – including his compatriot on the ERSC, the civil engineer Sir John Wolfe Barry – to investigate potential improvements to London's transport network.[67] Over the course of two years the committee held 112 meetings, interviewed 134 witnesses and visited the United States and 'various continental cities' to observe measures taken elsewhere to deal with the rapid growth in demand for urban transportation.[68]

Gibb's inclusion on such committees guaranteed that their deliberations received the input of a man with considerable experience in an industry that had faced the administrative challenges of large-scale organization – which had graphically taxed the extant military bureaucracy in their struggle against the Boers – and one that had 'led the way in developing relatively advanced techniques in business management'.[69] Yet Gibb's contribution to the enhancement of Britain's military administration was far from unique among his peers in the railway industry. At both an organizational and a

---

[65] T. R. Gourvish, 'A British business elite: the chief executive managers of the railway industry, 1850–1922', *Business History Review*, xlvii (1973), 289–316, at pp. 315–16.

[66] *Report of the Committee Appointed to Enquire into War Office Organisation* (Parl. Papers 1901 [Cd. 580], xl), p. 1. The companies consulted were the Army and Navy Co-operative Society; the Civil Service Co-operative Society; Armstrong, Whitworth and Company; the Midland Railway; the London and North-Western Railway; Vickers, Sons and Maxim; Rylands and Sons; the Great Northern Railway; and the Co-operative Wholesale Society. See *Minutes of Evidence Taken before the Committee Appointed to Inquire into War Office Organisation, Together with Appendices, Digest, and Index* (Parl. Papers 1901 [Cd. 581], xl), pp. 443–9.

[67] *Royal Commission on London Traffic* (8 vols., London, 1905–6).

[68] 'Royal commission on London traffic', *Commercial Motor*, 20 July 1905, p. 12.

[69] Gourvish, 'A British business elite', p. 290.

technical level, Britain's transport experts engaged in activities designed to improve the army's ability to operate efficiently in the years preceding the First World War. These links were particularly valuable in Britain as, unlike those of other European powers, the pre-war British army relied upon voluntary enlistment for its supply of officers and other ranks. The abolition of the purchase system in 1871 had been partly conceived to encourage men from middle-class backgrounds, 'by whose energy the industrial system' was maintained, to consider the army as a suitable profession.[70] However, the expanding influence and growing number of learned professions offered more attractive career prospects for the most talented public school and university graduates in the latter part of the nineteenth century. As the twentieth century began, the regular army remained dependent upon what Bidwell and Graham dubbed the 'left overs' of the landed classes for its supply of officers and unable to compel trained railwaymen to enlist in the ranks.[71] Consequently, while the continental armies enjoyed largely unfettered access to men with relevant technical and administrative skills – and could direct them into positions where their experience could be exploited most effectively – Britain's industrious and innovative young men were not drawn to the military life.

The effective use of military railways depended upon the planning of train schedules, an understanding of speeds and carrying capacities, the arrangement of loading and unloading facilities, the coordination of movements with the civil railway authorities and the knowledge of myriad technical details regarding locomotives, rolling stock and the infrastructure that comprised a railway.[72] To acquire the necessary expertise to handle such demands the Prussian army established a specialist section of railway troops in 1886, which comprised over 4,500 troops by 1900 and provided a cadre for the rapid (and vast) expansion of the railway section through the enlistment of civilian railwaymen upon the outbreak of war.[73] Following the shock of the Franco-Prussian War the French had comprehensively reorganized their railway administration. They created joint commissions of military and technical officers able to coordinate the national network

[70] *Report of the Commissioners Appointed to Inquire into the System of Purchase and Sale of Commissions in the Army* (Parl. Papers 1857 [2267], xviii), p. 293.

[71] S. Bidwell and D. Graham, *Coalitions, Politicians and Generals: Some Aspects of Command in Two World Wars* (London, 1993), p. 14.

[72] G. L. Herrera, 'Inventing the railroad and rifle revolution: information, military innovation and the rise of Germany', *Journal of Strategic Studies*, xxvii (2004), 243–71, at p. 253.

[73] M. Peschaud, *Politique et fonctionnement des transports par chemin de fer pendant la guerre* (Paris, 1926), pp. 37–41.

in the event of war, and drew their dedicated railway troops from among new recruits with experience on the railways, existing soldiers who wished to train in railway duties and from lists of employees supplied by the six largest railway companies in France.[74] The British army did not possess a regular force of railway troops until the formation of the 8th Company Royal Engineers in 1882, but did receive substantial voluntary support from the nation's largest railway company five years later. In 1887 the LNWR's locomotive works in Crewe provided 6,000 officers and other ranks for the 2nd Cheshire (Railway) Engineer Volunteers, the majority of whom were 'engine drivers, firemen, cleaners, boilermakers, riveters, fitters, smiths, platelayers, shunters, and pointsmen'.[75] Like all volunteers the railwaymen were only liable to be called out in case of invasion. Yet in addition to their ordinary infantry drill, the Crewe volunteers underwent a course of instruction in military engineering and – in recognition of their potential value to an army on campaign abroad – each man was encouraged to enlist in the Royal Engineers for a day before being placed on the army reserve for six years. As reservists the men received pay and made themselves available for service 'in case of need either at home or abroad'. By 1890 a 'considerable number of men' had taken advantage of the offer, and the Crewe volunteers served with distinction in South Africa between 1899 and 1902.[76]

However, the recruitment of volunteers did not provide the British army with a large, permanent pool of technically proficient railway personnel. Therefore, in addition to encouraging the enlistment of their men on the reserve list, the railway companies contributed to the practical education offered to regular soldiers. The Midland Railway provided both opportunities for officers to gain experience of the engineering work undertaken by one of Britain's largest private enterprises and the blueprint for a self-contained course organized by the Royal Engineers at the Chatham Dockyards.[77] At the Midland's gigantic locomotive works in Derby a voluntary course of

[74] Herrera, 'Inventing the railroad and rifle revolution'; F. P. Jacqmin, *Les chemins de fer pendant la guerre de 1870–1871: leçons faites en 1872 à l'École des Ponts-et-Chaussées*, (2nd. edn., Paris, 1874); Pratt, *Rise of Rail Power*, pp. 122–3, 154–5; V. Murray, 'Transportation in war', *Royal Engineers Journal*, lvi (1942), 202–32, at pp. 204–7.
[75] W. H. Chaloner, *The Social and Economic Development of Crewe, 1780–1923* (Manchester, 1950), p. 273.
[76] G. Findlay, *The Working and Management of an English Railway* (London, 1889), pp. 287–8; G. R. S. Darroch, *Deeds of a Great Railway: a Record of the Enterprise and Achievements of the London and North-Western Railway Company during the Great War* (London, 1920), pp. 22–3.
[77] TNA, WO 32/6164, instructions for officers while undergoing training in mechanical engineering at Chatham, instructions for officers while undergoing a course of instruction in mechanical engineering at H.M. dockyard, Chatham, May 1901.

instruction in mechanical engineering was offered to men who wished to enhance their knowledge of the machines deployed in support of a campaigning army. The course's existence demonstrated a growing awareness within the military of the importance of mechanical appliances to the fighting troops and the increasing complexity of the machines an army depended upon. The instructions issued to officers who took the course explained that it had been designed to give each man 'a thorough practical knowledge of machine design, the fitting, erection, and repair of machinery, and the care and working of boilers, such as will enable him to superintend work of these descriptions, and distinguish between good and bad material, workmanship, and design'.[78] Following completion of the course officers were employed 'upon machinery in the course of erection by the War Department' so that the army could make the best use of their newly acquired skills.[79]

In the case of Ralph Micklem, the Midland Railway provided the foundations upon which he constructed a successful military career. After eighteen months at the Royal Engineers' school of military engineering in Chatham, Micklem applied to specialize in the corps' railway section for 'no particular reason' in September 1904. Writing later in life, Micklem reasoned that his cousin Henry had 'gravitated towards the railway side in Sudan and South Africa and had done well in both countries'.[80] Micklem's initial experiences were less exotic than his cousin's, as his training with the Midland began with a 'fortnight at Brecon on a single line, then two or three months in London on goods working, then to Derby, where I did a month as a fireman, and then to various other places on civil engineering jobs. Altogether', he summarized, 'it was an enjoyable year'. Following a brief spell on the Royal Engineers' new instructional railway at Longmoor – opened in 1905 to provide soldiers with the opportunity to construct and operate railway lines in peacetime – Micklem followed in his cousin's footsteps and departed for Africa, where he joined the Egyptian army and participated in the survey of a possible line to link the Nile and Congo rivers. The railway was never built, but Micklem was involved in the construction of the Atbara to Port Sudan line in 1911 before he spent a 'pleasant' three

---

[78] TNA, WO 32/6164, instructions for officers, instructions for officers joining the Midland Railway Company's locomotive works for a course in mechanical engineering, 24 Aug. 1894, p. 1.

[79] TNA, WO 32/6164, instructions for officers, instructions for officers joining the Midland Railway Company's locomotive works, p. 2.

[80] IWM, private papers of Brigadier R. Micklem, 87/8/1, Ralph Micklem – an autobiography. Unless otherwise stated, all quotations in this passage are taken from this source.

years as assistant to the general manager at Atbara. In 1915 Micklem took charge of a company of the Egyptian Railway Battalion and was wounded at Cape Helles. Following recuperation in London he was passed fit for light duties and, thanks to his previous experience, was 'snapped up' by the movements directorate at the War Office. He spent the remainder of the war in London, where he was engaged in 'very technical work with the home railway companies'. The outcome of Micklem's seemingly impulsive decision to attend the Midland's training course in 1904 was that he became responsible for the railway arrangements connected to the defence of Britain and, despite having reached the compulsory retirement age of fifty-five in 1939, he played a prominent role in the directorate of transportation throughout the Second World War.[81]

The on-site vocational training offered by the Midland Railway, and later by the Royal Engineers themselves at Longmoor, were not the only examples of professional development available to military personnel who wanted to better understand the complexities of railway operations prior to the First World War. Alongside the practical experience delivered by a combination of civilian and military practitioners, the advent of the Liberal government in 1906 inspired the creation of an academic course that provided its students with a wider appreciation of the business methods that underpinned the railway industry. The incoming secretary of state for war, Richard Haldane, entered the War Office with the twin aims of promoting 'military efficiency' and reducing defence spending foremost in his mind.[82] A trip to Berlin during his first year in office gave him a chance to study the German general staff's organization in detail, and exposed him to an army that he considered to be 'as near perfection as possible, and at a cost proportionately much less than ours'.[83] Haldane's admiration for the German army remained high and was expressed once again in the infamous CID meeting of 23 August 1911, when the army and navy presented their plans for British intervention in the event of a European war. On that occasion he referred to the German military as 'a perfect machine'.[84] Haldane was

---

[81] Micklem was joined at the War Office in 1917 by his cousin Henry, who was responsible for 'the supply of material for railways, light railways and roads including the supply of special road-making, maintenance and repairing equipment, plant and materials'. According to their boss, Sir Sam Fay, Henry's workload was 'A heavy business!' See Fay, *The War Office at War*, pp. 46, 146.

[82] E. M. Spiers, *Haldane: an Army Reformer* (Edinburgh, 1980).

[83] Haldane's own account of this trip is given in R. B. Haldane, *Richard Burdon Haldane: an Autobiography* (London, 1929), pp. 200–9. Unless otherwise stated, all quotes in this passage are taken from this source.

[84] TNA, CAB 2/2, nos. 83–119, action to be taken in the event of intervention in a European war, 23 Aug. 1911, p. 7.

particularly struck by the degree of specialization in the German army. The general staff took no part in the administration and supply of the forces, a separation that left 'the army in the field free from the embarrassment of having to look after its transport and supplies'. The new secretary of state wished to implement the same partition within the British army, and envisaged a thorough reformation of the administrative staff tasked with providing logistical support to the fighting troops. As Hew Strachan noted, such an organization was particularly relevant to the British army due to the nature of Britain's imperial responsibilities – the planning for which predominantly required the fulfilment of tasks that were 'administrative and logistical' rather than the outcome of operational thought.[85]

To assist him in his goal of building an administrative organization composed of highly skilled experts – a 'thinking school' of officers – Haldane drew upon the knowledge and expertise of men from inside and outside the military profession. In January 1907, just a year after Haldane had become secretary of state for war, the first cohort of students enrolled on a course for the training of officers for the higher appointments in the administrative staff of the army at the London School of Economics (LSE).[86] The importance attached to such a training course is evident in the speed with which it was established, and owed much to the work of two men: Sir Edward Ward, the permanent under-secretary at the War Office and a former colonel in the Army Service Corps (ASC); and Halford Mackinder, the LSE's director.[87] The goal of the course they devised was the creation of a pool of officers who possessed a thorough knowledge of the principles required to run what Mackinder termed the 'greatest single business concern in the country'.[88] In time, as the officers who passed the course obtained promotions to senior positions within the logistics and supply departments of the army, Mackinder hoped that the course would

---

[85] Spiers, *Haldane*, p. 151; H. Strachan, 'The British army, its general staff and the continental commitment, 1904–1914', in *The British General Staff: Reform and Innovation, c.1890–1939*, ed. D. French and B. Holden Reid (London, 2002), pp. 75–94, at p. 87.

[86] G. Sloan, 'Haldane's Mackindergarten: a radical experiment in British military education?', *War in History*, ixx (2012), 322–52, at p. 328.

[87] P. Grant, 'Edward Ward, Halford Mackinder and the army administration course at the London School of Economics, 1907–1914', in *A Military Transformed? Adaptation and Innovation in the British Military, 1792–1945*, ed. M. LoCicero, R. Mahoney and S. Mitchell (Solihull, 2014), pp. 97–109.

[88] *Army. Report of the advisory board, LSE, on the first course at the LSE, January to July, 1907, for the training of officers for the higher appointments on the administrative staff of the army and for the charge of developmental services* (Parl. Papers 1907 [Cd. 3696], xlix), p. 11; S. Pelizza, 'Geopolitics, education, and empire: the political life of Sir Halford Mackinder, 1895–1925' (unpublished University of Leeds PhD thesis, 2013), pp. 117–18.

develop a tradition of its own – one that placed its graduates on a similar footing to those who passed through the staff college at Camberley. To ensure that the army accrued a long-term benefit from the material studied at the LSE, the age limit for entrants to the course was set at thirty-seven.[89]

The LSE course was designed to teach a new generation of officers the skills required to manage and operate a large-scale, data-intensive, complex organization. The first cohort of students studied subjects that disseminated lessons learned in the 'practical experience of recent campaigns, which had demonstrated the need for specialised administrative officers whose training should include financial, commercial and legal qualifications'.[90] The breadth of knowledge considered of importance to the British army's administrators can be deduced from the course's syllabus. Instruction was delivered in the following topics: accounting and business methods; commercial law; carriage by sea and land; economic theory; economic geography; and statistical method, and each class was taught by prominent academics or men with significant practical experience. Staff who contributed to the delivery of modules before the First World War included the statistician Arthur Bowley; the University of Birmingham's former professor of accounting, Lawrence Dicksee (who provided a colossal sixty lectures in the first year of the course's existence); the Allied Marine Assurance's Douglas Owen; and the railway expert and former North-Eastern Railway employee, Wilfred Tetley-Stephenson.[91] The teaching programme was supplemented by a sequence of informal after dinner 'smoking meetings', which included lectures provided by specially invited business leaders – referred to as 'practical men'. The guest lecturers in 1907 included Sidney Webb, who discussed the organization of trade unions; Thomas Brassey, who spoke about his role as the managing director of 'a group of distant mining and smelting works'; and T. H. Beckett, who explained the organizational systems that underpinned the Railway Clearing House. Mackinder reported in his survey of the course that their lectures had been 'greatly appreciated by the class', and in subsequent years the lectures were augmented with field trips to locations including the Railway Clearing House, the Great Western Railway's signalling school and Surrey Docks.[92] Emphasizing the

---

[89] C. W. Gwynn, 'The administrative course at the London School of Economics', *Royal Engineers Journal*, vi (1907), 229–35, at p. 229.

[90] W. Funnell, 'National efficiency, military accounting and the business of war', *Critical Perspectives on Accounting*, xvii (2006), 719–51, at p. 734.

[91] Gwynn, 'The administrative course', p. 231.

[92] Sloan, 'Haldane's Mackindergarten', p. 335; *Report of the advisory board, first course*, p. 5; Army. *Report of the advisory board, LSE, on the fourth course at the LSE, October, 1909, to March, 1910, for the training of officers for the higher appointments on the administrative staff of*

interaction of civilian and military figures prior to the war, the students on these observational visits were encouraged to ask questions and discuss matters with the academic staff to ensure that the course taught material of 'direct utility' to the forces.[93]

The syllabus taught at the LSE was not designed for men whose future career was expected to involve the administration of a so-called colonial gendarmerie. As Bertram Wilson, the leader of the business organization module explained, 'special attention [was] paid to the manufacturing industries, chiefly with regard to factory and office organization, arrangement of factory [sic] into departments for efficient control, methods to secure internal economy, storekeeping and checks on waste, systems of costkeeping, [and] systems of wage calculation'.[94] These were lessons that reflected Haldane's and Mackinder's shared belief in the coincidental intent of both military and civilian 'business', while the advisory board established to oversee the content and delivery of the course emphasized the complementary nature of the expertise possessed by those within the army and the private sector. Alongside Mackinder and Ward (who chaired the board), the LSE course was overseen by: the director of supplies, Brigadier-General Frederick Clayton; the director of staff duties, Lieutenant-General H. D. Hutchinson; the QMG, Major-General Herbert Miles; the director of fortifications and works, Brigadier-General R. M. Ruck; the commandant of the ordnance college, Woolwich, Colonel G. R. Townshend; the chairman of the institute of bankers, Sir Felix Schuster; and, in his position as the LSE's governor, by Sidney Webb. In his capacity as director of staff duties at the War Office, Sir Douglas Haig sat on the advisory board in 1908 and 1909. Unsurprisingly, given their size, the complexity of their operations and their pre-existing working relationships with the government and the army, the railway industry was also represented on the advisory board. Sir Hugh Bell, a North-Eastern director, sat on the board throughout the pre-war period, while Sir Frederick Harrison, the LNWR's general manager, contributed to the board's deliberations in 1907.[95]

The advisory board's conclusions on the course's first year were encouraging. They stated that:

> We desire to say that we are convinced that the results which have been achieved by this first class fully warrant the continuance of this experiment.

---

the army and for the charge of departmental services (Parl. Papers 1910 [Cd. 5213], ix), p. 5.

[93] Grant, 'Edward Ward, Halford Mackinder', p. 107; Sloan, 'Haldane's Mackindergarten', pp. 334–5.

[94] Report of the advisory board, fourth course, p. 5.

[95] Report of the advisory board, first course, p. 2.

The experience which has now been gained does not make it necessary to reorganise the scheme in any essential respects, but some minor changes and modifications in the original syllabus will be made.[96]

These modifications included the replacement of material on banking statistics, public administration and geography – perceived as being of 'less immediate practical bearing' – with lectures from Wilson on 'business organization'.[97] By 1909 the symbiotic process of military feedback and academic response had conceived a syllabus adjudged by the advisory board to be of such value to the army that they 'strongly recommend[ed] that the course be made a permanent annual institution, in order gradually to create a body of officers well fitted to undertake the varied administrative duties that may fall upon them'.[98] The only significant change to take place after 1909 saw an increased stress placed on Wilson's business organization module, which 'emphasised the importance of process and the elimination of waste' and incorporated the study of Frederick Winslow Taylor's pioneering *Principles of Scientific Management* following its publication in 1911.[99]

The so-called 'Mackindergarten' at the LSE created a forum for the dissemination of business knowledge that was otherwise absent from the professional training available to soldiers destined for administrative roles in the army. The advisory board recognized that the needs of a modern army reflected those of a supply-intensive industrial business, while the army acknowledged that the technical specialists responsible for maintaining a successful global trading empire could be tapped to improve its own knowledge base. However, it is important not to overstate the influence of the course's existence over the efficiency of the army's supply organization during the First World War. The conflict intervened before a substantial number of officers had participated in the course. Over the period 1907–14 only 243 officers successfully completed the course (see Table 1.1). In view of the army's administrative manpower requirements during the war such a small number of graduates meant that, by necessity, only a tiny minority of the army's supply duties were handled by men who had benefited from attendance at the LSE. Furthermore, the seven years between the course's inauguration and the outbreak of the war left insufficient time for the comparatively junior officers who attended the course to attain positions

[96] *Report of the advisory board, first course*, p. 6.

[97] *Report of the advisory board, fourth course*, p. 3.

[98] *Report of the advisory board, LSE, on the third course at the LSE, October, 1908, to March, 1909, for the training of officers for the higher appointments on the administrative staff of the army and for the charge of departmental services* (Parl. Papers 1909 [Cd. 4610], x), p. 3.

[99] Grant, 'Edward Ward, Halford Mackinder', p. 106; F. W. Taylor, *The Principles of Scientific Management* (New York, 1911).

Table 1.1. Number of officers to pass the administrative training
course at the London School of Economics, 1907–14.

Course	Dates run	Number of graduates
1	January–June 1907	31
2	October 1907–March 1908	30
3	October 1908–March 1909	31
4	October 1909–March 1910	29
5	October 1910–March 1911	31
6	October 1911–March 1912	30
7	October 1912–March 1913	29
8	October 1913–March 1914	32
	Total	243

Note: Number of officers from each rank to complete the course: 12 lieutenants; 162 captains; 64 majors; 4 lieutenant-colonels; 1 colonel.
Source: Various reports of the advisory board, 1907–14. For full details, see the bibliography.

of real influence at the army's highest levels of authority. Mackinder's vision for the long-term evolution of an 'administrative tradition', which reached to the highest positions of the supply branches of the army, was abruptly curtailed by the events of August 1914.[100]

Mackindergarten graduates were destined for roles that demanded proficiency in the execution of largely routine tasks, not those that involved planning the intricate network of inter-connected systems required to maintain the modern army in the field. At the outbreak of the war they were not in positions of sufficient seniority to influence the constitution of the arteries that directed the BEF's blood to its vital organs. None of the officers to occupy the principal supply positions of QMG, inspector-general of communications (IGC) or DGT on the western front during the First World War had attended the Mackindergarten. Only Frederick Clayton, a member of the advisory board, possessed any connection to the administrative course at the LSE whatsoever. Instead, in August 1914 the graduates maintained the blood flow around the body in junior management roles. Most of them remained in comparatively minor positions, where they followed orders rather than made policy decisions, for the duration of the war. However, the multitude of new vacancies created by the rapid expansion of the British army during the conflict did provide opportunities for a number of officers to apply the skills they had acquired

[100] *Report of the advisory board, first course*, p. 14.

at the LSE in positions of considerable influence. The appointment of Colonel E. E. Carter as director of supplies at GHQ in 1915 represented the pinnacle of achievement for a Mackindergarten graduate within the BEF's administrative hierarchy, and those who served in other theatres attained roles of similar responsibility. After a period at the base in Rouen, Major P. O. Hazelton – part of the first cohort to graduate in 1907 – became director of supplies and transport in East Africa in January 1916. Captain G. F. Davies and Major Wilfred Swabey, both from the class of 1908, occupied the same role in Egypt and Italy respectively. Two successive directors of supplies and transport for the British Salonika Force (BSF), Captain Oscar Streidinger and Major Philip Scott, graduated in 1909.

The syllabus developed by Ward and Mackinder represented an attempt to infuse mostly junior officers with business methods and mentalities, which were largely absent from the upbringings of such men. As noted above, the British army relied upon the landed classes – for whom the 'bourgeois ethic of business was anathema' – for its supply of officers throughout the pre-war period.[101] Such men did not typically arrive at the LSE with any grounding in the complex world of railway operations, and were almost entirely dependent upon Wilfred Tetley-Stephenson's module to provide them with an understanding of 'the conditions of railway work in relation to the army in times of peace and war'.[102] Their completion of the administrative course at the LSE provides a further example of the connections forged between Britain's civilian experts and military officers in the final years of peace. Yet these links were not restricted merely to lecture theatres, site visits and meeting rooms at Whitehall. As one young but precocious railway manager's early life demonstrated, the professional relationships between the army and the railways stretched far beyond British shores. For Eric Geddes, the permeability of civilian and soldier within the crucible of the empire proved invaluable in the conflict that followed.

The early career of Eric Geddes

Eric Geddes was not the most senior executive of a railway company in August 1914. Nor was he in the summer of 1916 when, as the battle of the Somme's voracious appetite threatened to paralyse northern France's transport network, he was despatched to the western front by Lloyd George to examine the BEF's supply organization. Rather than approach Geddes's superior, Alexander Kaye Butterworth, or the general manager of Britain's largest railway company, the LNWR's Guy Calthrop, the then secretary

---

[101] Funnell, 'National efficiency', p. 727.
[102] *Report of the advisory board, first course*, p. 9.

of state for war turned to the thirty-nine-year-old Geddes to conduct the investigation. The reasons behind Lloyd George's decision to send Geddes to GHQ in late August 1916 can be discerned from a study of the latter's early career, which demonstrates how Britain's senior transport experts were 'well known' within the walls of the War Office and Westminster ahead of the First World War.[103] Geddes's formative experiences reveal a 'remarkable man' on an unequivocal ascent to the peak of his profession,[104] and further entwines the strands of military, political and civilian expertise that have run throughout this chapter – and which underpinned the British response to events after 1914.

Born at Agra, India, on 26 September 1875, Eric Campbell Geddes was the eldest son of a Scottish civil engineer. Auckland Geddes had originally set sail for the east in 1857, and had established a private practice after undertaking railway survey and construction work on behalf of the Indian government – a clear example of the so-called 'diaspora of British engineering' in the nineteenth century.[105] The family moved to Edinburgh a year after Geddes's birth, and following a disruptive childhood in which he was 'asked to leave' a succession of public schools, he was eventually educated at the Oxford Military Academy. Geddes's studies were ultimately competent enough for him to pass the preliminary examination for entry into Woolwich. However, rather than follow in his father's footsteps and become an engineer (albeit along the military rather than civil path), the impetuous young Geddes 'set sail on a passenger liner to New York with ten pounds … and an introduction to family friends in Pittsburgh'.[106]

The army's short-term loss was its long-term gain. Over the following twenty years, Eric Geddes accumulated the breadth of knowledge and expertise required to fulfil the various roles he was asked to perform during the First World War. His professional education began in America, where he initially worked in occupations as diverse as theatrical agent, bar tender, typewriter salesman for Remington and labourer at Andrew Carnegie's

---

[103] Parliamentary Archives (PA), papers of David Lloyd George, LG/D/1/2/1 Butterworth to Lloyd George, 27 May 1915; Pratt, *British Railways and the Great War*, i. 45.

[104] The quotation refers to the impression left by Geddes upon the British military attaché in Paris, Colonel Herman Le-Roy Lewis, after their first meeting. See PA, Lloyd George papers, LG/E/3/14/29, Le-Roy Lewis to Lloyd George, 22 Nov. 1916.

[105] A. C. Geddes, *The Forging of a Family: a Family Story Studied in its Genetical, Cultural and Spiritual Aspects and a Testament of Personal Belief Founded Thereon* (London, 1952), pp. 89–104; R. A. Buchanan, 'The diaspora of British engineering', *Technology and Culture*, xxvii (1986), 501–24.

[106] K. Grieves, *Sir Eric Geddes: Business and Government in War and Peace* (Manchester, 1989), pp. 1–2; Geddes, *The Forging of a Family*, pp. 117–26.

steel works.[107] Both Remington and Carnegie were recognized innovators, and operated at the forefront of the systematic management ideology that spread across America and into Europe around the turn of the twentieth century.[108] Remington was among the first private enterprises to experiment with modern office equipment – such as the typewriters Geddes sold for them – and had been swift to adopt the card index as an organizational tool following its transmission from the library sector.[109] Carnegie's Pittsburgh steel works possessed a global reputation for the 'perfection' of its response to the challenges of modern big-business organization.[110] Whether the experience provided Geddes with similar insights into labour conditions as those espoused by Taylor and his disciples is unclear from the surviving records, but throughout his career Geddes extolled the virtues of manual labour for giving the budding manager 'sympathy with the point of view of the working man, the value of which cannot be exaggerated'.[111] By the time Geddes held high office in the railway industry, as Gourvish's research has demonstrated, managerial positions were increasingly held by men who benefited from the 'initial advantages of birth and education' rather than those who had climbed the internal ladder from the shop floor. Before 1890, nobody in the role of general manager at a prominent British railway company had attended university. After 1890 there were eight graduates appointed to the position, five of them among the eighteen appointments made after 1910.[112]

Geddes's first contact with the railway industry took place in America. Transport, he claimed after the war, soon became 'my religion. It interests me more than anything else. Transport contains elements that are not appreciated by the uninitiated'.[113] He clearly showed an aptitude for the sector, as he progressed swiftly from the position of station agent at a lumber-loading station in Virginia through to assistant yardmaster in a freight yard of the Baltimore and Ohio Railroad. With further promotions

[107] Grieves, *Sir Eric Geddes*, p. 2.

[108] D. Nelson, 'Scientific management, systematic management, and labor, 1880–1915', *Business History Review*, xlviii (1974), 479–500.

[109] M. Krajewski, *Paper Machines: About Cards and Catalogs, 1548–1929*, trans. P. Krapp (Cambridge, Mass., 2011), pp. 105–6.

[110] G. Brown, *Sabotage: a Study in Industrial Conflict* (Nottingham, 1977), pp. 121–2.

[111] Geddes to Ferguson, quoted in Grieves, *Sir Eric Geddes*, p. 2.

[112] Gourvish, 'A British business elite', pp. 293–7; T. R. Gourvish, 'The rise of the professions', in *Later Victorian Britain, 1867–1900*, ed. T .R. Gourvish and A. O'Day (Basingstoke, 1988), pp. 13–35, at p. 29.

[113] Geddes to Lord Riddell, 28 Aug. 1919, quoted in K. Grieves, 'Sir Eric Geddes, Lloyd George and the transport problem, 1918–21', *Journal of Transport History*, xiii (1992), 23–42, at p. 31.

Geddes became the car tracer for the southern group of railroads known as the big four. The American railways of the late nineteenth century were pioneers in modern management techniques, having faced the challenges associated with efficiently coordinating the energies and efforts of large numbers of employees earlier than the huge industrial concerns established by the likes of Carnegie and Henry Ford.[114] Illness impaired Geddes's ability to continue climbing the managerial ladder and to further absorb the methods and working practices of America's emerging corporations. He returned to Edinburgh in August 1895.

The United States provided Geddes with skills that proved invaluable the next time his 'volcanic energy' proved too large to be contained by the British Isles.[115] After his recovery in Scotland, Geddes's experiences in the railway and logging industries – assisted in good measure by his father's contacts – secured him a managerial role on a forest clearance project in the Himalayas. Part of the job called for the construction and operation of a light railway system, which was linked up to the Powayan Steam Tramway. Geddes oversaw the line's construction and managed the network, the efficiency of which so impressed an agent of the Rohilkund and Kumaon Railway (who happened to have been a former employee of Geddes's father) that the company assumed control of the line. Thence began Geddes's second rise in the railway industry. In 1901 he became the Rohilkund and Kumaon's traffic superintendent and moved to the prominent railway junction at Bareilly with his wife, Alice Stokes, whose brother Claude was an officer in the Indian army. His wife's ill health, exacerbated by the Indian climate, compelled Geddes to seek employment with a British railway company during a period of leave in 1903. His endeavours proved unsuccessful. However, upon his return to the sub-continent Geddes became reacquainted with the army he had decided not to join after he left school, and gained the opportunity to demonstrate his talents as a railway administrator to none other than Lord Kitchener.

The Russo-Japanese War, which broke out in February 1904, provided the catalyst for the meeting between Geddes and Kitchener. The Russians began to deploy troops to their frontiers upon the declaration of hostilities, to meet any force Britain may have decided to send north from India in support of its Japanese ally.[116] The build-up of soldiers on the Afghan border fed into longstanding British concerns over Russian intentions on

[114] A. D. Chandler, *The Visible Hand: the Managerial Revolution in American Business* (Cambridge, Mass., 1977).

[115] Geddes, *The Forging of a Family*, p. 202.

[116] P. Towle, 'The Russo-Japanese War and the defence of India', *Military Affairs*, xliv (1980), 111–17, at p. 112.

the north-west frontier, and led Kitchener to call for the conveyance of an all-arms force to the area as quickly as possible.[117] Several lines intersected in and around Bareilly, which made the junction a key component of any large-scale troop movements and placed a significant responsibility upon the Rohilkund and Kumaon to ensure a smooth concentration. Geddes devised the programme of movements with such efficiency that Kitchener personally congratulated him for its success.[118] It proved to be Geddes's final act in India. At the end of 1904 he was offered the post of claims agent at the North-Eastern Railway, then under the management of George Gibb. For the next decade the structure and working practices Gibb had introduced to the North-Eastern played a critical role in Geddes's maturation into the recognized transport expert he had become by August 1914.[119]

As the North-Eastern Railway provided the organizational culture within which Geddes obtained most of his pre-war experience, it is essential to establish both how the company operated and what Geddes learned from the North-Eastern's approach to the administration of a large-scale organization. The British railways had confronted increasingly difficult operating conditions from the 1870s onwards, caused by rising expenditure on resources and augmented by parliamentary controls designed to limit the companies' opportunities to shift price rises onto customers. The restrictive legislative situation produced an industrial environment in which efficient operating procedures became vital to the sustenance of profitability. However, contemporary observers such as William Acworth and George Paish believed that most British railway companies were unresponsive – and their managers too conservatively minded – to cope with the challenges that faced them. Acworth and Paish, although they stopped short of labelling Britain's railway managers 'donkeys', did compare their abilities unfavourably with those of their counterparts on the American railways.[120]

Thanks to Gibb's progressive attitude, contemporary observers did not consider the North-Eastern to be part of the conservative trend in late nineteenth-century British railway management. Instead, the North-

---

[117] National Library of Scotland (NLS), papers of Field Marshal Sir Douglas Haig, Acc. 3155/2D, diary entries, 13 June to 3 Oct. 1904 provide occasional references to Kitchener's concentration on mobilization questions during this period, alongside demonstrating Haig's own appreciation of the army's dependence on reliable transportation.

[118] R. J. Irving and R. P. T. Davenport-Hines, 'Geddes, Sir Eric Campbell (1875–1937)', in *Dictionary of Business Biography: a Biographical Dictionary of Business Leaders Active in Britain in the Period 1860–1980*, ed. D. J. Jeremy (5 vols., London, 1984), ii. 507–16, at pp. 507–8.

[119] R. Bell, *Twenty-Five Years of the North Eastern Railway, 1898–1922* (London, 1951), p. 30.

[120] W. M. Acworth, 'Railway economics', *Econ. Jour.*, ii (1892), 392–8; G. Paish, *The British Railway Position* (London, 1902), pp. 5–6, 14–15.

Eastern was held up as one of the too-few British companies to have revolutionized their working practices and organizational systems through the implementation of innovations developed across the Atlantic.[121] Gibb, upon becoming general manager in 1891, was convinced that the North-Eastern's extant managerial framework was defective and included 'few men in the higher grade of management who could give him a critical assessment of operating procedures which had remained basically unchanged' for over thirty years.[122] The by then traditional practice of promotion from within, coupled to an absence of managerial education opportunities, had created an executive branch that suffered from narrowness of vision and deficiencies in original thought. Subsequent historical analysis of the period has broadly accepted that Gibb's observations were applicable across the British railway industry.[123]

Gibb's response to such insularity of experience was the establishment of a traffic apprenticeship scheme, which provided Geddes with his introduction to the North-Eastern in 1904. The management development programme, created in 1897, focused on 'young blood, some of it not long out of the universities' and those from within the company who displayed the potential for higher appointments.[124] Ralph Wedgwood, the scheme's first graduate and the BEF's director of docks from 1917, typified the class of outsider Gibb sought to attract to the North-Eastern. A descendant of the famous pottery family with a degree in classics from Cambridge, Wedgwood possessed no experience in the railway industry prior to his arrival in York at the start of his apprenticeship. J. George Beharrell, who had entered the North-Eastern as a junior clerk in the secretary's office in 1888, was invited to participate in the scheme in 1902. By the time Geddes arrived in 1904 the traffic apprenticeship scheme had been refined into one that offered a carefully planned, comprehensive overview of the company's work. The programme was 'designed to allow the employee to move around the system experiencing the work of various grades of labour, as well as that of supervisory and management levels'.[125] Geddes, rather than being

---

[121] Paish, *British Railway Position*, p. 235; N. Crafts, T. Leunig and A. Mulatu, 'Were British railway companies well managed in the early twentieth century?', *Econ. Hist. Rev.*, lxi (2008), 842–66.

[122] R. J. Irving, *The North Eastern Railway Company, 1870–1914: an Economic History* (Leicester, 1976), pp. 214–15.

[123] For a synthesis of the existing literature, see D. A. Turner, 'Managing the "royal road": the London & South Western Railway 1870–1911' (unpublished University of York PhD thesis, 2013), pp. 14–18.

[124] Irving, *North Eastern Railway*, pp. 215–16.

[125] T. Strangleman, 'Railway and grade: the historical construction of contemporary identities' (unpublished Durham University PhD thesis, 1998), p. 45.

expected to learn 'on the job' through traditional but haphazard methods, received the benefits of a planned introduction to best practice upon his entry to the company.

The traffic apprenticeship scheme promoted the emergence of a unified managerial culture, which was diffused throughout the multitude of departments within which its graduates were employed. In 1907 the North-Eastern employed almost 48,000 workers, spread across the entire breadth of the company's network and engaged in myriad tasks that demanded close coordination. Geddes outlined the variety of tasks for which specialist working units had to be created in a 1910 lecture delivered to the York railway and lecture debating society:

> Sub-departments have been formed at headquarters to control the supply of wagons, the working of motor vehicles and the cartage of goods traffic. Advertising is the sole concern of a separate office. An inspector has been appointed to supervise the heating and lighting of the Company's premises. The inauguration of the commercial agency emphasised the distinction between the functions of the man who creates and obtains traffic and his operating colleague who is expert at moving traffic economically. Lastly, the development of the Continental business in recent years has led to the creation of an office where a wide knowledge of shipping and general business in indispensable. These examples by no means exhaust the list.[126]

The quantity of separate sections within the company reflected the increasing complexity of the railway industry and its corresponding demand for further specialization of duties. The traffic apprenticeship scheme reduced the need for overwhelming, time-consuming, and initiative-stifling central control of the North-Eastern's multiple business activities. Senior managers were relieved of administrative duties, which could be confidently devolved to talented junior executives 'on the spot'. The subsequent freedom from the burden of detail allowed those at the top to focus upon considerations of strategy and procedure, just as the existence of a competent staff – whose shared ethos increased the likelihood that predictable decisions would be made when required – liberated the army commander from the need to micro-manage his forces.

Geddes acquired a substantial appreciation of the challenges involved in freight rail operations as chief goods manager at the North-Eastern Railway between 1907 and 1912. His efforts in the goods department prepared him for the wartime challenge of supplying an army that demanded colossal amounts of work to be performed by limited pools of human and material resources. During the period 1899–1912 the North-Eastern improved its

---

[126] TNA, ZLIB 29/691, education and advancement of the railway clerk, pp. 5–6.

earnings per freight train by 87 per cent. In part these improvements were due to new loading practices implemented across the industry, but one historian has also suggested that the application of working methods based on statistical analysis played a considerable role in the North-Eastern's particularly notable rise. Data compiled by Beharrell – who became Geddes's assistant for the rest of their careers – was applied to measures that led to 'more work being done but [by] fewer trains, thus giving greater line capacity throughout the system … a smaller number of engines employed, economy in rolling stock, repairs, renewals, and … staff'.[127] In 1912 the North-Eastern's goods train mileage stood at roughly the same level as it had in 1906. However, over the same period the gross tonnage hauled over its lines increased substantially and its receipts per goods train mile rose from 75.2*d* in 1900 to 132.91*d* in 1912.[128]

Geddes's performance at the North-Eastern Railway marked him out as the 'coming man' in the railway industry. Both the LSWR and the Buenos Aires Southern and Western Combine attempted to lure Geddes away from York with promises of substantial wage increases and the title of general manager. The North-Eastern responded by promoting Geddes to the role of deputy general manager and renegotiating his salary. Upon taking up his new position Geddes became the highest paid railway official in Britain. According to Sir Hugh Bell, a North-Eastern director, it was a decision the company 'never regretted'.[129] With the incumbent general manager, Alexander Kaye Butterworth, scheduled to retire in 1916, Geddes's rise to the top of the company appeared to have its trajectory mapped out – he was considered the North-Eastern's general-manager-in-waiting by his colleagues. Yet while Butterworth's presence temporarily obstructed Geddes's path to the general manager's office, the former's religiosity acted to reconnect the latter to the institution he had almost joined after leaving school and assisted with aplomb while in India: the army.

Upon replacing Sir George Gibb as general manager of the North-Eastern in 1906, Butterworth received a commission into the ERSC. However, the quasi-military status evoked by his membership of the corps sat uneasily alongside Butterworth's faith and he resigned his commission in January 1907.[130] The North-Eastern was represented in the ERSC by the company's

---

[127] Irving, *North Eastern Railway*, pp. 241–9, 281; Crafts, Leunig and Mulatu, 'Were British railway companies well managed?', p. 853.

[128] Grieves, *Sir Eric Geddes*, pp. 6–7.

[129] PA, Lloyd George papers, LG/D/1/2/2, Bell to Lloyd George, 30 May 1915.

[130] Butterworth's father George was the vicar of St Mary's parish church in Deerhurst. Tragically, Butterworth's son, also named George, did not share his father's aversion to military service. A talented composer, and friend of Ralph Vaughan Williams, George

engineer, Charles Harrison (commissioned 1900), and from 1910 by its traffic superintendent, Henry Watson, but there was no representation of the North-Eastern's senior management in the corps for six years. The corps' rules of qualification were explicit: only general managers were permitted to hold commissions. Yet on Geddes's appointment as deputy – and in a further demonstration of the company's long-term ambitions for the man they had paid handsomely to retain – Butterworth began to lobby the ERSC to relax its entry criteria. Geddes's first recorded appearance at a meeting of the REC took place in December 1912, before – thanks to his superior's representations – he obtained his commission and became Lieutenant-Colonel Eric Geddes on 27 January 1913. He was the only deputy general manager of a railway company to gain admission to the ERSC before the First World War, and played an active role in the REC's deliberations in the final years of peace.[131]

Conclusion

Eric Geddes only briefly participated in the vibrant civil–military exchanges that characterized the peacetime relationship between Britain's army, government and prominent railway companies – and upon whose foundations a successful wartime partnership was constructed. The interactions of civilian and military figures assumed both practical and academic forms in the eighty years that preceded the First World War, and took place within a variety of domestic and imperial settings. As the railways spread across the British landscape almost every government department became invested in their efficient use, while army officers acquired a central position in the state's governance of the nascent railway industry. Further afield, the widespread construction of railways across the empire provided the Royal Engineers in particular with ample opportunities to construct and operate railways, and officers with experience gained in China, South Africa, Nigeria and on multiple Indian railways served in prominent roles on the western front between 1914 and 1918.[132] Similarly, Auckland Geddes and his eldest son represented just two of the many civil engineers and

---

Butterworth enlisted in August 1914 and was shot in the head by a German sniper on 5 Aug. 1916. His name is recorded on the Thiepval Memorial to the Missing of the Somme. See J. F. Addyman, 'G. S. Kaye Butterworth, M. C.', *The North Eastern Express*, xxxvii (1998), 64.

[131] Grieves, *Sir Eric Geddes*, p. 10; TNA, WO 32/9188, re-constitution of council. A meeting took place in the QMG's office on 10 Dec. 1912 to discuss the relationship between the REC and other permanent or temporary committees appointed to consider questions related to railway operations.

[132] *History of the Corps of Royal Engineers*, ed. H. L. Pritchard (11 vols., Chatham, 1952), v. 666–8.

railway officials who obtained valuable experience of railway construction and operation across the empire in the nineteenth and early twentieth centuries.

The creation and maintenance of a productive partnership between Britain's largest railway companies and the army during this period relied upon two complementary factors. On the one hand, the military proved keen to engage with and exploit the knowledge of those who emerged as specialists in the industry. Through the establishment of vocational courses at their vast workshops and their contributions of academic expertise to the administrative staff course offered to soldiers at the LSE, Britain's railway companies imparted skills and knowledge to those tasked with the military application of railway technology. In the creation of civil–military bodies such as the ERSC, the army underlined its respect for the opinions and observations of those whose transportation experience lay beyond the military sphere – in the operation and management of railways and docks, in civil engineering, and in the direction of the great contracting firms. On the other hand, the provision of transport for the army's annual manoeuvres, the development of timetables for the emergency movement of British forces around Britain and the identification of supply problems in the event of war each demanded substantial resource commitments from Britain's transport enterprises and ensured they were thoroughly conversant with the needs of a modern, industrial army.

Such support was provided willingly during the period before 1914, even when the army's interactions with civilian expertise were redeployed from domestic, purely defensive applications to those of a more blurred and potentially aggressive nature. Only Alexander Kaye Butterworth, the general manager of the North-Eastern Railway and a man of 'strong religious scruples',[133] chose to sever the connections that bound the higher echelons of the railway industry to the army before August 1914 – and even he was persuaded to take his place alongside colleagues from Britain's other strategically important lines following the outbreak of war. The REC, upon which Butterworth sat throughout the conflict, represented the ultimate manifestation of both sides' commitment to the development of fruitful and harmonious relations between civilian and military experts. From August 1914 onwards, at an organizational level, the committee provided the foundations upon which Britain's evolving domestic response to the war's transport demands was constructed. However, the REC did not provide the platform from which Britain's global response to transport requirements of an industrial war were met. Instead, the army's exploitation

[133] Townsend, *All Rank and No File*, p. 45.

of civilian knowledge and experience beyond British shores took place on an individual basis. For Geddes, the benefits of the tripartite pre-war relationship between army, government and the railways were most clearly evident from the summer of 1916 onwards. Yet for the BEF the interactions between Britain's transport experts, the military and the state bore fruit far sooner. In the first instance, the nature of their collaboration had profound implications for the nation's entry into the conflict.

2. A fruitful collaboration: Henry Wilson, the railways and the British Expeditionary Force's mobilization, 1910–14

A contributor to the North-Eastern Railway's staff magazine in 1912, inspired by the popular invasion literature of the time, mused upon the trauma that could potentially face the railway in the event of a German incursion on the Yorkshire coast:

> What an enormous strain would be thrown upon the NER and its officials! All ordinary traffic within the effected [*sic*] area would, for the time being, be suspended, and all resources taxed to the utmost … Supplies and all the necessary accoutrements, inseparable from an army on active service, would be rushed through in the wake of the troops. The railway line would have to be guarded throughout, together with all the bridges and tunnels – a most essential thing in time of war![1]

To meet such a challenge, the author argued, myriad details and orders had to be prepared in advance to ensure that the fluidity of the network was not compromised by the sudden onslaught of impromptu traffic. 'It is probably safe to assume', the author concluded, 'that the NER management have in their possession a secret timetable which could be put into operation at short notice in the event of mobilization'. In August 1914 the scenario was different, but the assumption was proven correct.

The evolution of modern, material-intensive industrial warfare engendered the development of armies that required quantities of manpower, munitions and equipment on scales incomparable in previous military experience. Britain's island status, global interests and command of the seas made it highly unlikely in the early twentieth century that a large military force would ever need to be deployed on the British mainland (Ireland was an entirely different story). Therefore, any mobilization scheme developed for the BEF following its creation in 1907 necessitated the provision of sufficient transport to move the force over land and sea. For its bulk transportation needs, the British army relied upon the railways and the Royal Navy.

---

[1] 'Our railways in time of war', *North-Eastern Railway Magazine*, ii (1912), 67. Unless otherwise stated, all quotations in this passage are taken from this source.

'A fruitful collaboration: Henry Wilson, the railways and the British Expeditionary Force's mobilization, 1910–14', in C. Phillips, *Civilian Specialists at War: Britain's Transport Experts and the First World War* (London, 2020), pp. 63–91. License: CC-BY-NC-ND 4.0.

Yet aside from a laudatory statement to the house of lords from the newly installed secretary of state for war, Lord Kitchener, which acknowledged that the railway companies had 'more than justified the complete confidence reposed in them by the War Office',[2] Britain's transport experts' contribution to the BEF's entry into the First World War received little recognition from contemporaneous military figures. Sir John French, the BEF's first commander-in-chief, reserved his plaudits for the 'Naval Transport Service and … all concerned in the embarking and disembarking of the Expeditionary Force'.[3] Elsewhere, recognition for the British army's successful mobilization was focused upon the QMG of the forces, Sir John Cowans,[4] and the director of military operations (DMO) at the War Office, Sir Henry Wilson. Lord Roberts, Wilson's friend and mentor, hailed the latter's importance to the army's entry into the war as early as 7 August. He wrote of Britain's 'indebtedness' to Wilson for all he had achieved at Whitehall.[5] Speaking shortly after the war, one of Wilson's subordinates in the directorate of military operations claimed that 'it was only the ardent spirit of Sir Henry Wilson, his tireless energy, wide vision and dauntless perseverance' that turned hypothetical projections into the practical arrangements that existed in August 1914.[6]

Consequently, the pre-war preparations made for the BEF's movement have been treated almost as if they were Wilson's personal possession. John Bourne, exemplifying the historical approach to the WF scheme, referred to it as Wilson's 'administrative Rolls-Royce'.[7] Robin Neillands, in the most thorough discussion of mobilization, transport and logistics in 1914 to have appeared to date, concluded his narrative with the observation that 'Henry Wilson's plan had worked to perfection'.[8] Such one-sided accounts imply that Britain's actions following the outbreak of war in August 1914 were a military-led response to the 'unaccountable disbelief of the authorities', which had retarded a comprehensive system of preparation for conflict.[9] Where the BEF's mobilization and concentration in France have not been ignored altogether,

[2]   Hansard, *Parliamentary Debates*, 5th ser., xvii (25 Aug. 1914), col. 503.
[3]   J. D. P. French, *1914* (London, 1919), p. 40.
[4]   D. Chapman-Huston and O. Rutter, *General Sir John Cowans, G.C.B., G.C.M.G.: the Quartermaster-General of the Great War* (2 vols., London, 1924), i. 287–8.
[5]   IWM, Wilson papers, HHW 2/73/45, Roberts to Wilson, 7 Aug. 1914.
[6]   TNA, WO 106/49A/1, history of the growth of the scheme; preparation of a plan for rendering military assistance to France, and notes on entrainment and embarkation, Address by Maj. Gen. Radcliffe on inception and working of scheme, p. 3.
[7]   J. M. Bourne, *Britain and the Great War, 1914–1918* (London, 1989), p. 17.
[8]   R. Neillands, *The Old Contemptibles: the British Expeditionary Force, 1914* (London, 2004), p. 96.
[9]   IWM, Wilson papers, HHW 2/73/45, Roberts to Wilson, 7 Aug. 1914.

references to them have been invariably brief and limited to affirmations that the processes 'proceeded remarkably well'.[10] Indeed, most of the available literature on Britain's entry into the war in 1914 tends to reinforce Julian Thompson's remark that logistics only predominate over the more glamorous and controversial themes of tactics and strategy when the logistics fail.[11] The presence of British troops at Mons on 23 August emphatically demonstrates that the BEF's logistical preparations did not fail in 1914.

The establishment, development and implementation of the WF scheme was not the result of one man's endeavours. Nor was it a spontaneous reaction to Belgian and French requests for assistance following Germany's invasion of their territory. Rather, Britain's mobilization programme was a thoroughly prepared example of civil–military cooperation, which depended upon the input of Britain's transport experts. Previous over-concentration on the political and military dimensions of the so-called July crisis – coupled with a desire to conceal the scheme's evolution from the public and parliament (and, indeed, much of the government) before the war – has created an imbalanced picture of British actions in the days immediately surrounding Britain's entry into the war. The success of the BEF's mobilization was in large part due to the existence of a sophisticated network fostered and managed by Wilson during his period as DMO. Wilson recognized, and consistently represented to his political superiors, that it was only through careful, detailed preparations – undertaken both with future allies and the technical specialists employed to operate Britain's largest transport companies – that the BEF could be rapidly and smoothly propelled into a European conflict. The nature of those preparations, and their influence over the decisions that governed Britain's initial contribution to what became the western front, are the focus of this chapter.

Henry Wilson and the development of the 'with France' scheme

Brigadier-General Henry Wilson became DMO at the War Office in August 1910. His small and isolated directorate, established in 1904, was principally responsible for the production and assessment of military intelligence and 'the development of strategic plans for the defence of Britain and the Empire'.[12] The duties assigned to the directorate included

[10] I. M. Brown, *British Logistics on the Western Front, 1914–1919* (London, 1998), p. 43. For other concise references, see H. Strachan, *The First World War: To Arms* (Oxford, 2001), p. 206; E. F. Carter, *Railways in Wartime* (London, 1964), pp. 80–1; J. N. Westwood, *Railways at War* (London, 1980), p. 138.

[11] J. Thompson, *The Lifeblood of War: Logistics in Armed Conflict* (Oxford, 1991), p. 3.

[12] T. G. Fergusson, *British Military Intelligence, 1870–1914: the Development of a Modern Intelligence Organization* (Frederick, Md., 1984), p. 203.

the collection of information about the military capabilities of the British empire, the collation of intelligence on Britain's possible opponents in a future war, and the preparation of mobilization schemes to meet potential threats.[13] Therefore, upon his appointment to the role of DMO, Wilson became intrinsically connected with two tasks: ensuring that Britain's political and military leaders knew the identity and strength of Britain's most likely adversaries; and preparing the army to respond to external dangers effectively.

Following the establishment of the directorate of military operations and the conclusion of the entente cordiale between Britain and France, Wilson's predecessors had developed schemes for the deployment of a force beyond the British Isles within the narrow confines of the War Office. The army's first two DMOs, Sir James Grierson and Spencer Ewart, had obtained government permission to establish contact with the French general staff and discussed the movement of British troops inland from the French coast to proposed concentration sites near the Belgian frontier. However, according to Wilson 'they had not had time' to investigate the question as to how the BEF was to be transported to the British coast from various locations across the country.[14] Wilson's remark was inaccurate. In fact, his predecessors had been explicitly forbidden from discussing the BEF's mobilization plans with anybody outside the War Office, including the railway companies whose infrastructure and resources were vital to the swift concentration and movement of troops and their supplies.[15] Within his first year in post, Wilson acknowledged that the 'old scheme' in place in August 1910 'had not been worked out in sufficient detail to admit of its being carried out'.[16] The mobilization scheme upon which the BEF's likely response to war in 1910 was founded consisted of hypothetical projections. It was not the product of a meticulous examination of the modern army's transport requirements undertaken in conjunction with the experts capable of assessing whether those requirements could be met. Prior to Wilson's arrival at the directorate of military operations, the arguments for diplomacy and national secrecy prevailed over the bureaucratic and technical realities that governed the mass movement of an industrial army.

---

[13] War Office (reconstitution) committee. Report of the War Office (reconstitution) committee. (Part II) (Parl. Papers 1904 [Cmnd. 1968], viii), p. 25.

[14] IWM, Wilson papers, HHW 3/7/2, minute to CIGS reporting progress on scheme of EF, Apr. 1913, p. 1.

[15] C. E. Callwell, Field Marshal Sir Henry Wilson Bart, G.C.B., D.S.O.: His Life and Diaries (2 vols., London, 1927), i. 91–2.

[16] IWM, Wilson papers, HHW 3/5/5, Wilson to Nicholson, 24 Apr. 1911.

The decision to preclude the railway companies from the planning process severely constrained both the quantity and quality of the work that the directorate could achieve in relation to the BEF's mobilization scheme. The British government's attitude contrasted sharply with circumstances in Germany, where the specialist railway section of the general staff – described by Mark Stoneman as a 'linchpin' of German war planning – cooperated closely with the civil railway administration throughout the pre-war period.[17] Yet the absence of an equivalent pool of technically proficient officers within the British army, which fed into a perception within the railway industry that Britain's military leaders consistently underestimated the railways' capacity,[18] did not mean that the army as an institution was ignorant of the technical aspects of railway operations. Each year the army made prolific use of the railways in conjunction with its annual manoeuvres. The transport demands for the army's exercises in the years preceding the First World War graphically illustrated the size and weight of the *impedimenta* attached to the modern fighting force. In 1910 the LSWR was responsible for the movement of 26,000 officers and men; 8,000 horses; seventy guns; and 1,200 transport vehicles to the manoeuvre area 'at the height of the holiday traffic' season. 'Between 9:55am on a certain Saturday, and 11:15am on the following Wednesday', the LSWR successfully arranged for 137 trains to be run under war conditions on the army's behalf.[19] Three years later, the LNWR used the pages of its staff magazine to proudly record the company's 'exceptional efforts' during the army's manoeuvres in East Anglia the previous September. The troops' concentration required the coordination of 209 trains, run over lines operated by the Great Western, Great Northern, Great Eastern and LSWR in addition to those handled by the LNWR. At the small and ill-equipped station of Potton alone, in 'practically 36 hours work' the railways delivered and unloaded trains containing 8,283 officers and men; 1,951 horses; forty guns and limbers; forty ammunition wagons; 251 four-wheeled wagons; eighty-four two-wheeled carts; and 124 bicycles. On the day after the successful completion of the movement, the LNWR's general manager received a telegram that conveyed the army's 'great appreciation of the remarkable and efficient and punctual manner in which the move of the ... army to the area of operations' had

---

[17] M. R. Stoneman, 'Wilhelm Groener, officering, and the Schlieffen plan' (unpublished Georgetown University PhD thesis, 2006), p. 153. On the strategic importance of railways to German war planning, see A. Bucholz, *Moltke, Schlieffen, and Prussian War Planning* (New York, 1991).

[18] 'Railways and military operations', *Railway Gazette*, 7 Aug. 1914, p. 174.

[19] 'Railways and military operations', p. 174.

been effected. 'There was', the telegram concluded, 'absolutely no hitch in the arrangements'.[20]

The movement of troops, both actual and potential, also engendered dialogue between the military and civilian experts concerned. Representatives of the War Office 'confided' with railway servants up to three months before the manoeuvres were scheduled to take place,[21] while senior railway executives made themselves available to the army both for consultation through the mechanisms of the REC and in informal meetings with individual soldiers. Wilson's papers from 1909 contain the notes he had taken during a discussion with the traffic managers of the South-Eastern and Chatham Railway (SECR) and the Great Eastern Railway, on the subject of a hypothetical mobilization of a division for a staff college exercise. The level of detail within Wilson's record of the meeting illustrates that, prior to his arrival at the War Office the following year, he was thoroughly conversant with the complexities that surrounded the army's use of railways.[22] That awareness undoubtedly contributed to his determination, upon becoming DMO, to overturn the restrictive governmental decree that forbade cooperation between the railways and the War Office for the purposes of planning the BEF's mobilization. On 9 January 1911 Wilson elaborated his reasons for seeking the railway companies' assistance with the process in a letter to the chief of the imperial general staff (CIGS), General Sir William Nicholson:

> As far as I am a judge no tables drawn up in this office are of practical value until they have been submitted to and worked out in detail by the Railway Companies concerned, and I submit that we have ample material on which to approach the railway companies as a preliminary to the detailed timetable being drawn up ... I am of course ready to discuss this question at any time, and to give any further information and assistance which it is in my power to give, but I hope no unnecessary delay may occur in having detailed timetables worked out by the W[ar] O[ffice] in conjunction with the railway companies, as until this has been done it is impossible to claim that our Expeditionary Force is ready to take the field.[23]

As the lengthy gaps between the ERSC's receipt of and response to the War Office's exercises in the latter part of the nineteenth century had

---

[20] W. E. Bradbury, 'Manoeuvres in East Anglia', *London and North-Western Railway Gazette*, Jan. 1913, pp. 6–9.

[21] J. F. Bradford, 'The war manoeuvres in the eastern counties—autumn 1912', *London and North-Western Railway Gazette*, March 1913, pp. 89–90.

[22] IWM, Wilson papers, HHW 3/3/11, appendix D – movement of troops by rail, Oct. 1909.

[23] IWM, Wilson papers, HHW 3/5/4, Wilson to Nicholson, 9 Jan. 1911.

demonstrated, the coordination of large-scale troop movements across the country required thorough, time-consuming preparations. The failure to undertake those preparations in peace, Wilson believed, reduced Britain's options in the event of a European war. The dominant military ideology of the period stressed the importance to an army of seizing the initiative through a rapid mobilization, followed by the application of that initiative to seek a decisive battle at a time and place that made success as likely as possible.[24] Without prepared railway timetables the BEF could not be mobilized rapidly, should the government decide to commence hostilities. Without the input of the railway companies, Wilson argued, the necessary timetables could not be prepared.

Wilson was by no means a lone voice in arguing for greater collaboration in the development of the BEF's mobilization scheme at the start of 1911. Sir Frederic Bolton's gloomy assessment of the railways' ability to cope with likely wartime demands elicited a politico-military response from the new under-secretary of state for war, Colonel Jack Seely. On 24 January he wrote to the prime minster, Herbert Asquith, about the '[e]specially valuable information' that could only be obtained from closer cooperation between the army and the nation's principal railway companies.[25] Asquith's reply raised no objections to the involvement of the 'General Managers of the principle [sic] railways', but emphasized 'that the conditions of secrecy which have hitherto prevailed should, so far as possible, be preserved'.[26] Unbeknownst to Seely, Wilson had obtained authorization to discuss matters with the railways from the secretary of state for war, Richard Haldane, and the foreign secretary, Sir Edward Grey, three days earlier.[27]

The government's decision to permit the War Office to discuss mobilization plans with the railway companies fundamentally shifted the relationship between the army and Britain's transport experts. From that point forwards, the latter's role in imperial military planning changed from one exclusively devoted to questions of national defence to one that was integral to the production of Wilson's offensive-minded WF scheme. And the DMO quickly set about incorporating the railway companies into the task of producing timetables for the BEF's concentration at the ports earmarked for its despatch to the continent. A schedule containing the details of every unit that required railway transport on mobilization was sent to the railway companies. The itemized list documented all of

---

[24] S. van Evera, 'The cult of the offensive and the origins of the First World War', *International Security*, ix (1984), 58–107.

[25] NCL, Mottistone papers, Mottistone 11/40, Seely to Asquith, 24 Jan. 1911.

[26] NCL, Mottistone papers, Mottistone 11/42, Asquith to Seely, 26 Jan. 1911.

[27] K. Jeffery, *Field Marshal Sir Henry Wilson: a Political Soldier* (Oxford, 2006), pp. 91–2.

the information required by the civilian specialists to make arrangements for the move: what the unit comprised in terms of men and equipment; the station(s) from which the unit was expected to commence its journey; and the time and date – after general mobilization – at which the unit was required to arrive at its designated port for embarkation. The companies arranged the technical details of the move in consultation with either the QMG's department or the individual home commands, and drew upon their experiences of planning the annual manoeuvres to pull together a workable programme. The companies were responsible for ensuring the provision of suitable rolling stock, for calculating the times when individual trains would pass through stations and junctions en route, for drawing up a complete timetable and for taking the necessary steps to guarantee that sufficient crews and engines would be available and ready for action when the need for them arose. Wherever potential clashes arose, the matter was referred up to Wilson's office where decisions as to the order of priority were made and communicated back to the railways.[28]

The LSWR became intimately connected to the evolution of the WF scheme after January 1911, thanks to its association with the principal departure point for the BEF. The port of Southampton, operated by the company since 1892, had a long history of military service and was earmarked to fulfil the same role at the outbreak of the next war as it had during the South African War.[29] The government covertly directed public funds to the LSWR, which were used to remove the only substantial bottleneck on Britain's dense railway network and to increase the length of track within the port to thirty-seven miles before 1914.[30] The LSWR became Wilson's 'secretary railway', handling correspondence between the War Office and the railway companies between 1911 and 1914. It installed bespoke diagram boards at the port to chart the specialist facilities required by certain units and allow the staff at Southampton to keep a visual record of the BEF's complex demands, and it was the only company to be entrusted with possession of the entire programme of movements.[31] The LSWR recruited

[28] TNA, WO 106/50, scheme for mobilization on a war footing – progress of scheme for despatch of forces (WH/1), memorandum by Captain H. O. Mance (staff captain, QMG 2) on the questions raised by the executive committee in their memorandum of 10 Dec. 1912, 23 Dec. 1912, pp. 2–3; E. A. Pratt, *British Railways and the Great War: Organisation, Efforts, Difficulties and Achievements* (2 vols., London, 1921), i. 27–8.

[29] I. F. W. Beckett, 'Going to war. Southampton and military embarkation', in *Southampton: Gateway to the British Empire*, ed. M. Taylor (London, 2007), pp. 133–46.

[30] D. Stevenson, 'War by timetable? The railway race before 1914', *Past & Present*, clxii (1999), 163–94, p. 174; Pratt, *British Railways and the Great War*, ii. 1008–9.

[31] Beckett, 'Going to war', p. 142; TNA, WO 106/49A/2, Wilson-Foch scheme – expeditionary force to France, outline of the scheme and details regarding mobilization and

a specialist clerk to work exclusively on the timetable, who received a list of the desired arrival times into Southampton for every train destined for the port upon mobilization. From that information each individual train was traced back to the point at which it was required to enter the LSWR's system. Thereafter, the company over whose lines the train passed immediately prior to its transfer onto the LSWR's network was notified of the time at which they were expected to pass the train over. Following the same method that company continued to plot the train's journey in reverse, either to its point of origin or the next handover point on its voyage across the patchwork of lines that made up the pre-war British railway network.[32] Once each journey had been traced back to its departure point an entrainment time was entered onto the corresponding unit's individual mobilization timetable.[33]

The process was not complete once the entrainment times for each unit were recorded, however. The Army Council amended the war establishments of certain components of the BEF every year, which meant that the timetables demanded constant revision. Changes to the departure time of individual trains had knock-on effects in terms of the journeys scheduled for other trains over the affected lines, while units' embarkation points could be switched to different ports as the army juggled with the varying capacities for troops, supplies and equipment at Avonmouth, Newhaven, Liverpool, Dublin, Glasgow, Belfast, Queenstown and the port of London. Given the numerous factors involved, the process of amending the timetables consumed a great deal of time and energy on the parts of both Wilson's directorate and the railway companies. Over the winter of 1912–13 the LNWR received such drastic alterations to their share of the programme that the company established a special department under W. E. Bradbury, chief of the timetable office in the company's southern division, which worked exclusively on ensuring that it would be ready to meet its obligations to the War Office.[34] Internal memoranda produced for the DMO indicate that the amendments handed down from the Army Council in December 1913 were only expected to be synthesized with the existing timetables in four months' time. The scheme for 1914 became operational on 1 April of that year, and provided the foundations for the mobilization programme followed by the BEF when war broke out four months later.[35]

---

staff arrangements, n.d., p. 12.

[32] Pratt, *British Railways and the Great War*, i. 112–14.

[33] TNA, WO 106/49A/1, history of the growth of the scheme, address by Radcliffe, p. 7.

[34] E. A. Pratt, *War Record of the London and North-Western Railway* (London, 1922), pp. 6–7.

[35] TNA, WO 106/49A/2, Wilson-Foch scheme, revision of programme, remarks of

Table 2.1. Numbers embarked at English and Irish ports
between 9 August and 21 September 1914.

Port	Officers	Other ranks	Horses	Nursing sisters and civilians
Southampton	5,028	171,708	51,434	1,389
Newhaven	66	409	—	9
Avonmouth	58	4,547	—	—
Liverpool	16	1,741	—	—
Devonport	30	844	421	—
Dublin				
Belfast	826	25,921	10,184	—
Queenstown				
Totals	6,024	205,040	62,039	1,398

Source: A. Hurd, *History of the Great War. The Merchant Navy* (3 vols., London, 1924), ii. 82.

In addition to moving the BEF's various units to their assigned ports of embarkation on the outbreak of war, the War Office also had to provide transport to ship the troops and their myriad accoutrements across the English Channel. The personnel figures recorded in Table 2.1 were supplemented by at least 93,364 tons of ammunition, stores, vehicles and other items during the BEF's initial movements. Such figures were comprehensively dwarfed later in the war, but they 'serve as a useful corrective' to the idea that military transport was ever 'a simple matter of embarking and disembarking personnel and horses'.[36] The pre-war army was thoroughly aware of that fact, and of the deceptive complexity involved in the embarkation and disembarkation of a military force. Exercises took place at Southampton in both 1912 and 1913 to test the army's embarkation procedures and identify the issues likely to arise when the mobilization took place for real. Henry Holmes, the LSWR's superintendent of the line, provided the War Office with guidance as to Southampton's capacity and underlined the importance of keeping the port's railway connections clear of obstructions during the latter exercise.[37]

While Wilson and the railway companies focused upon the challenges involved in the movement of the BEF on and away from British soil,

---

various directorates, unsigned memorandum, 4 Dec. 1913.

[36] A. Hurd, *History of the Great War. The Merchant Navy* (3 vols., London, 1924), ii. 83.

[37] TNA, WO 107/24, release of government personnel for active service: correspondence, Notes on the embarkation of the expeditionary force at Southampton; embarkation exercises, Southampton, 1913. Statement by L. S. W. Railway. On the 1912 exercise, see Chapman-Huston and Rutter, *Sir John Cowans*, i. 270; Beckett, 'Going to war', pp. 142–3.

Colonel Seely – who was appointed secretary of state for war in 1912 – focused on the requirements necessary for the effective disembarkation of the force on the other side of the Channel. He invited four shipping experts to investigate the complications the BEF was likely to face on the French coast: Sir Thomas Royden of the Cunard Company; Sir Lionel Fletcher of the White Star Line; Sir Richard Holt of Blue Funnel; and Sir Owen Philipps of the Royal Mail. Royden and Fletcher accepted Seely's invitation and, accompanied by officers from the military and naval staffs of both France and Britain, made a thorough reconnaissance of the French ports designated to receive the BEF.[38] The disembarkation of a modern army was just as complex a technical challenge as its embarkation, and one that required detailed examination of such questions as the availability of berthing facilities, tidal limitations, the number and power of the cranes at each port, and the amount of suitable storage facilities in the vicinity of the wharves. Royden and Fletcher 'gave up all their private work', and devoted an entire six months to the production of a comprehensive review of the BEF's shipping requirements. Their recommendations were handed over to the Admiralty in early 1913 and adopted as the foundations of the disembarkation instructions produced for issue to the troops.[39]

Royden's and Fletcher's expert investigations uncovered serious deficiencies in the infrastructure upon which the BEF's swift deployment depended. At Boulogne, Le Havre and Rouen – the ports destined to receive the BEF upon its arrival in France – they identified that insufficient crane facilities were available to handle the *impedimenta* that accompanied the troops. Therefore, to prevent backlogs and congestion the BEF's mechanical transport, which was projected at 950 lorries and 250 motor cars, was divided between all three ports rather than concentrated upon one facility.[40] Inevitably, such dislocations to the existing plans necessitated further revisions to the mobilization programme in Britain and created additional work for the timetabling staffs of the railway companies involved. A hand-written note on the surviving records confirms that the timetables for 1913 had been amended in response to the recommendations made in the Royden–Fletcher report.[41] However, the time and energy expended in peacetime to ensure that the BEF's preparations were solid reduced the risk

---

[38] J. E. B. Seely, *Adventure* (London, 1930), pp. 140–1.

[39] F. E. Smith, *Contemporary Personalities* (London, 1924), pp. 291–2; S. Cobb, *Preparing for Blockade 1885–1914: Naval Contingency for Economic Warfare* (Farnham, 2013), pp. 187–8; TNA, WO 106/49A/1, history of the growth of the scheme, address by Radcliffe, p. 4.

[40] TNA, WO 107/296, report of the British armies in France and Flanders, 17 March 1919, p. 38; WO 106/49A/2, Wilson-Foch scheme, i. Factors affecting plan of movement and staff work.

[41] TNA, WO 106/49A/2, Wilson-Foch scheme, v. Sea transport.

of delays in the time-sensitive period following the government's decision to mobilize in August 1914. As the QMG of the forces, Sir John Cowans, acknowledged in 1918, Royden and Fletcher 'rendered our Movements branch in the War Office ... enormous assistance before the war in drawing up the schemes for the despatch of the Expeditionary Force'.[42]

The WF scheme was not the only product of the collaborative environment of 1911–14. Wilson's tenure as DMO bore witness to an increase in inter-organizational coordination between institutions and departments that existed for martial purposes and those whose primary responsibilities lay in the government and administration of peacetime Britain. The work of Britain's transport experts took place concurrently with an acceptance within British strategic circles of the inapplicability of the term 'business as usual' to state affairs in wartime. As the CID's secretary noted in November 1910, many governmental departments had 'much important work to undertake – either consequent on, or contributory to, the naval and military mobilisations' of Britain's armed forces.[43] The government's response to the national and imperial requirements for coordinated action on the outbreak of war took physical form in the shape of the war book – a series of instructions to be followed by the appropriate government departments and industrial concerns, both upon the declaration of a precautionary period and following the order to mobilize. First produced in 1912, and updated in 1913 and 1914, the book acted as a step-by-step guide for officials whose responsibilities ranged from the provision of police officers to protect vulnerable railway junctions to the despatch of mobilization telegrams to the nation's soldiers and sailors. After 1913 the war book was arranged in chapters for each department, which allowed each to quickly obtain the instructions to guide their actions without having to concern themselves with material that only applied to others.[44] To establish their roles and responsibilities at the start of a war, employees at the post office and the customs and excise board, or in the general managers' offices of Britain's railway companies, simply consulted the relevant section of the war book. It represented the 'search for order and integration' that took place within Britain's largest businesses before the First World War on an imperial scale,[45] and guided Britain's entry into the conflict in August 1914.

---

[42] TNA, WO 107/16, inspector-general of communications, general correspondence, Cowans to Clarke, 23 Feb. 1918.

[43] TNA, CAB 15/2, memoranda, series K. 1–100, ote by the secretary, 4 Nov. 1910.

[44] Copies of all three war books are available at TNA, CAB 15/3–5, war book: summary of action taken by departments, 27 Feb. 1912 to 30 June 1914.

[45] J. Yates, 'Evolving information use in firms, 1850–1920: ideology and information techniques and technologies', in *Information Acumen: the Understanding and Use of Knowledge in Modern Business*, ed. L. Bud-Frierman (London, 1994), pp. 26–50, at pp. 29–30.

From his appointment as DMO through to the outbreak of the First World War, Henry Wilson never missed an opportunity to highlight how anxious he was to make his superiors aware of the BEF's state of readiness. His anxiety was frequently accompanied by a list of existing deficiencies that, in Wilson's view, rendered the BEF unprepared for war. In one of Wilson's regular letters to the CIGS he wrote that 'all the great powers and many of the smaller ones are straining every nerve to increase the numbers and the efficiency of their armies: we alone are doing nothing to increase our numbers and but little, and that slowly, to increase our efficiency'.[46] During the Agadir crisis, he argued that:

> There must be something radically wrong when a man in my position is forced to write, during a time of international strain, that he does not know when the E[xpeditionary] F[orce] can be made ready to take the field, nor even which of the larger units of that force could be made completely mobile, nor for how long the wastage of war can be made good; nor does he know if the Force will enter on the campaign with a serious deficiency of officers nor whether this deficiency will seriously increase. There must be something wrong when the officer responsible to you for the fighting efficiency of the E[xpeditionary] F[orce] in so far as plans of operation for that force are concerned is unaware that certain essentials in mobilization equipment are (or were) deficient; is unaware how long a time will elapse before the force is fully equipped with a resighted rifle and new ammunition; was unaware that there was a serious shortage in S[mall] A[rms] A[mmunition], or that the new howitzers would have to be fought in 4 gun batteries with a very inadequate supply of ammunition.
>
> I submit that such a state of affairs ought not to exist, and ought not to be allowed to exist.[47]

When Sir John French replaced Nicholson as CIGS in March 1912, Wilson placed on record his opinion that, 'as we stand today, we cannot claim that the E[xpeditionary] F[orce] is either ready to take the field, or capable of keeping the field as a thoroughly efficient fighting machine'.[48]

Yet through the collaborative efforts of the War Office and Britain's transport experts, by the summer of 1914 the WF scheme was complete. A full set of timetables, which recorded the peace stations, locations of equipment and places of mobilization for every component of the BEF had been copied, printed and issued to the relevant units. A series of tables indicating the day after general mobilization on which each of their units

---

[46] IWM, Wilson papers, HHW 3/5/21, Wilson to Nicholson, 26 Dec. 1911.
[47] IWM, Wilson papers, HHW 3/5/15, Wilson to Nicholson, 16 Aug. 1911.
[48] IWM, Wilson papers, HHW 3/5/22, Wilson to French, 3 Apr. 1912.

had to be ready to move had been delivered to each home command.[49] Every unit or part thereof had been assigned to a train, and the expected departure and arrival times for each train had been carefully recorded. At the ports of embarkation, troops and supplies had been allocated to a cross-Channel transport, and the serial numbers of each ship were ready to be telegrammed across the sea to inform the French authorities of their contents.[50] The 1st (Guards) Brigade's schedule illustrates the level of detail within the programme, even for a unit with a comparatively short journey to the coast. Each half-battalion of the brigade was assigned to one of eight trains, which were timetabled to leave Farnborough for Southampton between 2:27 a.m. and 2:31 p.m. on the fifth day of mobilization. Three battalions, labelled A to C, were allocated to ships that left Southampton on the same day and were expected to arrive at Le Havre before sundown. Following an enforced rest at a base camp outside the French port, on day seven the three battalions were to entrain in France for transportation inland to their destination stations at Ohis, Neuve Maison, La Capelle and Buironfosse. The fourth battalion, D, was to land at Le Havre before noon on day six and detrain two days later at the same stations as their comrades.[51]

The character of modern warfare among industrialized powers demanded that the intervention of the BEF on the continent was 'a diplomatic and military act too serious for its execution to be left to an eleventh-hour inspiration'.[52] The effective deployment of a British force relied upon thorough planning and detailed preparations. When the government made the decisions both to commence hostilities with Germany and to send the BEF to France, the British army was able to implement a programme of movements created by the combined efforts of civilian and military experts. The WF scheme was founded upon Britain's possession of a robust transport network and an abundance of technical specialists in the myriad professions required for the sustenance of a global trading empire. While the movement of the BEF in August 1914 represented a military manoeuvre more complex than any that had been previously attempted by a British force, the expertise that conceived and then oversaw the programme ensured that – unlike

---

[49] TNA, WO 106/49A/8, Expeditionary Force tables and details of the war establishments of units, mobilization dates by commands, Apr. 1914. The complete set of timetables is available at TNA, WO 106/49B/3, railway timetables, Expeditionary Force time tables, 1914.

[50] TNA, WO 106/49B/3, railway timetables, Expeditionary Force timetables, 1914.

[51] TNA, WO 106/49B/3, railway timetables, Expeditionary Force timetables, 1914; TNA, WO 106/49B/4, serial tables: Southampton for the 2nd–19th day details of train timetables: units and goods for entrainment, weight, etc., 1914; TNA, WO 106/49B/7, disembarkation tables, 1914.

[52] A. de Tarlé, 'The British army and a continental war', trans. H. Wylly, *Royal United Services Institution. Journal*, lvii (1913), 384–401, at p. 400.

Germany's so-called Schlieffen plan – it did not contain elements that were a gamble logistically. Martin van Creveld, in his pioneering assessment of the German commander's infamous scheme, concluded that 'Schlieffen does not appear to have devoted much attention to logistics when he evolved his great plan'.[53] Henry Wilson did, and he recognized the crucial need to involve Britain's transport experts in the process.

Wilson's role in the development of the WF scheme was that of a facilitator. In his biography of this 'political soldier', Keith Jeffery suggested that Wilson's 'larger than life persona' may have made him appear more of a driving force behind the scheme than he actually was.[54] The volume of work undertaken by the directorate of military operations under Wilson's direction provides ample evidence of his leadership skills, personal drive and energy in the role of project manager for the WF scheme. Yet it also demonstrates that Britain's mobilization planning before the First World War was a team effort. Wilson acknowledged the significant contributions of his subordinates, whose names and roles have been eclipsed by the theatrical and divisive personality of their director. Major Marr Johnson is one such figure. In the months immediately before the war began Johnson personally hand-wrote, typed and proofread the timetables for the 1914 edition of the programme before they were approved and printed on the War Office's secret press.[55] His work may have been largely forgotten by historians, but it was clearly appreciated by his pre-war chief. After the war Wilson attempted to gain Johnson a position in the newly established Ministry of Transport, writing to Sir Eric Geddes that:

> There is a Colonel Marr Johnson who used to work for me before the war and who did an immense amount of most detailed work on the railway side and the shipping side for getting the E[xpeditionary] F[orce] over to France. It is not too much to say that a great deal of the success of the initial moves of the troops from England to France was due to Colonel Marr Johnson.[56]

Johnson only completed his gargantuan task in July 1914, as Europe slid towards war.

Yet Johnson's timetables could not have been committed to paper without the substantial investment of time and resources made by the railway companies in the three years prior to the First World War. Wilson lobbied for access to the railway companies in 1911 because he understood

---

[53] M. Van Creveld, *Supplying War: Logistics from Wallenstein to Patton* (Cambridge, 1977), p. 138.

[54] Jeffery, *Henry Wilson*, p. 99.

[55] TNA, WO 106/49A/1, history of the growth of the scheme, address by Radcliffe, p. 7.

[56] IWM, Wilson papers, HHW 2/26/4, Wilson to Geddes, 3 Apr. 1919.

that a workable mobilization scheme required the support and input – in terms of knowledge, resources and time – of those who both maintained the flow of men and materials around the country on a daily basis and would be required to implement the programme when war began. Britain's transport experts were an integral part of the process that evolved from what 'would certainly have been a shambles' in 1910 to the thorough collection of instructions that existed by August 1914.[57] Thanks to the technical expertise possessed by a highly skilled, industrialized society, working in conjunction with an effectively managed directorate of military operations, the BEF's movements programme existed on paper. When the signal to mobilize was issued, the same combination of civilians and soldiers was responsible for converting the WF scheme from paper to practice.

Britain's mobilization and concentration in August 1914

Colonel Victor Huguet, France's former military attaché to Britain, wrote on 2 August 1914 to appraise Wilson of the day's events on the continent. Huguet informed Wilson that the French army's mobilization had begun, and that 'great hopes are entertained in France concerning British assistance. Should you not join us, it would be a great disappointment here'.[58] Wilson, a committed Francophile since his youth,[59] wished for nothing more than to see the BEF immediately mobilized and sent to France's aid. He had frequently stressed the importance of the BEF's swift mobilization in the event of war during his tenure as DMO, and claimed that the 'early intervention of our six divisions would be more effective than the tardy presence of double their numbers'. Therefore, he concluded, 'we must mobilise the same day as the French'.[60] However, as the French authorities began to mobilize in response to German activity on their eastern frontier on 1 August, the British government did not follow Wilson's advice. Furthermore, when the war council met for the first time on 5 August – following the expiration of Britain's ultimatum to Germany the previous night – it was unclear whether the timetables produced over the previous years were going to be put to their intended use.

In Britain the decision to go to war was far from constrained by the rigidity of railway timetables. In recent years David Stevenson has comprehensively debunked A. J. P. Taylor's provocative thesis, while William Philpott has

---

[57] B. Bond, *The Victorian Army and the Staff College, 1854–1914* (London, 1972), p. 258.

[58] IWM, Wilson papers, HHW 2/73/38, Huguet to Wilson, 2 Aug. 1914.

[59] Jeffery, *Henry Wilson*, p. 4.

[60] TNA, WO 106/47, defence and operational plans, conditions of a war between France and Germany (E2/25), 12 Aug. 1911.

even gone as far as to claim that 'the importance of Wilson's timetables has been overemphasised' in the historiography.[61] However, the existence of the WF scheme – with its interlocking transport schedules – did factor into the government's calculations in the days either side of the declaration of war.[62] On Sunday 2 August the QMG ordered Captain Henry Mance to 'bring all the mobilization programmes to 10 Downing Street to explain the railway situation' to the prime minister. Mance told Asquith that 'owing to the Territorials being scattered at that moment all over the country it would not be possible to make the following day the "First day of mobilization"[,] but that the programme would work if the "First day" was deferred' until Wednesday.[63] The army's use of the August bank holiday period as an opportunity to call out the territorials for summer manoeuvres meant that some 100,000 men were either far from their mobilization stations or in the middle of journeys to camp when the crisis in Europe deepened. All had to be returned before the mobilization programme could begin. Asquith ordered the immediate cancellation of the territorials' movements, before raising the question of whether the order for the BEF to mobilize could be detached from the order for it to embark for France. In response, Mance

> showed [Asquith] a diagram illustrating the different categories of moves to be carried out in connection with Home defence, mobilization, and the despatch of the Expeditionary Force, and how it was not possible to postpone the E.F.['s embarkation] to a later date without making the trains of the different programmes clash or disorganising the arrangements for rolling stock.[64]

Mance demonstrated that the various segments of the railway programme devised over the previous three years could not be operated separately without 'alterations to the orders to every unit and every railway'. Following consultation with the REC, Mance confirmed to the government that embarkation could begin on the fifth day of mobilization at the earliest.

The outcome of the meeting of 2 August was that both the size and destination of Britain's contribution to the war in Europe remained unsettled. Consequently, when the war council met three days later, those in the room

---

[61] A. J. P. Taylor, *War by Time-Table: How the First World War Began* (London, 1969); Stevenson, 'War by timetable?'; W. Philpott, 'The general staff and the paradoxes of continental war', in *The British General Staff: Reform and Innovation, c.1890–1939*, ed. D. French and B. Holden Reid (London, 2002), pp. 95–111, at p. 99.

[62] S. R. Williamson, Jr., *The Politics of Grand Strategy: Britain and France Prepare for War, 1904–1914* (Cambridge, Mass., 1969), pp. 336–7.

[63] TNA, PRO 30/66/9, correspondence and papers relating to the shipment of troops to Ireland and France, and the establishment and organization of the director-general of military railways, recollections of the first few days of mobilization, p. 3.

[64] TNA, PRO 30/66/9, correspondence and papers, recollections, pp. 3–4.

considered the BEF's deployment to be open for discussion. Sir John French, who had been issued command of the BEF, advocated that the WF scheme be rendered void because its mobilization had not been synchronized with that of the French army. In place of the proposed transportation of the British force to the French Channel ports, as coordinated by the directorate of military operations, Sir John suggested his troops be shipped to Antwerp to act in concert with the Belgian army it had ostensibly entered the war to protect. However, as Sir Charles Douglas, Sir John's replacement as CIGS pointed out, the arrangements that had been made for the BEF's despatch from Southampton and other ports had been drawn up with the journey times to and from France in mind. The extra distance to Antwerp demanded either the spontaneous sourcing of extra naval transports or the recalculation of the existing railway timetables for the delivery of troops and supplies to the British coast.[65] Wilson also believed that the waters of the Scheldt, the river upon whose banks the port of Antwerp sits, could be 'closed by a schoolboy'. In the event of a major European war, Wilson had warned Churchill three years earlier, Antwerp was likely to be 'cut off from all direct communication with the sea'.[66] Furthermore, the Royal Navy refused to guarantee the safety of naval transports north of the Dover Straits until the German fleet had been destroyed.[67] That the BEF's senior commander could raise such a logistically impractical suggestion augured ill for his appreciation of the role transportation was to play in the conflict.

The deliberations of the war council emphasized that the only practicable concentration scheme available to the British government in August 1914 was 'with France'. Wilson's carefully constructed plan boasted the benefits of interdepartmental cooperation, thorough logistical preparation and the input of suitably qualified transport experts. Once Sir Douglas Haig's suggestion to retain the BEF at home for two to three months – during which time the 'immense resources of the Empire' could be developed – had been rejected in favour of the immediate despatch of four divisions to the continent,[68] the programme of movements contained within the WF scheme dictated the location to which those troops would be sent. Sir John's attempts to reintroduce Antwerp as a possible base of British operations, coupled with Haig's desire to hold the BEF back from France, demonstrated the reluctance of Britain's two senior commanders to enter

[65] TNA, CAB 22/1, minutes of meetings, secretary's notes of a war council held at 10 Downing Street, 5 Aug. 1914, pp. 1–2.

[66] IWM, Wilson papers, HHW 3/5/16A, Wilson to Churchill, 29 Aug. 1911.

[67] P. Guinn, *British Strategy and Politics, 1914 to 1918* (Oxford, 1965), pp. 13–14.

[68] TNA, CAB 22/1, minutes of meetings, secretary's notes, 5 Aug. 1914, p. 2; secretary's notes of a war council held at 10 Downing Street, 6 Aug. 1914.

into too close a relationship with the French army.[69] The existence of Wilson's embarkation programme did not govern Britain's decision to enter the war. However, once that decision had been made – and the trains began to roll into Southampton on 8 August – the rigidity of railway timetables locked the British and French armies into a partnership, which materially diminished Britain's freedom of action for the duration of the conflict.

By the time the embarkation process began, Britain's transport experts and the directorate of military operations had already been preparing for war for more than a week. On 31 July Sir Sam Fay received a coded telephone call, which informed him that the 'precautionary stage' had begun. He set off for the LNWR offices that housed the REC and did not see his home again for a fortnight.[70] Mance was detained by the requirements of the WF scheme for even longer. He was in Worcestershire when the announcement of the precautionary stage reached him on 29 July, coordinating movements associated with the territorials' summer manoeuvres with the Great Western Railway. He returned to London that day and slept in the War Office every night from 31 July to 24 August.[71] Fay was joined in the capital by his REC colleagues, each of whom were connected to their home railways by the telephone network specially installed over the previous two years. On 4 August the War Office delivered a letter to 130 railway companies, which announced that the government had taken control of the railways and that they were to be managed on the government's behalf by the REC. It informed them that:

> Although the railway facilities for other than Naval and Military purposes may for a time be somewhat restricted, the effects of the use of the powers under [the Regulation of the Forces Act] will be to coordinate the demands on the railways of the civil community with those necessary to meet the special requirements of the Naval and Military Authorities. More normal conditions will in due course be restored, and it is hoped that the public will recognize the necessity for the special conditions and will in the general interests accommodate themselves to the inconvenience involved.[72]

---

[69] For a discussion of Sir John's strategic ideas before the First World War, see W. Philpott, 'The strategic ideas of Sir John French', *Journal of Strategic Studies*, xii (1989), 458–78.

[70] S. Fay, *The War Office at War* (London, 1937), pp. 20–1. The precautionary stage was the first of three phases of operations that took place at the outset of war, the others being mobilization (the movement of troops to their mobilization stations), and concentration (the delivery of troops to the theatre of operations). In the precautionary phase, the army communicated its railway requirements to the REC and guards were placed at tunnels and bridges to ensure the security of the network. See Stevenson, 'War by timetable?', p. 166.

[71] TNA, PRO 30/66/9, correspondence and papers, recollections, pp. 1–2.

[72] Quoted in J. A. B. Hamilton, *Britain's Railways in World War I* (London, 1947), p. 26.

Following the delivery of this letter the REC became responsible for 98 per cent of the railway mileage in Britain (but not Ireland, which was excluded from the Act), the instructions to general managers contained within the war book were brought into effect and the breadth of the secret plans developed over the previous three years was finally revealed to the majority of Britain's railway servants.[73]

The smoothness of the mobilization and concentration process depended upon the professionalism of the railways' employees and their military passengers. The annual peacetime manoeuvres had given both groups invaluable experience of the technical nature of railway transport before they were called upon to realize those movements in wartime: the army with regards to the loading and unloading of troops and their equipment, and the railways in terms of the coordination of the locomotives, crews and rolling stock required for the specialist moves.[74] The 12th Horse Transport Company's experience in August 1914 illustrates both the complications that arose from a mobilization during the summer bank holiday and the breadth of accoutrements that accompanied the industrial army. Based at Colewort barracks in Portsmouth, the company began August at their annual training camp on Salisbury Plain. Following receipt of the order to return to Portsmouth as quickly as possible, the unit undertook a march of fifty-five miles in just one day. Only one horse, which went badly lame, returned to the barracks by train. After medical inspections at Colewort the company marched out of the barracks to mobilize at Hilsea, where their ten-day mobilization schedule began. At Hilsea 'life became more strenuous than ever as stores of all description, but mainly wagons and harnesses, were drawn from Ordnance, and reservists and specially enlisted men began to roll in from all over the country'.[75] In addition to the extra men, the company had to incorporate between fifty and sixty horses per day as they arrived by rail according to pre-arranged impressment schedules devised before the war. In less than two weeks – and a day ahead of schedule – a pre-war transport company of around thirty men and forty horses emerged as No. 5 Reserve Park, which comprised seven officers, 289 men, 153 wagons and 358 horses. The unit 'shook down with local treks to accustom a variegated collection of

[73] Hamilton, *Britain's Railways in World War I*, pp. 26–9; TNA, PRO 30/66/9, correspondence and papers, recollections, p. 6.

[74] Norman Pattenden's series of articles describes the LSWR's experience of the manoeuvres in 1914, and highlights both the complexity of the work and the professionalism of the railway staff involved. See N. Pattenden, 'Armageddon? – No just practising', *The South Western Circular*, xii (2001).

[75] M. Young, *Army Service Corps, 1902–1918* (London, 2000), pp. 44–5.

82

men and horses to convoy duties and march discipline', and then headed to Southampton for embarkation.

The impressment of horses provides a further example of the depth of civil–military cooperation required for the WF scheme to function effectively. Wilson observed early in his tenure as DMO that 'there [would] be a difficulty about moving some 15,000 horses from the north of England to Aldershot'.[76] Wilson underestimated the size of the task. The BEF's peacetime establishment of horses was 19,000. Upon mobilization the BEF immediately required 55,000 horses and the territorials a further 86,000. A census of horses, compiled in each of the home commands from data provided by local police forces, confirmed that sufficient horses were available. However, a system for the identification, collection and transport of animals suitable for army requirements had to be created. Over the next two years the War Office used the information to draw up lists of horses available in various locations around the country, and trained around 1,400 'prominent local gentlemen of suitable knowledge and status' – usually landowners or experienced horsemen – to collect an assigned number of horses on mobilization.[77] By April 1913 the 'various horsebrows, slings and stores required by the home ports' for the embarkation of the animals had been purchased, and timetables for the movement of the horses were brought into operation on 1 April 1914 with the rest of the railway programme for the year.[78] The civilian collectors received their orders to commence the collection of horses on 3 August. Within twelve days the British army had successfully impressed 165,000 horses, an impressive number but one significantly smaller than the 615,000 initially mobilized by the German army.[79]

The timetable for the horses purchased on mobilization for the Aldershot command alone highlights the level of cooperation required for the scheme to work effectively. The civilian purchasers who provided horses for the first train to depart for Aldershot had to ensure their animals were at Worcester station by 8:40 p.m. on the first day of the programme, and the final departure for Aldershot was scheduled to leave Birmingham on the third day. During that period, trains for the Aldershot troops and Cavalry

[76] IWM, Wilson papers, HHW 3/6/4, note from a meeting in Major-General Heath's room, 27 July 1911.

[77] T. R. F. Bate, 'Horse mobilisation', *Royal United Services Institution. Journal*, lxvii (1922), 16–25, at pp. 18–19; J. Singleton, 'Britain's military use of horses 1914–1918', *Past & Present*, cxxxix (1993), 178–203, at p. 184; Chapman-Huston and Rutter, *Sir John Cowans*, i. 251–9.

[78] IWM, Wilson papers, HHW 3/7/3A, note containing '48 points' concerning the expeditionary force scheme prepared for the DMO, 3 Apr. 1913, p. 2.

[79] Singleton, 'Britain's military use of horses', p. 184.

Division were entrained at Worcester, Pershore, Coventry, Nuneaton, Amesbury, Basingstoke, High Wycombe, Abingdon, Reading, Aylesbury, Maidenhead, Chipping Norton, Oxford and the two Birmingham stations of Hockley and Small Heath. The stations of Bordon, Liphook and Farnborough received the animals, which had travelled along the lines of the Great Western, LSWR and LNWR on their journeys to Hampshire.[80] The surviving timetables for the eastern command present a similar story but on an even larger scale. The LNWR, Midland, SECR, Great Eastern, Great Northern and the London, Brighton and South Coast (LBSCR) railways all participated in the movement of horses from various sites across East Anglia and the south-east to fulfil the mobilization requirements of the Essex and Norfolk Yeomanry.[81]

The passages above demonstrate that the railways' work in August 1914 was not merely restricted to the delivery of the BEF and its stores to various ports in southern England. In the weeks that followed the government's decisions to enter the war and to send the BEF to France, Britain's railways were responsible for the mobilization and concentration of the BEF, the movement of the territorials, reserves and Royal Navy personnel, the supplies and equipment required by all of the above, and the maintenance of Britain's colossal passenger and freight traffic with as little dislocation to the rhythms of civil life as possible. Notices were pinned up at prominent stations across the country to warn commuters of potential disruption to regular passenger services by the demands of the armed forces. However, 'the business trains to and from London ran very much as usual, and the normal service was maintained on nearly all parts of the system'.[82] Britain's railways ran 1,408 specially timetabled trains for the carriage of over 334,500 troops during the first fortnight of the mobilization period. The Great Western alone handled 632 special troop trains, forty-one trains containing coal for the Admiralty and 149 trains containing petrol, oil and various stores. Aside from the suspension of excursion services, the ordinary goods and passenger traffic across the Great Western's system was maintained

---

[80] TNA, WO 33/657, mobilization railway time tables for southern command, section III: Aldershot command, table 4(G) – horses bought on mobilization (revised to July 1914). Under the army horse reserve agreement, approved by the Great Western's directors in March 1913, the railway agreed to provide the army with 221 horses on the outbreak of war. In August 1914 the army requisitioned a further 40 light draft horses from the railway, and 12 more animals were commandeered by the military authorities. See TNA, ZLIB 10/11, Great Western Railway: war reports of the general manager to the board of directors, 1914–1919, p. 16; S. Gittins, *The Great Western Railway in the First World War* (Stroud, 2010), p. 17.

[81] TNA, WO 33/676, eastern command mobilization railway timetables for horses, Category (G) – mobilisation horses, 1914.

[82] 'Railways and the war. Reduced passenger service', *Railway Gazette*, 7 Aug. 1914, p. 194.

throughout.[83] Welsh coal for the fleet began to arrive at Grangemouth on the Caledonian Railway's network on 10 August.[84] During that month the Caledonian handled 342 naval and military trains, most of which originated on their system and were composed of Caledonian stock. Thanks to the Admiralty's decision not to demobilize the fleet following test mobilizations in mid July, the human traffic connected to the Royal Navy's entry into the war was relatively small. Consequently, the comparatively tiny Highland Railway – the closest point on the railway network to the naval base at Scapa Flow – was not overloaded by the northward movement of reservists in early August. Those who had not been called up on 12 and 14 July headed towards Orkney from 2 August in a series of movements that were 'carried out with perfect smoothness'.[85]

It was at the opposite end of the country where the vast majority of the railways' attentions were directed in August 1914, however. All railroads may not have led to Southampton (see Figure 2.1), but at the start of the war a large volume of traffic was delivered into the BEF's principal port of departure. The first troops, charged with the establishment of supply bases and rest camps near the French Channel ports in readiness for the main body of the BEF, arrived on 7 August. Captain R. H. D Tompson was among them, and described arriving on a construction site as the railway access to the port was swiftly increased. The existing provision, described by Tompson as 'a strategic disgrace', was augmented by new construction undertaken by a 'very large group of navvies [who toiled] night and day'. Their exertions created 'a very fine piece of work', which 'seemed to grow almost as one watched' and was complete by 8 August when the bulk of the BEF started to filter through to the docks.[86] The programme demanded that the LSWR receive 350 trains – each composed of an average of thirty vehicles – into the port, unload their passengers and cargo, and remove them from the platforms in readiness for the next train within sixty hours. They did so within forty-eight. A train pulled into the docks every ten minutes in the first twenty-four hours of the concentration period, and for the following nineteen days in a row Southampton received ninety trains

---

[83] C. Hamilton Ellis, *British Railway History: an Outline from the Accession of William IV to the Nationalisation of Railways, 1877–1947* (2 vols., London, 1959), ii. 300–1; Gittins, *The Great Western Railway*, p. 13; C. Maggs, *A History of the Great Western Railway* (Stroud, 2015), p. 175.

[84] Pratt, *British Railways and the Great War*, ii. 546.

[85] Pratt, *British Railways and the Great War*, i. 110.

[86] Brotherton Library Special Collections (BLSC), Liddle collection, papers of Captain R. H. D. Tompson, LIDDLE/WW1/GS/1612, diary entries 7 and 8 Aug. 1914. My thanks to Mark Butterfield for alerting me to Captain Tompson's diary.

per day – each one loaded with men, horses, wagons, guns, ammunition and myriad other supplies. The flexibility and contingency that had been built into the programme resulted in the majority of trains arriving at Southampton between twenty-five and thirty minutes ahead of schedule. In the first twenty-four-hour period, just one train was recorded as having arrived late, and that by only five minutes.[87]

By 26 August, 65,814 officers and men had left Southampton aboard steamers sourced by the Admiralty from civilian firms – including, as the operators of a variety of waterborne services, many ships owned and operated by railway companies.[88] Second-Lieutenant Lyndall Urwick of the 3rd Battalion, Worcestershire Regiment, was among them. In an unpublished autobiography, he wrote that the scene on the evening he set sail was unforgettable:

> Pleasure steamers and ferry boats from every short sea crossing round the British Isles had been called in. For the Channel crossing to Le Havre and Rouen they could afford to pack the decks. Down both sides of Southampton Water as far as the eye could reach they were moored alongside the quays, stern to stern, and every ship was a mass of men. Yet still there were thousands more on the quays waiting their turn. In the middle of the Water, line ahead, going out into the sunset, were six more ships fully loaded. And every man Jack of 60,000 men was singing Tipperary.[89]

The accumulation of men and materials at the ports proved the only major concern for the QMG's staff during the concentration. Heavy fog in the English Channel meant that transports did not return to Britain as quickly as the timetable demanded. The weather, combined with an examination process at Southampton that Mance described as 'too rigorous', created a backlog of 'over one day's troops at the rest camps' around the port. Anxious at the build-up of men, Sir John Cowans went to Southampton to examine whether the railway programme should be postponed for a day to allow the port to be cleared. However, the fog dispersed, the examinations process was streamlined and the natural contingency built into the railway programme from its inception meant that the arrears were rapidly rectified.[90]

---

[87] Chapman-Huston and Rutter, *Sir John Cowans*, i. 281; Beckett, 'Going to war', p. 143; 'Railway administration in war', *Railway Gazette*, 20 Nov. 1914, pp. 529–30; TNA, ZPER 7/103, records of railway interests in the war, 1915, p. 18.

[88] A. J. Mullay, *For the King's Service: Railway Ships at War* (Easingwold, 2008).

[89] Greenlands Academic Resource Centre (GARC), papers of Colonel Lyndall Fownes Urwick, 8/3/2, management pilgrimage, p. 2.

[90] TNA, PRO 30/66/9, correspondence and papers, recollections, pp. 6–7; Chapman-Huston and Rutter, *Sir John Cowans*, i. 282.

Figure 2.1. Diagram showing how all (rail) roads lead to Southampton.

Source: E. A. Pratt, *British Railways and the Great War; Organisation, Efforts, Difficulties and Achievements* (2 vols., London, 1921), ii. 1008.

Once the troops had disembarked on French soil they passed beyond the limits of Britain's transport experts' influence on the WF scheme. From that point forward the successful concentration of the BEF depended upon the collaborative efforts of Wilson's staff and the French army. The planning for the final step of the BEF's concentration was complicated by the different train-loading methods employed on either side of the Channel.

In Britain trains 'of medium weight [were run] at a fairly high speed; on the continent the practice [was] to run very heavy trains at slow speed'. In Britain a battalion or battery was transported in two trains; in France they travelled in one.[91] Therefore, the units that disembarked in France did not entrain on vehicles of the same composition as those they had left behind in Britain. Fresh timetables had to be compiled between the British and French authorities to provide for the inland movement of the BEF. The French proved more than happy to collaborate in an activity that further cemented the growing alliance between the two nations, and in July 1914 Wilson ordered three staff officers to attend the French 11th Division's manoeuvres alongside Sir John French, Haig and other senior figures. The object of their visit was to become familiarized with the military workings of the French railways, and the lavishness of Huguet's praise is noteworthy:

> I also met at the same time [as Sir John, Haig, et al.] three of your officers, Radcliffe, Johnston [sic] and _____, and very glad to say they made a *very, very good impression*, first by themselves, their intelligence, their cleverness, their way of working, their seriousness … and also, I am glad to say, by the very good work which they had brought with them – our people were very gratified to see how well they work in the DMO department, how the thing has been seriously taken and carefully studied. In all this, I recognize the hand of my friend General Wilson, but all the same, it is really a pleasure to work with officers like those three whom you sent out.[92]

Unfortunately, the details of 'the very good work which they had brought with them' was not elaborated upon. However, the results of their endeavours became clear once the BEF arrived in France a month later. Each unit was issued with a manual that advised them of the procedures to be followed on their journey to the front, while officers who arrived in advanced parties received detailed instructions that delineated their responsibilities from those undertaken by the local authorities.[93] There were teething problems, particularly as the British troops struggled to entrain quickly onto unfamiliar rolling stock from rail level rather than from platform level. The first unit to entrain took five-and-a-half hours to load its vehicles; the equivalent

---

[91] A. M. Henniker, *History of the Great War: Transportation on the Western Front, 1914–1918* (London, 1937), p. 14.

[92] IWM, Wilson papers, HHW 2/73/27, Huguet to Wilson, 16 July 1914. Emphasis in original.

[93] TNA, WO 106/49A/7, instructions for First and Second army commanders and officers of advanced parties and railway transport establishment, pp. 1–3; WO 106/49B/1, instruction for entrainment and embarkation (short voyage) for units of the expeditionary force; BLSC, Tompson papers, LIDDLE/WW1/GS/1612, diary entry, 6 Aug. 1914.

French force was expected to complete the task in ninety minutes.[94] Yet the lack of standardization between the railway operations of the two countries, and the three-day gap between the mobilizations of their forces, did not materially obstruct the British concentration on the French army's left. The BEF entrained upon 361 trains from 15 August onwards, undertaking journeys from Boulogne, Le Havre and Rouen to the concentration area around Maubeuge, Busigny and Hirson that took up to seventeen hours to complete. Only thirty-six of the 343 trains that passed through Amiens during the period were more than thirty minutes late.[95] By 23 August, as the battle of the Frontiers raged to the south and took a horrific toll on the French and German forces involved, the BEF had crept forward to Mons where its own war began in earnest.

Conclusion

In an address to the American Luncheon Club on 13 November 1914, the acting chairman of the REC and general manager of the LSWR, Herbert Walker, remarked on the successful despatch of the BEF to the continent that:

> Magnificent and unprecedented as this feat was, we can pay the British railways no higher compliment than to say that it was expected of them, and that every man in the service knew the railways were equal to every demand that could be made on them, without it being necessary to dislocate ordinary traffic to one-quarter of the extent which mobilization involves abroad.[96]

Between 10 and 31 August Walker's own railway had carried 4,653 officers, 113,801 men, 314 guns, 5,221 vehicles, 1,807 cycles, 4,557 tons of stores and 37,469 horses on the army's behalf.[97] Its efforts – and those of its counterparts across the British Isles – pale by comparison to the railway efforts undertaken elsewhere across Europe as the gigantic conscript forces of the great powers readied for battle. In the first six days of mobilization alone the eastern border corps of the German army mobilized over 148,000

---

[94] Henniker, *History of the Great War*, p. 21.

[95] M. Peschaud, *Politique et fonctionnement des transports par chemin de fer pendant la guerre* (Paris, 1926), pp. 83–4; J. H. F. Le Hénaff and H. Bornecque, *Les chemins de fer français et la guerre* (Paris, 1922), pp. 215–22. Lieutenant Evelyn Needham's recollections of the 'interminable' journey from Le Havre to the front are quoted in P. Hart, *Fire and Movement: the British Expeditionary Force and the Campaign of 1914* (Oxford, 2015), pp. 63–4.

[96] 'Railway administration in war', p. 530.

[97] 'Modern armies and modern transport: the work of the London and South-Western Railway during the war', *Railway Gazette*, 31 Jan. 1919, p. 160.

men, mainly drawn from Berlin and the Rhine area. At the same time, on the other side of the country, the XV, XVI and XXI Army corps received almost 112,000 men and 23,000 horses by rail. In total, the German and French armies ran 20,800 and around 10,000 trains respectively in support of their mobilization programmes; far higher than the 1,408 trains operated on the British rail network. The 361 trains that delivered the BEF to its concentration area from the Channel ports were a mere fraction of the 4,278 trains that traversed the French rail network in the French army's fourteen-day concentration period – during which up to 380 trains per day delivered some 1,300,000 French troops into position.[98] The British railways' effort was 'magnificent and unprecedented', but it was undertaken on a far smaller and less conspicuous scale than those on the European mainland in August 1914.

However, the substantial investments of time and resources made by Walker, his colleagues in the REC and the thousands of unnamed railway servants to the British mobilization should not be reduced to a mere footnote in the history of the First World War.[99] For over three years Britain's most prominent railway companies provided labour 'greatly in excess of what had previously been necessary' in response to the army's complex and changeable transport demands.[100] They participated in an exchange of ideas and expertise with Henry Wilson and his staff within the directorate of military operations. Their joint endeavours underline both the applicability of industrial knowledge and skills to military operations and the existence of a far closer working relationship between the army and industry than has hitherto been acknowledged. The British army, 'contemptible' in size as it may have been, was absolutely reliant upon privately owned locomotives, rolling stock and track for its propulsion to the front. Henry Wilson understood this, and turned a professional relationship that had been focused upon defensive and educational activities into one that was integral to the swift deployment of Britain's 'strike force'. By the time Europe's armies began to roll out their gargantuan mobilization schemes – drawn and redrawn by large-scale, state-sanctioned military organizations – Britain possessed a thoroughly researched and mapped out scheme for the deployment of the BEF.

---

[98] Reichsarchiv, *Der Weltkrieg, 1914 Bis 1918: Das Deutsche Feldeisenbahnwesen* (Berlin, 1928), p. 12; *Les armées françaises dans la grande guerre: la direction de l'arrière* (Paris, 1937), p. 24; Le Hénaff and Bornecque, *Les chemins de fer français*, p. 28; Stevenson, 'War by timetable?', p. 167.

[99] Robin Neillands, in an otherwise broad account of the mobilization process, only referred to one railway company's contribution to the WF scheme: the LSWR. See Neillands, *The Old Contemptibles*, p. 90.

[100] Pratt, *British Railways and the Great War*, i. 16.

Railway timetables did not impose the First World War upon the statesmen of Britain or any other of the European powers in 1914, as A. J. P. Taylor famously asserted in the 1960s.[101] Henry Wilson's vituperative diary entries in early August, as his carefully coordinated plans for a simultaneous mobilization and embarkation were cast aside by Asquith's government, illustrate the limited influence of logistics over the state's decision-making processes.[102] Yet once it had taken the decision to commence hostilities, the lack of alternatives to the WF scheme severely constrained the government's freedom of action. No other programme of movements existed in the detail required for it to be carried out swiftly in August 1914. From the very outset the options available to British civil and military leaders were limited by the transport factor. For the four years that followed, the roads, railways and shipping lanes that had delivered the BEF to the western front became part of a global transport infrastructure that sustained and shaped the fighting that took place during the First World War. The contribution of Britain's transport experts to the processes and procedures of industrial warfare did not end once the firing started.

---

[101] A. J. P. Taylor, *The First World War: an Illustrated History* (London, 1963), p. 20.
[102] IWM, Wilson papers, HHW 1/23, diary entries, July–Aug. 1914. The period is summarized in Jeffery, *Henry Wilson*, pp. 128–9.

II. Expansion

3. Stepping into their places: Britain's transport experts and the expanding war, 1914–16

In August 1915, the *Great Western Railway Magazine* carried photographs and brief biographies of twenty-four of the company's employees, each of whom had recently died while in the service of their nation. The men's details illustrate the great diversity of work undertaken and the geographical sprawl of Britain's largest pre-war railway companies. Porters, clerks, relayers, lamp men, carpenters, packers and labourers were among the two dozen men featured in the magazine, who had enlisted from such places as the company's works at Swindon, its London terminus at Paddington, the engine shed at Newton Abbot and the engineering department at Plymouth. The locations of their deaths highlight how Britain's commitments to the First World War had grown beyond France and Flanders in the first twelve months of the war. While Emmanuel Rowland died of pneumonia in Exeter, and Lance-Corporal F. Hammond was killed in action at Hill 60 near Ypres, both James Gully and Walter Lamacroft were lost on 12 May 1915 when HMS *Goliath* was torpedoed and sank at Cape Helles.[1] As the magazine went to print the 5th Battalion, Wiltshire Regiment – whose recruits included many with pre-war attachments to the Great Western's Swindon works – were overwhelmed by Turkish forces at Chunuk Bair. At least nineteen of those who died in the assault were railwaymen. The Great Western, Great Eastern and LNWR all lost former employees on the slopes of the Sari Bahr ridge. In the first year of the First World War the railwaymen of Britain found themselves at the forefront of Britain's expanding war effort.[2]

Alongside their contributions to the fighting forces on various front lines, employees from some of Britain's largest railway companies helped underwrite the organizational effort required to gear the nation towards the prosecution of a war effort on an unprecedented scale. The creation, maintenance and sustenance of a mass army capable of rivalling the conscripted forces of the other European powers presented Britain's civil and military authorities with a series of colossal challenges, which revealed

[1] 'G.W.R. men who have lost their lives in the war', *Great Western Railway Magazine*, Aug. 1915, pp. 206–7.
[2] J. Higgins, *Great War Railwaymen: Britain's Railway Company Workers at War 1914–1918* (London, 2014), pp. 100–1.

themselves gradually from the moment the decision to raise such a force was taken. 'Issues of management and logistics', claimed two British scholars, 'were the primary concerns of senior commanders for the first three years of the war'. The army, they continued, underwent a 'conceptual change' that involved the mobilization of businessmen who brought 'their knowledge of forecasting and economies of scale to military logistical supply'.[3] This process is viewed almost exclusively through the prism of Sir Eric Geddes and the creation of the directorate-general of transportation in the autumn of 1916. Even Elizabeth Greenhalgh's *Victory through Coalition*, which more than any previous history analysed the bureaucratic machinations behind the inter-allied management apparatus, devoted little more than two pages to the logistical complexities and frailties that afflicted the Franco-British coalition before the battle of the Somme.[4]

Yet prior to Geddes's arrival on the western front, as this and the following chapters will demonstrate, the British army actively sought out and engaged with the skills and expertise possessed by Britain's transport experts. Individuals and companies from across the country, the empire and beyond, were tasked with identifying solutions to the multiple conundrums posed by modern industrial warfare. Both at home and in the multiple theatres that emerged as the war spread from Europe, those who possessed technical and managerial experience in the railway industry found profitable employment in the production and distribution networks that underpinned the British war effort. However, the army's initial attempts to utilize civilians to grapple with the implications of industrialized warfare were relatively small in scale and limited in scope. Furthermore, they were hamstrung by the constant presence of the two factors that inhibited the British army's development before 1916: a continued belief in the impermanence of trench warfare and subsequent disinclination to prepare for a static war; and the French army's reluctance to relinquish control over the shared transport infrastructure upon which the coalition's forces depended.

Transport organization and the clash of arms

The War Office, like its compatriots across Europe, entered the First World War with no proposals for how to deal with the consequences of an indecisive opening engagement. 'Whether the General Staffs really expected a war to

[3] D. Todman and G. Sheffield, 'Command and control in the British army on the western front', in *Command and Control on the Western Front: the British Army's Experience, 1914–18*, ed. G. Sheffield and D. Todman (Staplehurst, 2004), pp. 1–11, at p. 6.
[4] E. Greenhalgh, *Victory through Coalition: Britain and France during the First World War* (Cambridge, 2005), pp. 33–5.

end by Christmas', argued David Stevenson, 'all they planned in detail was the opening campaign'.[5] The pre-war conversations between the French and British general staffs had resulted in an arrangement whereby the BEF's transport requirements were to be 'manned and controlled by the French'. The BEF's hosts undertook to complete 'the work of construction, repair, maintenance, traffic management and protection' necessary to maintain the British troops on the western front.[6] In a sign of the French army's confident approach to the impending hostilities – and its assumption that the BEF was unlikely to substantially increase in size over the course of the war – the French also committed to provide for all the BEF's transport needs after the allies had pushed the Germans back beyond the French-Belgian frontier. The terms of the coalition arrangement were such that the duties nominally assigned to a British director of railway transport (DRT) were almost entirely assigned to the French army. Consequently, the British officer earmarked for the position, Colonel John Twiss, remained in London when the BEF set sail. Only a small staff of liaison officers, the railway transport establishment, crossed the Channel to act as intermediaries between the BEF and the French railway authorities in early August 1914.[7]

The twenty-nine officers who constituted the initial railway transport establishment – twelve of whom were students at the staff college when war was declared – were thrown into action immediately. Captain R. H. D. Tompson, one of the staff college students, recorded in his diary the disappointment he felt at being given such a 'poor job'. 'It is very hard after all this sweat of getting to the Staff College', he wrote on 6 August, 'to find oneself in a rotten job like this'. Tompson, a veteran of the South African War, embarked on the LSWR ship SS *Vera* on 9 August, and went onshore at Le Havre at 6:30 a.m. the following morning. Before he embarked for France, the relationship between the British officers and the French railway authorities was clearly explained. The French were to control operations within the stations, and Tompson's role was to ensure that disembarking troops complied with the instructions issued by the station staff. For each train that arrived at his allocated station of Vaux (and later Busigny), Tompson worked alongside the 'excellent fellows' of the *commission de la gare* to ensure the swift unloading of troops and equipment from the trains, their removal from the station's surroundings and the preparation of the

[5] D. Stevenson, 'War by timetable? The railway race before 1914', *Past & Present*, clxii (1999), 163–94, at p. 166; I. M. Brown, *British Logistics on the Western Front, 1914–1919* (London, 1998), pp. 75–6.

[6] TNA, WO 33/686, instructions for the IGC, part II, section 1, 1914; A. M. Henniker, *History of the Great War: Transportation on the Western Front, 1914–1918* (London, 1937), p. 13.

[7] Henniker, *History of the Great War*, pp. 16–17.

area to receive the next arrival. The work was incessant. The steady stream of trains meant that he could only snatch occasional periods of rest, and as the concentration period climaxed Tompson recorded that he 'had not slept in three days and two whole nights'. By 24 August, when the BEF had just commenced its infamous retreat from Mons, Tompson's boots had already worn through.[8]

The French authorities' desire to retain control inside the railway stations near the concentration area was not matched by the ability to honour their agreement with the BEF at the coast. The pre-war arrangement at the ports was broken before the bulk of troops had begun to arrive. Before he set sail to take up his post, the IGC Sir Frederick Robb was dismayed to discover that the French 'had not kept their promises about the dock employees, they can only furnish 1000 stevedores out of the 3000 [and] they propose not to work at night. I have had to be very firm about this, they have now promised to try and get some more'.[9] This inauspicious beginning to the practical operations of the Franco-British coalition set a pattern that continued once the fighting commenced in earnest.

The outbreak of war sparked a colossal volume of traffic on the French railways. According to one American correspondent, 'no fewer than 1,800,000 troops were gotten to the front, and each of these soldiers were handled three times, so that in reality 5,400,000 troops were delivered at the required points … while possibly 5,000,000 of the civil population were also travelling' between 1 and 24 August.[10] The mass of refugees provided a noteworthy concern for the allies' supply officers. Around 100,000 people 'threw themselves' at any trains heading west from Laon in the final days of August, while at Busigny Tompson witnessed the 'poor haunted creatures' who choked the station in search of transport away from the front and complicated the movements of troops.[11] In the less sympathetic view of the QMG, Sir William Robertson, the refugees were 'an awful nuisance, blocking our roads, and even our fire' during the retreat from Mons. In a letter to the king's private secretary he described 'colonies of perhaps 200 or 300 families', who in some cases had travelled from central Belgium to the outskirts of Paris. 'The selection they have made of their belongings', Robertson noted, 'has amused me more than anything. It includes in some cases of a flock of about 1,000 sheep. Two or three wagons of what looks

[8] All quotations in this passage are taken from BLSC, Tompson papers, LIDDLE/WW1/GS/1612, diary entries, 6–24 Aug. 1914.

[9] IWM, Wilson papers, HHW 2/73/49, Robb to Wilson, 10 Aug. 1914.

[10] 'France saved by her railroad men', *Railway Gazette*, 16 July 1915, p. 58.

[11] J. H. F. Le Hénaff and H. Bornecque, *Les chemins de fer français et la guerre* (Paris, 1922), p. 37; BLSC, Tompson papers, LIDDLE/WW1/GS/1612, diary entries, 24 and 25 Aug. 1914.

like straw or hay, but which really consist of furniture and clothing, hidden under the straw. Bicycles, mattresses, perambulators, boxes, cocks and hens, turkeys and so on' comprised all the refugees had been able to remove in the face of the German advance.[12]

As QMG, it was Robertson's responsibility to ensure that his troops were supplied with the items they required. Yet as the location of the front constantly shifted, the BEF's administrative departments behind the lines could not be certain of the troops' locations from day to day.[13] By the time rendezvous points had been selected by GHQ and communicated to Robb's headquarters on the lines of communications, there was no guarantee that British troops would be in position to receive the supplies once they had been shifted forwards. Closer to the front the quartermasters of individual formations struggled to maintain contact with their troops as the road network became increasingly congested with troops, guns, refugees and abandoned supplies.[14] Lyndall Urwick recalled that 'only once or twice during the retreat and the Battle of the Marne had our regimental transport caught up with us'. Consequently, the food Urwick and his comrades received on the retreat 'had been uncertain but monotonous, consisting, when we got any, almost entirely of bully beef and biscuit' or whatever the enterprising soldier could scrounge.[15] With the distribution system insufficient to meet the needs of an army on the move, Robertson arranged for food and ammunition to be 'dumped' at busy crossroads for the men to take as they passed.[16] Inevitably, he recalled, such a system led to 'excessive waste' and huge volumes were left behind. 'But when the troops are fighting hard', he reasoned, 'one does not like to worry them too much about administrative matters. The chief thing is to beat the enemy' rather than obsess over red-tape.[17]

[12] Liddell Hart Centre for Military Archives (LHCMA), papers of Field Marshal Sir William Robertson, 7/1/1 Robertson to Wigram, 1 Sept. 1914.

[13] Brown, *British Logistics*, p. 61.

[14] A. Whitty, *A Quartermaster at the Front: the Diary of Lieutenant-Colonel Allen Whitty, Worcestershire Regiment, 1914–1919*, ed. E. Astill (Eastbourne, 2011), pp. 22–31; H. A. Stewart, *From Mons to Loos: Being the Diary of a Supply Officer* (Edinburgh, 1916), pp. 54–72.

[15] GARC, Urwick papers, 8/4, apprenticeship to management, pp. 34, 47; 8/3/2, management pilgrimage, p. 3; F. Richards, *Old Soldiers Never Die* (London, 1933), p. 27.

[16] J. Spencer, "'The big brain in the army": Sir William Robertson as quartermaster-general', in *Stemming the Tide: Officers and Leadership in the British Expeditionary Force 1914*, ed. S. Jones (Solihull, 2013), pp. 89–107, at p. 97. Urwick recalled the scene at one roadside dump where, had it not been for the posting of guards with fixed bayonets, the Royal Irish Rifles 'would have looted the lot'. See GARC, Urwick papers, 8/4, apprenticeship to management, p. 37.

[17] TNA, WO 95/27, branches and services: quarter-master general, Robertson to Maxwell, 23 Oct. 1914; W. R. Robertson, *From Private to Field-Marshal* (London, 1921), pp. 208–10.

The initial campaigns of the First World War underlined the inaccuracy of Lord Kitchener's insistence that Sir John French's command was 'an entirely independent one'.[18] The arrangements by which control of the railway network remained in the hands of *Grand Quartier Général* (*GQG*), and the relative strengths of the French and British forces, ensured that priority was consistently given to the demands of the former during the emergency of the war's opening months. All orders for railway transport had to be made through the French railway authorities. Therefore, the BEF was entirely reliant upon their host's willingness to run trains filled with British supplies.[19] By the end of September Robertson grumbled that he had 'always doubted the possibility of our obtaining much, if any, transport from French sources'.[20] Furthermore, *GQG* allotted the best railheads to French troops, which forced the BEF to rely on inferior facilities, and when French supply trains blocked the lines ahead of British railheads the BEF simply had to wait. As battle raged around Ypres on 23 October, Robertson observed with mounting frustration that the troops were struggling to obtain sufficient ammunition:

> Some of the ammunition trains yesterday were within a few miles of our railheads but we could not get them there. It seems ridiculous that it should take some eighteen hours from Boulogne [to] here but it does, and the greater part of that time is probably spent near where we are … *If anything* goes wrong with the ammunition train there may be a shortage, of which there can be no greater QMG's offence. Besides, it is exceedingly wearing and worrying for one every day to be wondering whether the ammunition required will be forthcoming.[21]

The French army was engaged heavily in a struggle for national survival alongside a small, untested force, perceived in Paris as hesitant and unreliable. Unsurprisingly, the BEF's requests were persistently subordinated by *GQG* to the demands of their own troops and civilian population.

The discussions surrounding the BEF's transfer to Flanders ahead of the first Ypres battle neatly illustrate the reality of the Franco-British relationship in 1914. In late September Sir John conceived a plan to unite his forces and undertake a huge enveloping manoeuvre against the Germans concentrated

---

[18] W. Philpott, *Anglo-French Relations and Strategy on the Western Front, 1914–18* (Basingstoke, 1996), pp. 15–16.

[19] TNA, WO 95/3949, headquarters branches and services. Inspector general, Robertson to Maxwell, 24 Oct. 1914.

[20] TNA, WO 95/3949, IGC war diary, Robertson to Maxwell, 29 Sept. 1914.

[21] TNA, WO 95/27, QMG war diary, Robertson to Maxwell, 23 Oct. 1914. Emphasis in original.

on Lille. The plan would take 'a week or nine days' to execute, and if successful would put an end to the German invasion of France.[22] Sir John wrote to his opposite number, the French commander-in-chief General Joseph Joffre, to forcefully request that British troops be moved north to put his ambitious plan into action. 'Both from strategical reasons and tactical reasons', Sir John argued, 'it is desirable that the British Army should regain its position on the left of the line. There remains the question of *when* this move should take place. I submit that *now* is the time'.[23] In response, Joffre stated that he would 'endeavour to satisfy this request', but warned that 'the movement of the British troops can only be carried out in succession'. Joffre's letter went on to 'assure Marshal French' that 'the greatest efforts' would be made to concentrate the whole of the BEF in the northern sector of the front, but noted that to immediately comply with Sir John's wishes would severely delay the French army's intended operations. Consequently, the British troops moved not as a whole but in small groups, and they travelled according to a schedule devised at *GQG* rather than GHQ.[24]

The BEF's movement to Flanders and the onset – although not perceived as such at the time – of static warfare, provided Robertson and his colleagues with an opportunity to reflect on the efficacy of the transport arrangements on the western front. To maintain a regular supply to their forces, the British needed a thoroughly staffed traffic organization able to coordinate the BEF's transport needs with those of the French. The first steps to providing such an organization had already been taken. Between 18 September and 1 October, the BEF's railway transport establishment more than doubled thanks to the arrival of thirty-two new officers in France, while several officers with experience on the Indian railways – including Major Henry Freeland and Lieutenant-Colonel Valentine Murray – took up positions on the IGC's staff to coordinate their activities. By 10 October the newly established traffic office, under Freeland's command, had seventy-five men at its disposal to facilitate the BEF's movements on the French railways.[25]

---

[22] S. Badsey, 'Sir John French and command of the BEF', in *Stemming the Tide: Officers and Leadership in the British Expeditionary Force 1914*, ed. S. Jones (Solihull, 2013), pp. 27–50, at p. 48.

[23] IWM, Wilson papers, HHW 2/73/62, note (signed by Sir John French), 29 Sept. 1914. Emphasis in original.

[24] LHCMA, Robertson papers, 2/2/85, Joffre to French, 5 Oct. 1914; TNA, WO 95/27, QMG war diary, Railway transport for the British army, 12 Oct. 1914; Le Hénaff and Bornecque, *Les chemins de fer français*, p. 223; Greenhalgh, *Victory through Coalition*, p. 19. See Henniker, *History of the Great War*, p. 73 for a schedule of the rail movements involved in the transfer of British troops from the Aisne, and Indian divisions from Marseille, to the northern flank of the French army.

[25] Henniker, *History of the Great War*, p. 41.

Further forward, the BEF's contribution to the construction and repair of railways was also dealt with. Colonel Twiss took up his duties as DRT on 16 September, and attended a meeting at *GQG* to discuss the subject of railway construction the following day. The 8th (Railway) Company, Royal Engineers, had been in France for a month. Since their arrival the company had laid a short length of siding in the port of Le Havre and shadowed the work of French railway engineers, but by mid September the company had achieved little of value to the allies. Indeed, such was the unit's perceived redundancy that in late August a proposal had been floated that men without knowledge of specific railway trades should be withdrawn to replace losses among the front-line troops.[26] Following Twiss's meeting at *GQG* the French agreed to modify the pre-war arrangements, and accepted the offer of British railway troops to assist with repairs on the Chemins de Fer du Nord – provided that the French retained overall control of the work undertaken. The 8th (Railway) Company was immediately set to work repairing the Pont de Metz, 'a lofty brick bridge of two spans' to the south-west of Amiens that had been destroyed by the Germans during their retreat from the Marne. Upon completion of the heavy timbering work required for the 'semi-permanent repairs' to the bridge the company headed to St Omer en Chaussée and Gamaches, where they worked alongside 300 employees of the Nord railway to install short connections that eliminated the need for engines to be reversed on single lines.[27]

With the BEF's available transport assets fully engaged in the task of keeping the troops supplied and the French army aware of their requirements, a comprehensive examination of the BEF's supply arrangements could not be handled from within. Therefore, when Colonel Twiss and Lieutenant-Colonel Murray arrived at the IGC's office to commence an investigation into the BEF's transport organization, they were accompanied by a transport expert sent to France by Kitchener to oversee the task. Despite the rank, Brigadier-General Sir Édouard Percy Cranwill Girouard was not a serving officer in October 1914. However, his role cannot be considered equivalent to that of the purely civilian railway experts that contributed to the army's operations later in the conflict. Girouard, a French-Canadian from Montreal, had acquired railway construction and operation experience – both in war and peace – across three continents. After graduation from the Royal Military College at Kingston he had spent two years on the engineering staff of the Canadian Pacific Railway, before he accepted a commission in

---

[26] Henniker, *History of the Great War*, pp. 53–4.

[27] TNA, WO 95/4052, lines of communication troops. 8 Railway Company Royal Engineers, diary entries, 28 Aug. to 21 Oct. 1914.

the Royal Engineers to become the Royal Arsenal at Woolwich's first traffic manager.[28]

At Woolwich Girouard observed the 'confusion and waste' that emerged when competing departments attempted to exert influence over a shared transport system. Before his arrival no central administration existed to oversee traffic flows around the 824-acre site, and each factory had arranged their own train schedules on the arsenal's narrow-gauge railway. Girouard took control of the rolling stock and motive power – thirty-six engines and 1,000 carriages, vans and trucks – and centralized all traffic requests within the arsenal. Under his authority, the narrow gauge became an integral component of operations at Woolwich, provided a 'valuable link between office and shop, storehouse and magazine', and even acted as a passenger service for the employees of various factories.[29]

If Woolwich had taught Girouard that competitive behaviour between units that were working towards a shared goal had to be avoided – whether in the south-west of London or the north-east of France – his next role exposed him to the practical advantages to be gained from the use of railways for military purposes. In 1896 he was seconded to the Egyptian army, and was responsible for overseeing the construction of a railway across the Nubian Desert early the following year.[30] By July 1898 the 'cholera-decimated' engineers under his charge had extended the railway to Atbara, and the line was of sufficient quality to sustain Kitchener's 22,000-strong force as it triumphed at Omdurman. According to Edward Spiers, 'this remarkable victory in which 10,200 Mahdists had been killed, and possibly another 16,000 wounded in a morning, derived from the transformative' effect of the Sudan Military Railway'.[31] Girouard received the Distinguished Service Order for his efforts on the project and was appointed president of the Egyptian railway and telegraph administration. However, less than a year later his services were required in a theatre of war at the other end of Africa. Girouard became director

[28] R. P. T. Davenport-Hines, 'Girouard, Sir Édouard Percy Cranwill', in *Dictionary of Business Biography: a Biographical Dictionary of Business Leaders Active in Britain in the Period 1860–1980*, ed. D. J. Jeremy (5 vols., London, 1984), ii. 570–4; J. Flint, 'Girouard, Sir (Édouard) Percy Cranwill (1867–1932)', *ODNB* <https://doi.org/10.1093/ref:odnb/33415> [accessed 14 Sept. 2014]; O. F. G. Hogg, *The Royal Arsenal: its Background, Origin and Subsequent History* (2 vols., Oxford, 1963), ii. 878, 1292; M. Smithers, *The Royal Arsenal Railways: the Rise and Fall of a Military Railway Network* (Barnsley, 2016).

[29] Hogg, *The Royal Arsenal*, pp. 878, 1309–10.

[30] W. Baker Brown, *History of the Corps of Royal Engineers* (11 vols., Chatham, 1952), iv. 256–66; J. N. Westwood, *Railways at War* (London, 1980), p. 94.

[31] E. M. Spiers, *Engines for Empire: the Victorian Army and its Use of Railways* (Manchester, 2015), pp. 96–111.

of railways in South Africa following the outbreak of hostilities with the Boers in 1899.

The war in South Africa presented Girouard and his staff with a range of operational challenges. The British captured some 1,100 miles of damaged railways from the Boers and utilized thousands of miles of railway to supply an army of 250,000 troops.[32] As director of railways he assembled a 'special military staff' to act as the sole channel of communications between the army and the technical railway personnel. The directorate provided a centralized organization for the management and maintenance of the network, and ensured that the overall efficiency of railway operations was not jeopardized by the localized concerns of commanders ignorant of the force's overall transport priorities. Following the war, Girouard catalogued the lessons that had emerged from the British army's largest undertaking prior to 1914. Alongside delivering a lecture, subsequently published in the *Royal Engineers Journal*, he wrote the first volume of *History of the Railways during the War in South Africa, 1899–1902* in 1903.[33] The text was quickly recognized as a valuable educational resource for officers tasked with understanding the role of railways in modern warfare and, despite post-war fiscal retrenchment that precluded the publication of subsequent volumes in the series through official channels, the Royal Engineers chose to publish the works themselves.[34]

The publication of Girouard's histories emphasized their author's mastery of the details and the high regard with which the army valued his expertise on railway matters.[35] His specialist background, coupled with his fluency in the French language, made him an obvious choice to undertake the transport investigation in October 1914. He met first with Sir Ronald Maxwell, who had taken over from Sir Frederick Robb as IGC on 19 September, before he travelled to Paris to discuss matters with the military commission responsible for running the Chemins de Fer du Nord and the French director of railways. Prior to his departure from France he met with Robertson at GHQ and visited the port of Boulogne to assess its suitability as an army base.[36] The report he submitted after his inspection outlined

---

[32] For a brief overview of the British army's operation of railways in the South African war, see C. M. Watson, *History of the Corps of Royal Engineers* (11 vols., Chatham, 1914), iii. 104–8.

[33] É. P. C. Girouard, *History of the Railways during the War in South Africa, 1899–1902* (London, 1903); É. P. C. Girouard, 'Railways in war', *Royal Engineers Journal*, ii (1905), 16–27.

[34] 'Detailed history of the railways in the South African War, 1899–1902', *Royal Engineers Journal*, i (1905), 133–5.

[35] A. H. M. Kirk-Greene, 'Canada in Africa: Sir Percy Girouard, neglected colonial governor', *African Affairs*, lxxxiii (1984), 207–39, at p. 237.

[36] TNA, WO 32/5144, report on rail transport arrangements for British army on the continent by General Sir E. Girouard, Girouard to Cowans, 24 Oct. 1914.

the French system of railway organization, and compared it both with the practice recommended in the pre-war British *Field Service Regulations* (*FSR*) and the system in place at the time of his visit. The French railways – quite apart from the efforts required to mobilize and concentrate the French army, transport the BEF inland from the ports and remove hundreds of thousands of civilians from the areas of France and Belgium under threat of German invasion – had been called upon to handle large numbers of locomotives and rolling stock removed from the Belgian railways and the supply of a multitude of entangled fighting units both in retreat and advance.[37] Robertson's frustrations with the lack of priority afforded to the BEF's needs notwithstanding, Girouard concluded that the French had managed their numerous tasks with a remarkable degree of success.[38] Their organization resembled Girouard's system from the Royal Arsenal on a far larger scale, with control over the entire network located at *GQG*. The ability to direct railway operations from the principal information centre of the French army ensured that they both deployed their resources in response to the latest intelligence and could react to Joffre's strategic designs immediately.

The French system of organization compared favourably to that laid down in *FSR* part two. The British divided responsibility for transport and supply between two officers. Maxwell, the IGC, maintained stocks at the bases and controlled traffic on the lines of communications from his headquarters at the BEF's advanced base.[39] Robertson described Maxwell's role as 'something like the managing directors of Harrods' Stores and Carter Paterson rolled into one'.[40] Robertson, the QMG, worked alongside Sir John French at GHQ and took charge of the administrative arrangements between Maxwell and the units at the front. Under the guidelines laid down before August 1914 the general staff was to identify priorities for movement, Robertson to issue instructions to the relevant units and Maxwell to coordinate the move.[41] The system collapsed on contact with the enemy. Between 25 August and 1 September GHQ changed location five times. The frequency of GHQ's movements afforded little opportunity for adequate communications to be established at each site, which made contact between Maxwell and Robertson almost impossible to sustain.

[37] Le Hénaff and Bornecque, *Les chemins de fer français*, pp. 31–52.
[38] TNA, WO 32/5144, Girouard report, pp. 1–2.
[39] TNA, WO 33/686, instructions for the IGC, Part II, sections 5–6.
[40] Robertson, *From Private*, p. 199. Carter Paterson was a road haulage firm that sold many vehicles and horses to the War Office in 1914.
[41] R. G. Miller, 'The logistics of the British Expeditionary Force: 4 August to 5 September 1914', *Military Affairs*, xliii (1979), 133–8, at p. 133.

Messages and orders from GHQ regularly failed to reach their destination or were wholly inapplicable to the prevailing circumstances by the time they arrived. As Captain Tompson reflected, 'the situation altered so rapidly that railheads required to be settled finally only a few hours before the arrival of the daily supply trains'.[42] With the communications system unable to maintain reliable connections between the base depots, GHQ and the front, Robertson advocated a temporary abandonment of the principles laid down in *FSR* part two. In its place Robertson favoured the guidelines recommended in *FSR* part one, which emphasized that the 'man on the spot' use his initiative when the circumstances compelled such an action. As Robertson was located at GHQ, the spot upon which the most up-to-date information on the dispositions of troops and the military situation as a whole was to be found, he argued that he was best placed to identify the BEF's priorities and respond to urgent requests.[43]

To discharge the duties he had taken on, Robertson sought out an expert in French railway practices who could coordinate the movements he ordered from GHQ. Major Marr Johnson, who had created the railway timetables for the WF scheme and worked alongside the French to develop the BEF's concentration inland from the ports, was well equipped for the role. He was fully conversant with the technical aspects that governed French railway operations, and had been on the IGC's staff since his arrival in France. Johnson arrived at GHQ – supposedly on a temporary basis – on 26 August, and began to act as an expert conduit between Robertson and the French authorities immediately. The sheer volume of railway questions that required his expert consideration meant that Johnson remained attached to Robertson's staff, where he was engaged in a process of streamlining the BEF's transport arrangements when Girouard's examination took place.[44]

The existence of a solitary British officer with expert knowledge of the French railways and their military applications contrasted sharply with the expertise available to the French army. Following the defeat of 1870–1, in which uncoordinated military command of the railways contributed to confusion and congestion on the lines, French efforts had been channelled

---

[42] BLSC, Tompson papers, LIDDLE/WW1/GS/1612, short account of the introduction of an 'advanced regulating station' into the French traffic system and the subsequent control (British) of the railhead area, p. 1.

[43] TNA, WO 95/3950, Headquarters branches and services. Inspector general, French to Kitchener, 20 Nov. 1914; Spencer, '"The big brain in the army"', pp. 95–6.

[44] LHCMA, Robertson papers, 2/2/83, Robertson to Maxwell, 3 Oct. 1914; TNA, WO 95/27, QMG war diary, Robertson to Maxwell, 21 Oct. 1914; Henniker, *History of the Great War*, pp. 26–8. The difficulties experienced at Maxwell's headquarters by Johnson's prolonged absence are discussed in Henniker, *History of the Great War*, pp. 39–41.

into the establishment of a unified civil–military command system to operate the network during wartime.[45] As the French director of railways admitted to Maxwell in September 1914, the absence of a recognized command structure had 'resulted in most serious failures in the working of our railways during the war of 1870'.[46] Orders and counter-orders had been issued direct to the civilian operating staff by the general staff, the administrative staff, individual departments and even the minister of war.[47] The French were keen to avoid a repeat, and established an administrative system that reduced the possibility of contradictory and conflicting orders being issued by local commanders unaware of the army's wider requirements. Upon mobilization in 1914 the entire network came under the control of a single railway authority, which consisted of commissions made up of senior military officers and professional railwaymen with an encyclopaedic knowledge of their portion of the system. Each of the commissions existed in peacetime, and took over their designated section of the network upon mobilization.[48] Therefore, the staff on each line possessed an intimate knowledge of the system's limitations, and were thoroughly aware of individual stations' capacities. The combination of military and civilian expertise minimized the prospect of trains being despatched to stations incapable of handling them, and the location of the railway directorate at *GQG* ensured that orders were sent from the source best placed to take a holistic view of the situation.

The methodical structure of the French organization contrasted sharply with Girouard's assessment of the BEF's arrangements. Johnson's transfer to GHQ symbolized the collapse of *FSR* part two's guidelines, which were themselves an abandonment of the principles Girouard had advocated a decade earlier. Maxwell was unable to comply with the directive that all communications with the French railway authorities were to be made through the IGC's office because the officer responsible for those communications no longer worked there.[49] However, Johnson's relocation had not engendered a thorough re-examination of the BEF's administrative structure. The various directorates concerned with supply and movement continued to be split between the IGC's headquarters and GHQ. The

[45] F. P. Jacqmin, *Les chemins de fer pendant la guerre de 1870–1871: leçons faites en 1872 à l'École des Ponts-et-Chaussées* (2nd edn.,Paris, 1874), pp. 55–87.

[46] The directeur des chemins de fer to Maxwell, 19 Sept. 1914, quoted in Henniker, *History of the Great War*, p. 5.

[47] Girouard, 'Railways in war', pp. 17–18.

[48] PA, Lloyd George papers, LG/F/18/2/8, Geddes to Lloyd George, 8 Aug. 1918.

[49] TNA, WO 33/686, instructions for the IGC, Part II, section 6; WO 32/5144, Girouard report, pp. 6–8.

director of works, for example, reported to Robertson. However, his office space was located at Maxwell's base on the lines of communications.[50] Effective liaison between the two staffs, particularly given the unsatisfactory state of communications during the war of movement, had proven largely impracticable.

Following the tumult of defensive action, retreat, advance, offensive operation and movement around the front, Girouard's investigation represented both the BEF's first attempt to assess the applicability of pre-war regulations to wartime conditions and its first opportunity to replace reactive, ad hoc adjustments to the system with an effective, long-term solution. Girouard's recommendations consisted of abandoning the guidelines laid down in *FSR* part two and replicating the French system of organization. He argued that the BEF had to establish coordination between the French and British railway staffs at all levels of authority, right up to the executive branch of the transport hierarchy, upon which the British point of view was absent.[51] A modification on these lines, Girouard suggested, would provide the BEF with a forum within which it could influence the allies' future transport policy. The French had begun to request that British troops be provided to repair lines behind the BEF in the event of a general advance, and Girouard deemed it highly desirable that any organization established to oversee the reconstruction and operation of the Belgian railways 'should have a considerable [British] voice'.[52]

Both Robertson and Maxwell had already recognized the need for greater liaison between the French and British staffs. Their attempts to provide better facilities for inter-allied contact included Twiss's arrival in France and the increased establishment of railway transport officers on the lines of communications. However, Girouard believed that Twiss had 'not [yet] been encouraged to take his proper place in the field and assume control' of the BEF's railway staff.[53] The former's report codified the latter's role as DRT in France. Twiss's directorate became responsible for collecting the 'various demands for railway transport' that arose across the BEF, and for coordinating them 'in the manner best suited to meet the organization of the British Army, while putting as little strain on the French railways, which were being worked under very high pressure'.[54] In the final five months

---

[50] Brown, *British Logistics*, pp. 48–55; J. E. Edmonds, *History of the Great War: Military Operations, France and Belgium, 1914* (2 vols., London, 1928), i. 415–16.

[51] TNA, WO 32/5144, Girouard report, pp. 5–6, 12.

[52] TNA, WO 32/5144, Girouard report, p. 12; Girouard to Cowans, 24 Oct. 1914.

[53] TNA, WO 32/5144, Girouard report, pp. 6–7.

[54] TNA, WO 107/69, work of the QMG's branch of the staff: and directorates controlled, British armies in France and Flanders 1914–1918: Report, p. 15.

of 1914 alone, the northern and eastern networks in France carried 12,000 supply trains for the French and British armies – far higher than the volume of peacetime traffic but, unsurprisingly, well below the traffic figures recorded as the war expanded in scale in subsequent years.[55] The success with which Twiss felt he had achieved harmonious relations with the French railway staff was summed up in a letter that he sent to his counterpart in June 1915. Twiss celebrated the 'confidence and good feeling between your railway staff and mine … and may I also say my dear Colonel Le Hénaff, between you and me, [which] is a matter of the greatest satisfaction to me'.[56] Twiss's retention of Marr Johnson – well known among the French staff – as assistant DRT undoubtedly contributed to the 'good feeling' that accompanied Franco-British collaboration on railway matters after October 1914.

The whereabouts of Twiss's department also solved the question of which senior officer, Robertson or Maxwell, should take responsibility for traffic coordination on the western front. Maxwell, identified as the authority in the pre-war instructions, had proven unable to exercise effective control over the railways. It was impossible for Maxwell to retain responsibility for the coordination of traffic unless the supply departments were placed under his direct control, which meant their relocation from GHQ and an inevitable reduction in their access to the latest intelligence on the army's requirements. These disadvantages produced agreement between Maxwell and Robertson that Girouard's recommendations should be adhered to, the French hierarchy mirrored and the DRT accommodated at GHQ. Consequently, Robertson accepted overall responsibility for the coordination of the BEF's traffic arrangements.[57]

Maxwell and Robertson's adaptability in October 1914 confirms Ian M. Brown's argument that a working environment fostered by staff college training, pragmatism and professionalism existed within the BEF's administrative echelons.[58] Robertson's actions during the retreat from Mons were evidence of his belief that regulations and procedures were 'hand-rails to guide decision-making rather than barriers to creativity',

---

[55] Le Hénaff and Bornecque, *Les chemins de fer français*, p. 63. In 1915 the figure rose to 65,000, or 180 trains per day. In 1916 it rose still further, to 84,500 (an average of 231 per day). The figure decreased in 1917 to 72,000 trains before it rose again in the first half of 1918 to 45,000 trains.

[56] TNA, WO 95/64, branches and services: director of railway transport, Twiss to Le Hénaff, 19 June 1915.

[57] TNA, WO 32/5144, Girouard report, Maxwell to Robertson, 23 Oct. 1914; WO 95/3949, IGC war diary, Robertson to Maxwell, 24 Oct. 1914.

[58] Brown, *British Logistics*, p. 61.

while Maxwell's admission that 'the French system is likely to give the best results' highlights the broadminded response to Girouard's investigation from the officer most directly affected by the proposed changes.[59] However, Maxwell's and Robertson's approach was not universally accepted within the army, as the War Office's responses to Girouard's report demonstrate. The former IGC, Sir Frederick Robb, denounced Girouard's proposals as 'nothing new' and criticized the 'absurdity' of holding one man responsible for all transport requirements in the theatre of war. Furthermore, while Robb noted correctly that the system Girouard had reviewed was not that recommended in the pre-war guidelines, he blamed the 'co-efficient of human nature' for the modifications to those instructions that had taken place between August and October 1914. The thinly veiled implication of Robb's statement was that Robertson's actions, rather than a genuine response to inadequate communications between the BEF's administrative departments, had been an attempt to centralize authority under himself and reduce the IGC's influence.[60] An even more condemnatory reaction to Girouard's report emerged from the QMG of the forces, Sir John Cowans. In a note written three days after Girouard submitted his observations – and with the fighting for the town of Ypres increasing in intensity – Cowans argued that Girouard had 'far exceeded his instructions. He was not told to produce a scheme for uprooting organizations deliberately laid down after deep deliberation … The Regulations have been issued and acted upon and it is no time in the middle of a campaign to tinker with them'. For Cowans, despite his personal misgivings about the 'anomalies' in the structure of the BEF's original supply system, the short-term exigency of ensuring the troops engaged around Ypres remained fed and equipped superseded the rearrangement of rearward services decided upon before the war.[61]

The contents of Cowans's note flatly contradict the commentary on Girouard's report that appeared in a hagiographic biography of the former published after his death. Cowans's biographers claimed that the report had been 'shelved' by the BEF, 'most probably because the authorities in France were not ready for any change and because they … resented anything that looked even faintly like interference from home'.[62] This statement is not borne out by the evidence, such as Sir John's confirmation

---

[59] Spencer, '"The big brain in the army"', p. 106; TNA, WO 32/5144, Girouard report, Maxwell to Robertson, 23 Oct. 1914.

[60] TNA, WO 32/5144, Girouard report, note by Major-General Sir F. S. Robb on Sir Percy Girouard's proposals, n.d., pp. 1–3.

[61] TNA, WO 32/5144, Girouard report, note on memo. by Sir P. Girouard, 27 Oct. 1914.

[62] D. Chapman-Huston and O. Rutter, General Sir John Cowans, G.C.B., G.C.M.G.: the Quartermaster-General of the Great War (2 vols., London, 1924), ii. 102.

to Kitchener that the centralization of responsibility under the QMG was 'working to the satisfaction of all concerned' after a month,[63] or Robertson's correspondence with Cowans following the reorganization. Far from exhibiting lingering resent over Girouard's contribution, Robertson wrote to enquire whether the Canadian was going to return to France to address the many 'important questions' on the operation of the Belgian railways that required consideration.[64] French and Belgian representatives had already engaged in bilateral discussions and, in an echo of Girouard's observations, Robertson underlined the need for the BEF to have a 'voice' in any formal agreements reached between their coalition partners. For reasons that have not been established, Colonel Henniker, rather than Girouard, was chosen to participate in what became known as the Calais commissions. However, as Robertson's letter makes clear, the decision over Girouard's non-participation was made in London rather than at GHQ.[65]

Girouard's report contained only one recommendation that the BEF could be accused of having 'shelved'. To guarantee Britain's possession of sufficient dock accommodation to offload the troops and supplies (and to store the latter) required to maintain the BEF over the winter and beyond, Girouard advised that the pre-war agreements with the French needed to be revisited.[66] In early December 1914 Maxwell amplified Girouard's concerns in letters to Robertson, which voiced the IGC's concern that the limit for traffic allotted to the BEF was about to be reached. He warned that only two additional infantry divisions and a cavalry division could be supplied through the existing port space allocated to the BEF, and urged GHQ to consider the establishment of a second line of communications and request immediate access to Dieppe.[67] However, in an early demonstration that the BEF was never the sole focus of British military attention during the war, Sir John was reluctant to commence negotiations with the French over port space until the outcome of developments 'in the Eastern theatre of operations' became clearer. Consequently, the matter was dropped for the rest of the year.[68]

---

[63] TNA, WO 95/3950, IGC war diary, French to Kitchener, 20 Nov. 1914.

[64] LHCMA, Robertson papers, 2/2/24, Robertson to Cowans, 28 Nov. 1914.

[65] Henniker recorded the outcome of the commission's deliberations in his volume of the official history. See Henniker, *History of the Great War*, pp. 93–101.

[66] TNA, WO 32/5144, Girouard report, p. 15.

[67] TNA, WO 95/3951, headquarters branches and services. Inspector general, Maxwell to Robertson, 1 and 5 Dec. 1914.

[68] TNA, WO 95/3951, IGC war diary, Robertson to Maxwell, 3 Dec. 1914; Brown, *British Logistics*, p. 76.

Sir John's 'wait and see' approach exemplified a desire not to commence potentially longwinded negotiations over organizations, allocations and responsibilities that – should the fighting formations achieve success – may never be required. Throughout the First World War, as will be demonstrated further below, the allies had to balance the requirements of their immediate demands with preparations to ensure sufficient resources for the continuation of a war of indeterminate length. His views reflected a widespread tendency to regard the trench-bound stalemate of late 1914 as a temporary anomaly, and were founded on the hope that the BEF would soon be operating again on Belgian rather than French soil.

This optimistic outlook, combined with the French army's continued desire to adhere to the pre-war agreement, thwarted a second attempt to re-evaluate the BEF's transport arrangements in the war's first six months. In December 1914 Kitchener summoned Eric Geddes, the railway organizer he had been impressed by a decade earlier in India, to a meeting at the War Office. In the Geddes family chronicle, written by Geddes's younger brother Auckland, what happened next was presented as an example of the insular and protective military family closing ranks to avoid the criticisms of an expert from outside the profession. Auckland Geddes claimed that Kitchener wished to send Eric to France to 'see what was wrong' with the BEF's transport services, but that the proposition was vetoed by Cowans:

> Eric realised … that such a mission would be hopeless unless he had the good will of the soldiers; and, from the way in which Lord Kitchener, in Eric's presence, sprang the proposal on a totally unprepared QMG, it was obvious that the officer must think Eric had already passed adverse judgment on his department's handling of railway transport. In such circumstances good will would inevitably be lacking.[69]

It was not the first time that Geddes had been in contact with the War Office since the start of the conflict. In August 1914 he had approached the DOM, Brigadier-General Richard Montagu Stuart-Wortley, with the idea of raising a battalion of skilled railwaymen of all grades for service in France. Stuart-Wortley rebuffed the approach, and Geddes wrote later that he had been told 'that the military railway personnel were competent to deal with the situation in France and that railway units were not wanted'. Doubtless coloured by that incident, Geddes reflected that Stuart-Wortley's rejection of his offer was the result of the latter's membership of a 'military machine' that was not prepared to accept civilian specialists in its ranks.[70] The legacy

---

[69] A. C. Geddes, *The Forging of a Family: a Family Story Studied in its Genetical, Cultural and Spiritual Aspects and a Testament of Personal Belief Founded Thereon* (London, 1952), p. 222.

[70] Geddes's introduction in J. Shakespear, *A Record of the 17th and 32nd Service Battalions*

of this misunderstanding, as demonstrated by Geddes's recollection of the event twelve years later, needlessly politicized the transportation mission in 1916 and infused Geddes's memory of his second attempt to work alongside the army at Kitchener's behest in December 1914.

Cowans was a fellow officer of the Rifle Brigade and close friend of Stuart-Wortley's, and Geddes later suggested to Lloyd George that it was personal jealousy and professional 'demarcation' that curtailed any possible transportation mission in January 1915.[71] Cowans's biographers made no mention of the meeting in their celebratory account of the QMG's career, perhaps as a result of Geddes's subsequent contributions to the British war effort, while Peter Cline's alternative proposal – that the North-Eastern Railway's reluctance to release Geddes contributed to the scheme's abandonment – appears unlikely.[72] The North-Eastern's chairman did write that the 'board of the … railway were loath to dispense with Mr Geddes's services, even for a short period' when Lloyd George recruited Geddes for the Ministry of Munitions the following summer,[73] but the company conceded that its deputy general manager was 'not very fully employed' by the railway's wartime operations. Under such circumstances, Alexander Kaye Butterworth observed, 'when every man of ability should be utilised to the best advantage of the State, I feel it to be one's duty to let those in authority know of any good men who are "spare"'.[74]

Yet it would be unfair to portray the aborted transportation mission, and the earlier dismissal of Geddes's offer of a specialist railway battalion, as purely the result of rhadamanthine military attitudes to civilian assistance in Whitehall. Cowans, for example, was a prominent early advocate of civil–military collaboration in the field of army sustenance. At Deptford Cattle Market, rented from the City of London Corporation for military use at the start of the war, Cowans 'realised from the outset' that the depots required 'men with business, rather than military experience to run them and to maintain their efficiency once they were organised'. Each depot was

---

Northumberland Fusiliers, N.E.R. Pioneers, 1914–1919, ed. H. Shenton Cole (Newcastle upon Tyne, 1926), p. xiii.

[71] TNA, MUN 9/35, Geddes. A handwritten, undated note in this file suggests that Kitchener's project fell through when the QMG's department claimed responsibility over railway organization in France; K. Grieves, 'The transportation mission to GHQ, 1916', in *'Look to Your Front!' Studies in the First World War by the British Commission for Military History*, ed. B. Bond et al. (Staplehurst, 1999), pp. 63–78, at p. 71.

[72] Chapman-Huston and Rutter, *Sir John Cowans*; P. K. Cline, 'Eric Geddes and the "experiment" with businessmen in government, 1915–22', in *Essays in Anti-Labour History*, ed. K. D. Brown (London, 1974), pp. 74–104, at p. 77.

[73] PA, Lloyd George papers, LG/D/1/2/6, Knaresborough to Lloyd George, 5 June 1915.

[74] PA, Lloyd George papers, LG/D/1/2/1, Butterworth to Lloyd George, 27 May 1915.

managed by a businessman, supported by 'junior officers with thorough knowledge of railway work, shipping, labour, accounting, etc., so that the whole of the work was carried out, as far as possible, on business lines'.[75] Furthermore, less than a month prior to the uncomfortable meeting between Kitchener, Geddes and Cowans, the latter had been instrumental in the recruitment of Commander Gerald Holland to create an inland water transport (IWT) department on the western front (see Chapter 5).

These examples illustrate that the British army was receptive to specialist, non-military advice in early 1915. However, the differences between the Deptford Cattle Market on the one hand, and the proposed Geddes mission on the other hand, were considerable. Deptford Cattle Market was a British base, located in the centre of London and free of any requirement to consider the wishes of Britain's coalition partners. The French railway network remained firmly under the control of the French rail authorities, who had shown little inclination to materially alter the arrangements both nations had agreed to before the war. Under such circumstances – and considering Girouard's recently completed investigation, the comparatively miniscule size of the BEF's demands, the continued applicability of the pre-war agreement and the apathy of his subordinates towards the proposal – it is perhaps understandable that Kitchener did not consider another transportation mission to be a high priority at the start of 1915.

The allies' failure to produce a substantial battlefield success, rather than military intransigence, was responsible for GHQ's inability to act upon the most prominent of Girouard's recommendations. The division of responsibility between the French and British forces for the repair and operation of the Belgian State Railways in the event of an allied advance consumed much of Girouard's discussions with the French railway authorities. Girouard recognized that there was 'little doubt that any retirement of the enemy [would] be accompanied by very grave damage to the railway lines and structures. Much damage', he observed, had 'already been effected both by the Belgian authorities and the enemy'. The Belgian public works department, responsible for repairs to the railway network in wartime, had 'practically disappeared', and the staff of the Belgian State Railways were scattered across occupied, allied and neutral European territory.[76] Therefore, Girouard argued, it was important that the coalition partners agreed their duties with regard to the repair and operation of the railway lines before they were liberated from German occupation (the ports

---

[75] Chapman-Huston and Rutter, *Sir John Cowans*, ii. 55.
[76] TNA, WO 32/5144, Girouard report, p. 10.

of Ostend and Zeebrugge were included in Girouard's considerations for the same reason). For Girouard, the effective exploitation of the Belgian railways provided the most important reason why the British needed to possess a voice in inter-allied transport discussions. While those conversations remained hypothetical until the final months of the war, elsewhere Britain's transport experts became embedded in the very real development of the nation's expanding continental commitment.

Materials, manpower and the continental commitment

The anticipated German retirement failed to materialize in 1915. The western front's relative stability during the year ensured that the BEF's transport experiences were comparatively unremarkable. In the week commencing 10 March, 25,000 reinforcements were concentrated around Neuve Chapelle, while the roads behind the First Army were improved by working parties drawn from the 125th Rifles and the 34th Sikh Pioneers, Lahore Division. Wooden tramlines 'greatly facilitated' the accumulation of ammunition in readiness for the attack, as the average number of ammunition wagons sent forward rose from fifteen per day to fifty-five in the period 11–15 March.[77] At 7:30 a.m. on 10 March the artillery opened a thirty-five-minute bombardment that expended more munitions than the British had fired in the entirety of the South African War. Yet Colonel Henniker's only comment on the preparations for the battle claimed that, from a transport perspective, they were 'insignificant'; traffic around the battle zone was 'heavy but not more than the railways could cope with'.[78]

The French transport network had proven itself able to deliver to the BEF's artillery firepower on a scale hitherto unprecedented in British military history. Yet as the subsequent fighting in Artois demonstrated, a colossal weight of fire at the outset of the attack was not enough to propel the attacking troops through the German positions into open country. Furthermore, the First Army's munitions expenditure of 10 March could not be sustained without jeopardizing the stockpiles available to the rest of the BEF. The First Army had fired the equivalent of 132 rounds per 18-pounder field gun on 10 March. The rate of fire dropped to sixty-four rounds per gun the following day and to forty-nine rounds per gun on the 12th. By 16 March the BEF's supplies of field gun ammunition were almost exhausted and, at a replenishment rate of just 7.5 rounds per gun per day, it took seventeen days to replace the stocks fired indecisively at Neuve

[77] J. E. Edmonds and G. C. Wynne, *History of the Great War: Military Operations, France and Belgium, 1915* (2 vols., 1927), i. 82–3.
[78] Henniker, *History of the Great War*, p. 118.

Chapelle. The BEF was unable to maintain pressure on the enemy under such circumstances, a fact recognized by Sir John when he announced to Kitchener on 15 March that future offensives had been abandoned 'until sufficient reserves are accumulated'.[79]

The availability and deployment of munitions has been central to many debates on the quality of British generalship during the First World War.[80] The South African War had emphasized the 'absolute necessity' of a good supply of ammunition being made available 'at the required place at the required time', but it was between 1914 and 1918 that artillery became the 'great destructive force' in modern warfare.[81] Artillery fire was responsible for perhaps 60 per cent of combatant casualties during the First World War, and for much of the devastation of the landscape over which the war was fought. Yet in 1915, for a variety of reasons, the available artillery was unlikely to hit German targets concealed from view. Consequently, the BEF required sufficient volumes of shells to effectively 'drench' a target area with fire in support of an infantry assault. Under such conditions the strictly limited 'per diem' allocation of shells reduced the likelihood that enemy positions could be suppressed by artillery activity.[82] The call for 'more shells' and 'more heavy howitzers' from *The Times*'s military correspondent in May 1915 both exposed the paucity of pre-war preparations for a war of extended duration and over-simplified a complex technical and organizational challenge.[83] The heavy howitzers from which the shells were delivered to their ultimate recipients in the German lines were just the final step in a vast wartime production and distribution network.

The creation of the Ministry of Munitions signified an official recognition that the war had outgrown the War Office's capacity to handle it. As the British army had been a comparatively small purchaser of arms before the war, the British arms industry was not adapted to the demands of a large

[79] D. French, 'The military background to the "shell crisis" of May 1915', *Journal of Strategic Studies*, ii (1979), 192–205, at p. 200; Edmonds and Wynne, *History of the Great War*, i. 149–50.

[80] See, e.g., J. B. A. Bailey, *The First World War and the Birth of the Modern Style of Warfare* (Camberley, 1996); R. Prior and T. Wilson, *Command on the Western Front: the Military Career of Sir Henry Rawlinson, 1914–18* (Oxford, 1991); P. Harris and S. Marble, 'The "step-by-step" approach: British military thought and operational method on the western front, 1915–1917', *War in History*, xv (2008), 17–42.

[81] R. Prior and T. Wilson, *Passchendaele: the Untold Story* (New Haven, Conn., 1996), p. 17.

[82] See Prior and Wilson, *Command on the Western Front*, pp. 36–43 for a discussion of the technical framework within which the BEF's artillery operated in 1915.

[83] 'Shells and the great battle', *The Times*, 14 May 1915, p. 9.

army engaged in protracted operations.[84] Furthermore, the arms industry and associated sectors had not been subject to government protection at the outbreak of war. By mid 1915, as Britain's industrial labour force participated in the so-called rush to the colours, the two government factories and sixteen private firms engaged on munitions work were short of 14,000 workers. Almost 24 per cent of male employees from the chemical and explosives industry, 19.5 per cent from the engineering trades, 18.8 per cent from the iron and steel industry and 16.8 per cent from small arms manufacturers had enlisted in the army by June 1915.[85] The new ministry commenced its work from a less than opportune position. However, within one month of taking up the newly created post of minister of munitions, David Lloyd George had appointed almost one hundred men of 'first class business experience' to execute the decisions made at the ministry's headquarters in Whitehall Gardens.[86] Their principal goal was to coordinate the output of the nation's factories, and to ensure that future battles on the western front were not hamstrung by insufficient firepower.

Eric Geddes was among the first railwaymen to apply his business skills to the challenge of munitions production, following what Lloyd George later referred to as 'one of the luckiest discoveries in my life'.[87] According to the Geddes family chronicle, Lloyd George interviewed Geddes with a view to utilizing his talents in the newly formed ministry after he had received a glowing account of Geddes's abilities from the foreign secretary, Sir Edward Grey (a director and former chairman of the North-Eastern Railway). Geddes admitted to knowing nothing about munitions production at his interview, but professed a 'faculty for getting things done'. This conviction was supposedly enough for Lloyd George to make Geddes the head of a department in the nascent ministry.[88] However, as Keith Grieves has demonstrated, the accounts provided by Lloyd George and the Geddes family of the former's 'discovery' of Eric Geddes were 'largely fictional'.[89] In fact, Geddes was first interviewed by Christopher Addison, the ministry's parliamentary secretary, as part of the 'man-grabbing' process that took

[84] H. Strachan, *The First World War: To Arms* (Oxford, 2001), p. 1066.

[85] Strachan, *To Arms*, p. 1071; P. Simkins, *Kitchener's Army: the Raising of Britain's New Armies, 1914–1916* (Manchester, 1988), p. 111.

[86] D. Lloyd George, *War Memoirs of David Lloyd George* (2 vols., London, 1938), i. 254.

[87] PA, Lloyd George papers, LG/F/18/4/36, Lloyd George to Geddes, 24 Feb. 1922.

[88] Geddes, *The Forging of a Family*, p. 223; K. Grieves, *Sir Eric Geddes: Business and Government in War and Peace* (Manchester, 1989), p. 12; TNA, MUN 9/35, 'Sir Eric Geddes', undated note.

[89] K. Grieves, 'Improvising the British war effort: Eric Geddes and Lloyd George, 1915–18', *War & Society*, vii (1989), 40–55, at p. 40.

place immediately after the ministry was established.[90] Addison's first impression – that Geddes appeared to be 'first rate' – was supplemented by positive references forwarded to Lloyd George by Grey, Alexander Kaye Butterworth, Sir Hugh Bell, Sir Percy Girouard and the North-Eastern's chairman, Lord Knaresborough.[91]

The positive reports Lloyd George received confirmed what Geddes's pre-war career had demonstrated in detail. His referees described Geddes as a successful administrator of complex, large-scale organizations, a man of energy, efficiency and drive, and a manager with the capacity to think big and work outside the constraints of established routine.[92] Geddes had successfully managed a vast, geographically dispersed workforce at the North-Eastern Railway, a proficiency that suited the national distribution of the ministry's production facilities.

From July 1915 onwards Geddes acted as a trouble-shooter on Lloyd George's behalf. He dealt first with the slow delivery of rifles, then moved on to the task of reducing congestion at Woolwich. In December 1915 he became head of the national filling factories and component distribution organization. Lloyd George's aversion to questions of detail provided Geddes and the ministry's other directors with executive freedom, which facilitated administrative innovation. Over the winter and spring of 1915–16 Geddes and the other 'men of push and go' that were drawn into the ministry's operations infused the manufacture of munitions with the latest managerial methods: the so-called American practice of statistical analysis adapted from the North-Eastern Railway; the scientific management techniques popularized by Frederick Winslow Taylor; and the motion studies developed by Frank and Lilian Gilbreth were all incorporated within the Ministry of Munitions' approach.[93]

The progressive, analytical managerial style that Geddes had been introduced to in the United States – and applied to the operations of the North-Eastern Railway for a decade before the war – combined with the

[90] C. Addison, *British Workshops and the War* (London, 1917), p. 5; R. J. Q. Adams, *Arms and the Wizard: Lloyd George and the Ministry of Munitions, 1915–1916* (London, 1978), pp. 38–55.

[91] Cline, 'Eric Geddes and the "experiment"', p. 78; Grieves, *Sir Eric Geddes*, pp. 12–13; PA, Lloyd George papers, LG/D/1/2/1, Butterworth to Lloyd George, 27 May 1915; LG/D/1/2/2, Bell to Lloyd George, 30 May 1915; LG/D/1/2/6, Knaresborough to Lloyd George, 5 June 1915.

[92] Adams, *Arms and the Wizard*, pp. 45–8.

[93] *History of the Ministry of Munitions: General Organisation for Munitions Supply* (12 vols., London, 1922), ii. 19; C. Wrigley, 'The Ministry of Munitions: an innovatory department', in *War and the State: the Transformation of British Government, 1914–1919*, ed. K. Burk (London, 1982), pp. 32–56.

pioneering methods of some of Britain's leading industrial figures to raise the nation's munitions output ahead of the battle of the Somme. Through a management technique Geddes referred to as 'intelligent control' – founded upon statistics devised by J. George Beharrell, which allowed the team to compare outputs, identify available capacities and create more accurate forecasts of production – the department established a more efficient method for the use of available labour supplies and raw materials.[94] Improvements in output were substantial, despite the complications involved in the production of modern armaments. A single 18-pounder shell contained sixty-four components; a complete round of 4.5-inch ammunition comprised fifty-seven individual pieces, all of which had to be drawn together, assembled, and despatched to the front in an organized, efficient flow.[95] By 1 June 1915 only 1,992,000 of the 5,573,000 shell bodies ordered by the War Office had been delivered. Yet by July 1916 'the weekly deliveries to the War Office averaged just over a million rounds … of which rather over 50 per cent were high explosive shell, as compared with a weekly average of 166,500 rounds in June 1915, of which only 23 per cent were high explosive'.[96] Geddes received a knighthood in recognition of his contribution to the vast improvement in munitions production that had taken place in the ministry's first year of existence.

Geddes was by no means the only figure from the transport industry to influence the operations of the Ministry of Munitions. Nor were his pre-war compatriots employed by the ministry purely for their managerial qualities. Both Henry Fowler and Vincent Raven, the chief mechanical engineers of the Midland and North-Eastern railways respectively, brought technical expertise and supervisory abilities to various facets of munitions production. Fowler was initially appointed to direct the national projectile factories department, before becoming both deputy controller of shell manufacture and superintendent of the Royal Aircraft Factory at Farnborough in September 1916. He became assistant director-general of aircraft production in December 1917, and chaired the inter-allied conference on the standardization of aircraft components in 1918.[97] Raven initially joined Geddes in the organization of efforts at the Royal Arsenal, and acted as chief superintendent at Woolwich from December 1915. When the incumbent of the post, Brigadier-General Sir Hay Frederick Donaldson

---

[94] PA, Lloyd George papers, LG/D/3/1/6, Geddes to Lloyd George, 15 March 1916.

[95] Grieves, 'Improvising the British war effort', p. 44; I. F. Marcosson, *The Business of War* (New York, 1918), pp. 269–70.

[96] *General Organisation for Munitions Supply*, pp. 29, 30.

[97] *General Organisation for Munitions Supply*, p. 265; G. W. Carpenter, 'Fowler, Sir Henry (1870–1938)', *ODNB* <https://doi.org/10.1093/ref:odnb/37427> [accessed 29 Sept. 2017].

– himself a mechanical engineer who had worked for the LNWR, in Goa, on the Manchester Ship Canal and at London's India Docks in a peripatetic career – went to the United States and Canada to arrange new sources of supply for the British forces, Raven took over his responsibilities. Following Donaldson's death aboard HMS *Hampshire* in June 1916 (he had been selected to act as a technical advisor on Kitchener's trip to Russia), Raven's temporary appointment was made permanent.[98]

Raven's technical background and familiarity with the North-Eastern's statistical methods made him a perfect fit for the task at Woolwich, which principally involved the coordination of 88,000 workers. In the House of Commons in December 1915, Lloyd George celebrated Raven's contribution:

> The manufacture and filling of various articles has increased since he took it in hand in some cases by 60 per cent, and in others by as much as 80 per cent, whereas the staff has only increased 23 per cent. One of the reforms he initiated are statistical records of the output. These records were not compiled prior to his assumption of control. Now they are having, and will continue to have, a potent effect not only upon the output, but upon the cost of output. As an illustration of the use to which such figures can be put, I will mention that when the output of a certain shop or section of a shop is noted the following morning it is possible for the superintendent or the works manager to immediately put their finger upon the fact that perhaps the flow of raw material fails, or that owing to congestion of the arsenal railways the output cannot be got rid of; and, therefore, the inefficiency can be checked. Such hitches in the daily work of a factory can only be avoided and minimised by a most complete system of statistical control, and that has been instituted at Woolwich.[99]

Both Raven and Fowler received knighthoods for their work in the higher administration of the ministry, and demonstrated the transferability of skills they had applied successfully to the management of large departments within their peacetime professions. As the Ministry of Munitions' official history attests, they were accompanied throughout the nascent organization by railway servants from across the British railway industry.[100]

Yet the contributions of Britain's transport experts to the widening war were not made solely by the employees of British railway companies. Other men with valuable experience, such as Alfred Collinson and Follett Holt, were drawn into the munitions production effort from much farther afield.

---

[98] *General Organisation for Munitions Supply*, p. 272; A. Everett, *Visionary Pragmatist: Sir Vincent Raven: North Eastern Railway Locomotive Engineer* (Stroud, 2006), pp. 131–5; J. Pollock, *Kitchener* (London, 2002), p. 479.

[99] Hansard, *Parliamentary Debates*, 5th ser., lxxvii (20 Dec. 1915), col. 105.

[100] See the list of some of the principal officers employed in the Ministry of Munitions during the war in *General Organisation for Munitions Supply*, pp. 260–75.

Collinson's and Holt's careers emphasize both the global dispersal of British expertise prior to the First World War and the extent of an international talent pool open to Britain's political and military authorities during the conflict. Following a career that had begun on the Great Northern Railway in the late 1880s, the decade before 1914 saw Collinson acting as engineer-in-chief on the construction of several railways in China and as a consulting engineer to the Chinese Government Railways. Between 1914 and 1918 he was responsible for the inspection of munitions outside London, and travelled across the Atlantic to organize the munitions inspection process in the United States. Holt's engineering career began in the LSWR's locomotive superintendent's office, and took in service in India and South America prior to his retirement in 1910. In April 1915 he became one of the first transport experts to acquire a position in the Ministry of Munitions, when he entered the shell production organization.[101]

Holt was by no means unique in possessing an attachment to munitions production that pre-dated Lloyd George's famed recruitment of 'men of push and go'. The pre-war relationship between the military and the railway companies facilitated contacts between the two groups, and helped place the skills and productive capacity of Britain's transport experts at the War Office's disposal from the very outset of the war. Sir George Gibb, Sir Percy Girouard and George Booth (a ship owner and director of the Bank of England), alongside representatives of the machine tools industry and private armaments firms, offered their services to the government in the conflict's opening months.[102] The royal ordnance factories were first to take advantage of the adaptability of the railway companies' plant, when they requested the REC's assistance in the manufacture of 4.5-inch howitzer carriages on 13 August 1914.[103] A second request followed in early September, as the higher-than-anticipated number of casualties created an urgent demand for ambulance stretchers. Eleven companies agreed to construct 12,250 stretchers between them (see Table 3.1), and the first deliveries were made as early as 12 September. A fortnight later twenty-two companies from across England, Scotland and Ireland agreed to produce 5,000 general service wagons for the Royal Artillery, a figure which ultimately rose to

---

[101] 'Alfred Howe Collinson', *Grace's Guide to British Industrial History*, 2017 <http://www.gracesguide.co.uk/Alfred_Howe_Collinson> [accessed 29 Sept. 2017]; '1922 who's who in engineering: name H', *Grace's Guide to British Industrial History*, 2017 <http://www.gracesguide.co.uk/1922_Who's_Who_In_Engineering:_Name_H> [accessed 29 Sept. 2017].

[102] *General Organisation for Munitions Supply*, p. 17; D. Crow, *A Man of Push and Go: the Life of George Macaulay Booth* (London, 1965), pp. 68–113.

[103] *History of the Ministry of Munitions: Industrial Mobilisation, 1914–1915* (12 vols., London, 1922), i. 106.

Table 3.1. Railway companies' manufactures of ambulance stretchers, September 1914.

Railway company	Total number to be made	Number to be made per week
London and North-Western	2,000	300
Great Western	1,500	200
Midland	1,500	200–250
Great Central	1,000	200
Great Eastern	1,000	500
Lancashire and Yorkshire	1,000	200
London, Brighton and South Coast	1,000	200
North-Eastern	1,000	200
South-Eastern and Chatham	1,000	100
Great Northern	750	100
London and South-Western	500	100
Total	12,250	

Note: The Midland Railway increased production from 200 to 250 stretchers in the third week of the schedule.
Source: E. A. Pratt, *British Railways and the Great War; Organisation, Efforts, Difficulties and Achievements* (2 vols., 1921), ii. 584.

9,300 wagons as the war progressed. The productive facilities available to the railway companies were the subject of further enquiries from the Royal Arsenal in October 1914, and by the end of the month the REC had recognized the necessity for a civil–military sub-committee to coordinate the 'various requests made by or through the War Office to the railway companies to assist in the manufacture of war-like stores and equipment'.[104]

The creation of a large-scale industrial production system drew heavily upon the manufacturing capacity of the railway companies and the organizational abilities of their managers, whose professional careers were dominated by challenges that involved the coordination of dispersed but interlinked activities. Following the constitution of the railway war manufactures sub-committee all applications for work to be undertaken on behalf of the War Office, or its subcontractors, were submitted to the REC through one of two military members of the sub-committee: Captain Henry Mance, the pre-war liaison between the army and the railway companies;

[104] E. A. Pratt, *British Railways and the Great War: Organisation, Efforts, Difficulties and Achievements* (2 vols., London, 1921), ii. 583–88.

and Brigadier-General Herbert Guthrie Smith, the director of artillery. The railway members – who included the general managers of two railways, the chief mechanical engineers of four others and the Midland Railway's carriage and wagon superintendent (see Table 3.2) – decided whether the railway companies could fulfil the request and devised a programme for completion when the project had been accepted. The sub-committee met for the first time on 2 November 1914. By the end of the year the great railway workshops at Derby and Swindon had erected and despatched twenty-three gun carriages for 8-inch howitzers. Further requests, for products that varied from water tanks, miners' trucks and heavy-capacity wagons to picketing posts, artillery wheels and brake blocks arrived from the armed forces and private firms between November 1914 and April 1915. The railway companies also accepted responsibility for the manufacture of 'two armoured trains for home defence, gun carriages and limbers, fittings for 60-pounder guns, sets of elevating gear, drop forgings, limber hooks, wagon hooks and flanges, [and] mountings for 6-pounder Nordenfeldt guns'. Each order was divided between as many as twenty-four companies, while four companies committed to work on behalf of the Admiralty in addition to their work on the construction of ambulance trains for use at home and overseas.[105]

Yet for the secretary of state for war these efforts did not prove ambitious enough. Kitchener believed that further use could be made of the railway companies' facilities. In response, the REC expanded the membership of the railway war manufactures sub-committee and inverted the procedure by which the companies' workshops became involved in war-related production. The sub-committee was provided with a further representative from the War Office, Lieutenant-Colonel A. S. Redman, and augmented by an extra eleven members from railways across Britain. With further input from the Admiralty and – following its creation – the Ministry of Munitions, the sub-committee provided a link between the government, the railways and the manufacturers of Britain for the remainder of the war. From April 1915 the committee took a more active role in the allocation of work to the railway workshops. Rather than passively await enquiries from the War Office or armament firms, the sub-committee divided the country into six districts and despatched members to visit manufacturers in their area to ascertain what assistance the railways could offer in the realization of their government contracts.[106] After just one month the new procedures had generated a diverse range of demands for the railways' plant. The first list of

[105] Pratt, *British Railways and the Great War*, ii. 586–7.
[106] Pratt, *British Railways and the Great War*, ii. 589–90.

Table 3.2. Members of the railway war manufactures sub-committee, October 1914 and April 1915.

	Name	Organization	Date of appointment
Military representatives	H. Guthrie Smith	War Office	October 1914
	H. O. Mance	War Office	October 1914
	A. S. Redman	War Office	April 1915
Railway company managers	A. Beasley	Taff Vale Railway	April 1915
	G. Calthrop*	London and North-Western Railway	October 1914
	C. H. Dent*	Great Northern Railway	October 1914
	D. A. Matheson*	Caledonian Railway	April 1915
Railway superintendent officers	D. Bain	Midland Railway	October 1914
	C. J. B. Cooke	London and North-Western Railway	October 1914
	J. Cameron	Taff Vale Railway	April 1915
	G. J. Churchward	Great Western Railway	October 1914
	P. Drummond	Glasgow and South-Western Railway	April 1915
	H. D. Earl	London and North-Western Railway	April 1915
	H. Fowler	Midland Railway	October 1914
	H. N. Gresley	Great Northern Railway	October 1914
	A. J. Hill	Great Eastern Railway	April 1915
	G. Hughes	Lancashire and Yorkshire Railway	October 1914
	W. Pickersgill	Caledonian Railway	April 1915
	V. L. Raven	North-Eastern Railway	April 1915
	J. G. Robinson	Great Central Railway	April 1915
	A. R. Trevithick	London and North-Western Railway	April 1915
	E. A. Watson	Great Southern and Western Railway, Ireland	April 1915

Note: * denotes members of the Railway Executive Committee.
Source: E. A. Pratt, *British Railways and the Great War; Organisation, Efforts, Difficulties and Achievements* (2 vols., London, 1921), ii. 586, 589–90.

items requested of the railway companies ran to eleven pages. It included lethal items such as shells, bombs, steel forgings for guns and 6-pounder Hotchkiss guns alongside more mundane but morale-boosting equipment such as travelling kitchens, kettles and drinking cups. By November 1918 the railway companies had sub-contracted work from over one hundred firms, and the list of items produced at the railway workshops stretched to 121 pages.[107]

The pre-war composition of the British railways played a crucial role in allowing the railway companies to respond to the army's needs. The plethora of privately owned companies across the country meant that, 'in important respects, the British railways were over-resourced and over-manned, with substantially more locomotives and twice the carriages per mile of Germany and three times that of France. This surplus capacity became invaluable in war time'.[108] British railway companies could afford to redirect their plant into the production of war-like stores to a degree unobtainable in France and Germany without profound implications for the efficiency of the railway network – every man employed on the manufacture of gun carriages was unavailable for the restoration of the passenger carriages or goods wagons that carried the regular and military traffic upon which the nation depended.

The application of civilian expertise to the conundrum of munitions production greatly increased the BEF's capacity to wage intensive warfare. By 1 July 1916 the BEF possessed 714 heavy guns, more than double the 324 that had been in France on 1 January that year. By the time the battle of the Somme had ground to an inconclusive end in November the number of heavies in France had risen to 1,127. As the new divisions of Kitchener's army continued to pour into France – and correspondingly increased the number of trains required to service the force's appetite for food, ammunition and other supplies – the formerly insignificant logistical challenges posed by the presence of British troops on the continental mainland assumed imposing proportions. For the daily provisions of the new army divisions alone the French railway network was tasked to accommodate an extra 59.5 divisionally packed supply trains per week.[109]

The demand that the French railway authorities provide sufficient transport capacity to supply the enlarged BEF in addition to its own

---

[107] Pratt, *British Railways and the Great War*, ii. 594–95, 602.

[108] A. Gregory, 'Railway stations: gateways and termini', in *Capital Cities at War: Paris, London, Berlin 1914–1919*, ed. J. Winter and J. L. Robert (2 vols., Cambridge, 2007), ii. 23–56, at p. 27; J. A. B. Hamilton, *Britain's Railways in World War I* (London, 1947), pp. 29–30.

[109] Brown, *British Logistics*, p. 112.

army's needs further exposed the inadequacy of the pre-war arrangements agreed between the coalition partners. As early as November 1914 the 10th (Railway) Company and three special reserve companies were despatched to France to join the 8th (Railway) Company on railway construction and repair duties. However, events around Ypres meant that the companies were immediately employed on defensive rather than railway works until March 1915.[110] The War Office, unlike in August 1914 when presented with Geddes's proposal for the recruitment of a skilled railway battalion, perceived that five companies were insufficient for the volume of work likely to be generated by the fighting in 1915 – particularly should the much hoped for advance into Belgium take place. They requested that the REC recruit a large force of construction and railway operating troops, to provide a pool of skilled labour for the army's use. The REC's railway recruitment sub-committee was established under the stewardship of William Forbes, general manager of the LBSCR and the ERSC's commandant. Forbes was assisted by Francis Dent, general manager of the SECR, and Arthur Watson, general manager of the Lancashire and Yorkshire Railway. Dent took responsibility for the recruitment of railway operating personnel – the chrysalis of the Railway Operating Division (ROD) – and initially also interviewed all applicants for commissions and non-commissioned officers' roles within the new construction and operating companies. He recommended men for those positions based upon their 'railway qualifications', while Watson identified men suitable for enlistment as railway traffic officers. Forbes was responsible for enlisting the construction personnel that comprised the first four companies to be despatched to France, among whom were to be found platelayers, carpenters, timbermen, blacksmiths and telegraphists.[111]

The railway recruitment sub-committee's activities were complicated both by the pre-existing number of army reservists employed by the railway companies and the uncoordinated nature of British military recruitment at the beginning of the war. Within eleven days of the outbreak of hostilities 27,600 railway servants had left their peacetime occupations for service with the armed forces as reservists, territorials or volunteers. The War Office swiftly realized the importance of efficiently operated railways to the sustenance of Britain's war effort, and issued instructions to recruiting agents not to enlist railwaymen without the written approval of the man's head of department. However, by 23 February 1915 around 17 per cent of

[110] *History of the Corps of Royal Engineers*, ed. H. L. Pritchard (11 vols., Chatham, 1952), v. 596–97.
[111] ICE, Original communications, O.C./4277, H. A. Ryott, 'The provision of personnel for military railways in the war of 1914–1918', p. 1; Pratt, *British Railways and the Great War*, ii. 613–14.

the companies' pre-war male staff had signed up. The haemorrhage of men was slowed, but not entirely stopped, by the introduction of 'badging' in July 1915. By then, significant numbers of men who might have been more effectively employed in railway-related activities later in the war had already been recruited for the fighting formations.[112] The enlistment of skilled railwaymen elsewhere in the forces notwithstanding, the first railway construction company arrived in France just before Christmas 1914, and was engaged on the unloading of stores and the construction of sidings at Arques.[113] They were joined by ten further companies by the summer of 1915, and their duties included doubling the Hazebrouck–Poperinghe line, the construction of a new line between Candas and Acheux, the building of a railway store at Audruicq and the laying of sidings at Abbeville, Calais and Blargies. By December 1915 British construction companies had laid 105 miles of track on the western front, and by the end of the war forty-five companies had been raised from among the employees of British or dominion railways.[114]

Recruitment for the companies initially followed the blueprint provided by the 'Pals battalions', with individual railways raising complete sections that allowed their employees to serve together. In 1914 Canada possessed a far higher proportion of workers who were suitable for railway construction companies than did Britain, as increased migration into the prairie west had stimulated the expansion of Canada's railway network in the 1900s. The first two railway construction companies from Canada were formed of '500 picked men from the construction forces of the Canadian Pacific Railway', who arrived in France as the Canadian Overseas Railway Construction Corps in August 1915. They were the nucleus of a corps that eventually attained a strength of 16,000 men on the western front.[115]

The arrival of the Canadians in France was emblematic of the British empire's deepening commitment to the transport infrastructure on the western front, and emphasized Britain's access to global supplies of suitable manpower. The appointment of Lieutenant-Colonel William Waghorn as

[112] Pratt, *British Railways and the Great War*, i. 348; Simkins, *Kitchener's Army*, p. 111; Gregory, 'Railway stations: gateways and termini', pp. 34–5; TNA, ZPER 7/103, records of railway interests, pp. 23, 27.
[113] TNA, WO, 95/4053, lines of communication troops. 109 Railway Company Royal Engineers, war diary.
[114] Pritchard, *History of the Royal Engineers*, v. 596; C. Baker, 'The RE railway construction companies', *The Long, Long Trail: the British Army of 1914–1918 – for Family Historians* <http://www.1914-1918.net/re_rlwy_cos.htm> [accessed 20 September 2017].
[115] A. E. Kemp, *Report of the Ministry, Overseas Military Forces of Canada, 1918* (London, 1919), p. 355; G. W. Taylor, *The Railway Contractors: the Story of John W. Stewart, His Enterprises and Associates* (Victoria, BC, 1988), pp. 107–8.

chief railway construction engineer in the spring of 1915 underlined both the transnational reach of the BEF's call for suitable officers and the extent to which pre-war imperial service imbued men with the skills required to prosecute an industrial war. Waghorn had entered the Royal Engineers at the age of twenty, and seen engineering service in India and South Africa before he became acting manager of the North-Western Railway in India in June 1910. By 1914 he had established himself as one of the leading experts in the construction and operation of railways in the army – and, indeed, the empire – and as he was on leave in England when war was declared he was immediately appointed to the role of deputy DRT.[116] When Waghorn took on the role of chief railway construction engineer the majority of his subordinates were drawn from Argentina and Brazil. In South America, as in Canada, ample opportunities for ambitious railway builders to develop their careers had existed before the outbreak of war.[117]

Britain's transport experts and the global war

The arrival of skilled men from the Americas and Asia into the major European theatre of war in 1915 demonstrated Britain's growing military commitment to the western front. In contrast, the deployment of the first railway construction companies raised in Britain – drawn from the ranks of the LNWR and Great Western's huge workforces – illustrates how Britain's military focus had widened since August 1914. The 115th and 116th railway construction companies arrived in Egypt between December 1915 and March 1916, where they were employed to strengthen Britain's military position in the Middle East against the Ottoman empire.[118] The British had acted to secure their interests in the Persian Gulf at the beginning of the war, and soon turned their attentions to the defence of Egypt and the Suez Canal. Egypt's location, climate, and resources made it an ideal centre for the concentration and training of troops from India and Australasia, while

---

[116] M. Kaye Kerr, 'Waghorn, Brigadier-General Sir William Danvers', in *Biographical Dictionary of Civil Engineers in Great Britain and Ireland: 1890–1920*, ed. M. M. Chrimes et al. (London, 2014), p. 626; C. Messenger, *Call-to-Arms: the British Army, 1914–18* (London, 2005), p. 224; Pritchard, *History of the Royal Engineers*, v. 591; J. Bourne, 'William Danvers Waghorn', *Lions Led by Donkeys* <http://www.birmingham.ac.uk/research/activity/warstudies/research/projects/lionsdonkeys/t.aspx> [accessed 20 Sept. 2017].

[117] Pritchard, *History of the Royal Engineers*, v, pp. 35–6. On Britain's wider economic interests in South America before 1914, see P. A. Dehne, *On the Far Western Front: Britain's First World War in South America* (Manchester, 2009), pp. 8–39; C. Emmerson, *1913: the World Before the Great War* (2013), pp. 252–66.

[118] For details of the two companies' activities during the war, see the relevant war diaries in TNA, WO 95/4410, general headquarters troops; WO 95/4718, line of communication troops.

the canal formed a 'vital artery for war and merchant ships'. At least 376 transport ships, carrying 163,700 troops, passed through the canal between August and December 1914. Its continued accessibility to allied tonnage saved time and fuel, and allowed Britain to bring the weight of its global empire to bear on the fighting.[119]

Egypt, although legally part of the Ottoman empire, was heavily under British influence when Sultan Mehmed V declared 'holy war' on 11 November 1914. The higher tax officials and senior officers of the police and army were British, and the country had been under British administration since 1882. Commercial life in Egypt was also dominated by the British – alongside French, Italian and Greek expatriates – and the Suez Canal Company and the Egyptian State Railways were run by French and British officers respectively.[120] The three senior managers of the Egyptian State Railways were all retired Royal Engineers. Major Sir George Macauley (general manager), Major R. B. D. Blakeney (deputy general manager), and Captain C. M. Hall (traffic manager) represent further examples of the diaspora of British engineering knowledge in the pre-war period, which played a significant role in the creation and maintenance of the imperial trading network. The peacetime experience acquired by such individuals became invaluable to the sustenance of the empire's global supply system after 1914. Macauley was appointed director of railways in the Egyptian Expeditionary Force (EEF), with Blakeney and Hall named among his deputies.[121]

Cooperation between the military and civil authorities was, as in Europe, of 'utmost value' to British operations in Egypt. By the end of 1914 the pre-war garrison of 5,000 soldiers had been replaced by a force of 70,000 men drawn from India, Australia and New Zealand. The rapid expansion of Britain's military presence in Egypt outpaced the construction of accommodation and highlighted the poor state of communications in the canal's vicinity. There were no metalled roads alongside the canal and only a single-track railway linked it to the main line terminus at Zagazig. The railway company built sufficient sidings and stations to supply eight defence posts along the length of the canal during the autumn of 1914, and provided the Royal Engineers with temporary officers to supervise the works required to improve the canal's defences. When the much-anticipated Turkish attack upon the canal was finally launched on the night of 2–3 February 1915 it was

[119] E. Rogan, *The Fall of the Ottomans: the Great War in the Middle East, 1914–1920* (London, 2015), p. 116.

[120] R. Johnson, *The Great War and the Middle East: a Strategic Study* (Oxford, 2016), pp. 60–3; Pritchard, *History of the Royal Engineers*, vi. 161.

[121] Pritchard, *History of the Royal Engineers*, vi. 175–6.

repulsed, and the assaulting troops withdrew having lost over 1,000 men. The Ottoman forces retreated to Beersheba and made no further organized attacks upon the canal during the war. Consequently, the British developed Egypt as a base for the expeditions to Gallipoli and Salonika during 1915 with only minor disruptions from Turkish raiding parties.[122]

The expansion of Egypt's capacity as a training area and base for military operations drew heavily upon the resources and expertise of its foreign-run institutions. As Kristian Coates Ulrichsen noted, the EEF 'recognized and acted upon the need to introduce civilian expertise into military matters' long before Sir Eric Geddes's feted transport mission in August 1916.[123] After some hesitation the Suez Canal Company placed its engineering resources and a range of craft at the army's disposal, and in December 1915 a civilian-operated IWT organization was inaugurated to supplement the railway network. The public works ministry provided the EEF with ninety-seven tugs, steam barges, and lighters, 260 sailing craft, and all the canal and river craft of its inspection fleet. They were used for the distribution of road metal, coal, tools and machinery to the troops engaged on construction work around the canal.[124] The renewed threat of a Turkish attack on the canal in 1916 – following the allied evacuation of the Gallipoli peninsula – acted as a catalyst for increased engineering activity in Egypt over the winter months. A five-mile-long light railway constructed by the 116th Railway Company comprised just a fraction of the ninety miles of track laid to service the defensive positions built six to seven miles east of the canal during this period.

The Egyptian State Railways were integral to the infrastructure improvements on the eastern shore of the canal. The company provided most of the labour and resources utilized in the 'forward railway policy' implemented by the EEF,[125] including 15,000 local labourers deployed in support of the two British railway construction companies in the country. The Egyptian State Railways doubled the line between Zagazig and Ismailia, laid out large stations for camps in the Nile Delta and on the canal and erected new sidings from the mainline to the banks of the canal. Further north the company oversaw 'a considerable programme of improvements' to the

---

[122] Johnson, *The Great War and the Middle East*, pp. 64–7; Pritchard, *History of the Royal Engineers*, vi. 163–9.

[123] K. C. Ulrichsen, *The Logistics and Politics of the British Campaigns in the Middle East, 1914–22* (Basingstoke, 2011), pp. 59, 60–1.

[124] Pritchard, *History of the Royal Engineers*, vi. 166, 177.

[125] TNA, WO 106/712, defence of Suez Canal and railway policy, notes on the forward railway policy in Egypt, outlined in War Office telegram, 3 Dec. 1915; estimate by the general staff of forces required for defence of Egypt, 11 Dec. 1915.

single line along the canal from Port Said, which included the development of existing sidings, the addition of nearly eight-and-a-half miles of branch lines, passing loops, extensions to the dock facilities and eight stations at camp sites. The doubling of the line between Zagazig and Ismailia alone consumed almost one hundred miles of track, and the commencement of a further line east into Sinai in February 1916 left the company with no reserve stocks for maintenance.[126] The situation became increasingly critical once the EEF decided to advance on Palestine after the battle of Romani in August. However, following a slow, methodical progression across the desert, by December 1916 the railway from Qantara was of sufficient quality to maintain a force of 200,000 men on the Palestinian border.[127] Both the size of the force and the length of the line expanded further in the second half of the conflict.

Whereas the British possessed a free hand to undertake construction and improvements to the transport infrastructure in Egypt, circumstances in southern Europe bore closer resemblance to those in France. Forces from both Britain and France participated in the operations that took place following the landing of troops at Salonika. However, in western Europe the chain of command and the limits of British authority had been established and agreed upon during the pre-war conversations between the French and British general staffs. No such agreements existed to guide the transport organization in Macedonia, and to complicate matters still further both nations' forces disembarked on the soil of a third party. In the words of the Royal Engineers' history of the campaign,

> The allied Balkan campaign of 1915–18 was conceived in haste in the autumn of 1915 in fundamental disagreement between the British and French governments, and even more so between their military advisers. There was also disagreement in the Greek government upon whether they should be neutral or belligerent, and if the latter then on which side.[128]

As Asquith pointed out in a heated exchange with France's political and military leaders in June 1916, 'the British Government ... had sent their

[126] 'The Palestine campaign', *Railway Gazette: Special War Transportation Number*, 21 Sept. 1920, pp. 119–28.

[127] For a description of the 'Herculean British efforts that were made to construct the railway', the battle of Romani, and the EEF's systematic advance across the desert, see Johnson, *The Great War and the Middle East*, pp. 116–20; Rogan, *The Fall of the Ottomans*, pp. 312–18.

[128] Pritchard, *History of the Royal Engineers*, vi. 95. On the development of a Franco-British presence at Salonika, see TNA, CAB 28/1, papers I.C. 0–12, Balkans (Salonica expedition), July 1918, pp. 2–4.

troops [to Salonika] reluctantly at the request of Greece, at the insistence of France, and had always tried to get them away'.[129] The expedition to assist Serbia was viewed in Britain as a 'useless dissipation of effort and resources', a conviction founded upon Britain's role as the principal provider of shipping for the expedition and unmodified by the Greek railway authorities' 'uncooperative attitude' towards the British Salonika Force's (BSF) advance from the coast.[130]

Salonika was far from an ideal location for the disembarkation of a modern army and its accoutrements, and the Admiralty strongly disputed French claims that 2,000 tons per day could be landed once suitable piers had been constructed. Only three full-depth berths existed at the port, and the facilities required for the unloading and removal of stores from the water were virtually absent. These difficulties were reflected in the Admiralty's estimation that the capacity of the port was just 500 tons per day, a quarter of the French assessment.[131] The infrastructure to support the movement of men and materials inland was equally poor. The BSF possessed 350 lorries, but there were few roads upon which to use them. Locomotives and rolling stock were scarce and in bad repair. Even before the first troops arrived, on 1 October a small party of officers had landed at Salonika with the intention of establishing contact with the Greek authorities. Led by Colonel Maurice Sowerby of the Sudan Government Railways, the party proved unable to provide their hosts with information on the BSF's requirements for four days. The Royal Engineers' history records the officers' reception as having been 'so frigid and obstructive, not to say hostile', that Sowerby was advised by the War Office to 'go to ground until the situation became clearer'. Relations gradually thawed as the allied presence around the town grew, and the British railway staff eventually managed to 'cajole the Greeks into providing the absolute minimum' of the force's railway requirements.[132] The BSF, unlike their Egyptian counterpart, was unable to make effective use of the transport expertise provided to it by an imperial railway. In December 1915 Sowerby headed back to North Africa, where he made a valuable contribution to the construction of the EEF's railway line to El Arish.[133]

[129] TNA, CAB 28/1, papers I.C. 0–12, proceedings of a conference held at 10 Downing Street, London, S.W., on Friday, 9 June 1916, at 11:30 a.m., pp. 30–2.

[130] D. J. Dutton, 'The Calais conference of December 1915', *Hist. Jour.*, xxi (1978), 143–56, at p. 144; A. Palmer, *The Gardeners of Salonika* (London, 1965), p. 40.

[131] TNA, CAB 28/1, papers I.C. 0–12, notes of a conference held at 4:30 p.m. on Friday 29 Oct. 1915, at 10 Downing Street, p. 2; C. Falls, *History of the Great War: Military Operations, Macedonia* (2 vols., London, 1934), i. 44.

[132] Pritchard, *History of the Royal Engineers*, vi. 101–2.

[133] Pritchard, *History of the Royal Engineers*, vi. 229–31.

Just as Sir John French had found his desired railway movements superseded by the demands of the French army on the western front, the BSF was assigned a lower priority than the French forces in Salonika when it came to the allocation of transport in Macedonia. On 4 November 1915 the movement of elements of the 10th Division to Doiran was postponed for a week as the French demanded use of the trains. When the departure of the troops was rescheduled for 12 November, it was again aborted as no trucks were available. Journeys originally arranged for 6 November were not completed until nine days later.[134] The arrival of the 117th Railway Construction Company in the first week of February 1916 permitted the construction of new sidings in the base depots and passing places on the single-track railway that headed north from the port, but civilian traffic took top priority and the local railway staff permitted only one or two military trains to run each day. The Greek authorities remained in control of the railways between the port and the Serbian frontier until June 1916, when the French commanding officer General Maurice Sarrail appointed a Franco-British administration to take over operations.[135] Like the Gallipoli campaign before it, the Macedonian sideshow was incapable of replacing the western front as the British army's primary theatre of operations. However, unlike around the Dardanelles, French pressure to maintain an allied presence at Salonika ensured that British troops remained stationed in the theatre for the rest of the war.

Conclusion

Britain's position at the heart of a global empire profoundly influenced its approach to the First World War's expansion. The acquisition, development and maintenance of imperial possessions in Asia and Africa had provided the Royal Engineers with ample opportunities to augment their theoretical training with practical experience in the nineteenth and early twentieth centuries. Conditions in India had demanded the construction of roads and offered soldiers the chance to build railways before the war, while the Egyptian and Sudanese state railways provided the wartime army with a pool of experts capable of working comfortably under both civilian and military conditions. Beyond the theatres of war vast railway construction and civil engineering projects across the globe created demands for a pool of skilled engineers, which the British war effort took advantage of both at home and abroad.

[134] LHCMA, papers of Field Marshal Sir George Milne, general staff (operations) Army of the Black Sea, diary entries, 4–15 Nov. 1915.
[135] 'Railways and the Salonica campaign', *Railway Gazette: Special War Transportation Number*, 21 Sept. 1920, pp. 110–18, at p. 110.

Sir George Macauley and Follett Holt exemplified both the global experience acquired by Britain's transport experts before 1914 and the diversity of applications found for such expertise within the expanding British war effort. Macauley, a Royal Engineers officer, had resigned his commission after service in the Sudan to take up the posts of general manager on the Sudan Military Railway and subsequently the Egyptian State Railways prior to the war.[136] His knowledge of the Egyptian railways made him the obvious candidate to take charge of the EEF's railway requirements. Holt entered the Ministry of Munitions at the end of a career that had seen him work on the London and South-Western, the Indian State, the Buenos Aires and Rosario, the Great Western of Brazil and the Entre Ríos railways. He was joined at the ministry by established engineers from the British railway industry, such as the Midland's Henry Fowler and the North-Eastern's Vincent Raven, and by managerial figures like Eric Geddes. Together with their compatriots on myriad REC sub-committees, these men laid the groundwork for the material-intensive war fought by the British army in the second half of the war.

At home and in Egypt, where Britain possessed a free hand, Britain's transport experts exerted a substantial influence over the empire's approach to the First World War from the very outset. Macauley and his colleagues from the Egyptian State Railways coordinated transport operations and constructed significant lengths of railway on behalf of the troops stationed either side of the Suez Canal. Elsewhere, Britain's position within an international coalition restricted the extent to which the army could immediately make use of the empire's transport experts. The BEF and BSF, in France and Salonika respectively, lacked the freedom of action enjoyed by the EEF. Control of the ports and railways in the former remained firmly in the hands of the French authorities, while inter-allied relationships in Macedonia were complicated by the concerns and priorities of a neutral – and by no means friendly – host nation.

The shadow of the Franco-British pre-war arrangements and the enticing prospect of an imminent return to manoeuvre warfare loomed large over organizational developments on the western front. In France the BEF could not call upon the services of an experienced, civil–military transportation service, liberated from the requirement to consider the needs of a strong-willed and fiercely independent ally. While the BEF's expansion was encouraged by the French military authorities, it was not mirrored by a corresponding growth in the facilities and authority offered to the British.

---

[136] A. Fox, *Learning to Fight: Military Innovation and Change in the British Army, 1914–1918* (Cambridge, 2017), p. 176.

As Valentine Murray later reflected on the BEF's predicament:

> From first to last, and whatever its size, so far as the railway situation was concerned, the British Army was only a unit of the French Army – and in accordance with the principles of centralisation … its railway requirements had of necessity to be centralised at French GHQ – so that no British railway move could take place without the authority of the French. Consequently, all British railway requirements had to be carried out under the authority and orders of the French Railway Commissions and in this respect from the outset, the British Army was completely under the orders of the French.[137]

The stability of the line in front of the BEF made the lines of communications more recognizably civilian in appearance. The supply task that faced the British command equated to the provision of a small, rapidly growing town, rather than the supply of a small, rapidly growing *and* moving town.[138] As such, trench warfare created a situation for which Britain's transport experts' civilian occupations provided excellent apprenticeships. As the following chapters demonstrate, the BEF's position as the junior partner in the land-based coalition in France and Flanders constrained the exploitation of British transport expertise in the war's principal theatre between 1914 and 1916. However, that did not prevent the BEF from accessing and engaging with the employees of some of Britain's largest companies as it sought to perfect its continental commitment.

---

[137] V. Murray, 'Transportation in war', *Royal Engineers Journal*, lvi (1942), 202–32, at p. 207.

[138] Brown, *British Logistics*, p. 65.

4. Commitment and constraint I: the South-Eastern and Chatham Railway and the port of Boulogne

Britain's commitment to the First World War increased substantially during 1915. On the western front alone the BEF more than trebled in size. Between January and October over 650,000 men were added to the ration strength in France and Flanders, as the BEF rapidly grew into 'a force outside the bounds of British tradition and experience'. To maintain the growing requirements for food, fodder, munitions, equipment and raw materials generated by the BEF – alongside the supply of Britain's forces operating on the fringes of Europe and beyond – the British army and state 'needed a sound, coherent support infrastructure'.[1] Yet in the historiography of the British war experience of 1915 – dominated by the dismal failure of the Gallipoli campaign, General Charles Townshend's retreat to Kut Al Amara, the inconclusive results of the BEF's operations at Neuve Chapelle and Loos and the ongoing transformation of Kitchener's volunteers into the citizen army that fought the battle of the Somme – the administrative achievement that ensured Britain's vast manpower and resource commitments were not accompanied by starvation has been eclipsed. The dominant narrative centres upon shortages and insufficiency, the so-called shells scandal that underlined the soldiers' inability to prosecute the war effort and engendered the establishment of the Ministry of Munitions. In the words of its first minister, David Lloyd George, the 'war lords' were 'surly, suspicious, and hostile' towards the new ministry – a by-product of the 'ingrained distrust, misunderstanding and contempt of all businessmen ... which was traditionally prevalent in the Services'.[2]

This chapter builds upon the previous, and demonstrates how the British army actively sought out and engaged with civilian specialists during this year of rapid growth. It was aware of the potential benefits to be gained from the application of civilian knowledge to the challenges of the modern battlefield,

---

[1] I. M. Brown, 'Growing pains: supplying the British Expeditionary Force, 1914–1915', in *Battles Near and Far: a Century of Operational Deployment*, ed. P. Dennis and J. Grey (Canberra, 2004), pp. 33–47, at p. 35.

[2] D. Lloyd George, *War Memoirs of David Lloyd George* (2 vols., London, 1938), i. 83, 144.

and attempted to use non-military skills to improve the efficiency of the working methods employed upon its lines of communications. However, as Spencer Jones argued, in 1915 the BEF 'was thrust into a style of warfare for which it was conceptually and materially ill-prepared'.[3] The British army – and the nation as a whole – had yet to comprehend the magnitude of the war in which it had become embedded by 1915, and failed to appreciate the scale and character of the commitment that would ultimately be required to successfully prosecute it. 'The stationary character of the warfare of the first two years', the QMG's post-war report stated, 'placed no undue strain upon the QMG's branch' to maintain communications between the coast and the front.[4] The quantities of material demanded by front line commanders had not yet attained sufficient scale to severely tax the transportation system behind the growing BEF. Consequently, the early experiments in the application of Britain's transport experts to the logistical challenges of the First World War were governed by localized responses to short-term issues rather than a clear understanding of the Franco-British coalition's long-term priorities.

The wartime contribution of Francis Dent, general manager of the SECR, exemplifies the British approach to transportation during 1915. Dent's experience of the railway industry and his pre-war working relationship with the army made him an exceptionally useful figure, a man upon whom the government and military could rely to provide specialist technical guidance and leadership on a range of organizational problems. As this chapter illustrates, his personal war effort was diverse in subject and global in scope. Yet Dent's influence over the BEF's transportation operations in France was restricted by three factors: a British army that had not accepted that its governing structures were ill-suited to the type of conflict that emerged from the stalemate of 1914; a French army and state unwilling to relinquish command over the foreign forces engaged on its soil; and his own insufficient understanding of the scale and complexities inherent in modern military logistics. Taken together, these constraints severely curtailed the influence that a man of Francis Dent's undoubted abilities was able to have over the direction and character of the war prior to the battle of the Somme.

Creating capacity at the Bassin Loubet

The port space available on the French Channel coast played a fundamental role in determining the upper limits of the British force that could be

---

[3] S. Jones, '"To make war as we must, and not as we should like": the British army and the problem of the western front, 1915', in *Courage Without Glory: the British Army on the Western Front 1915*, ed. S. Jones (Solihull, 2015), pp. 31–55, at p. 31.

[4] TNA, WO 107/69, work of the QMG's branch, p. 1.

deployed on the western front during the First World War. The size and composition of modern armies – made up of millions of men and horses with constant demands for food and fodder, and supported by complex machinery dependent upon manufactured spare parts and a steady supply of raw materials – demanded that the belligerents provided their forces with reliable connections between centres of industrial production and the theatres of military operations. In simple terms, the amount of port space that could be occupied and worked efficiently by the allies had a direct correlation to the scale of war effort that the Franco-British coalition could sustain between 1914 and 1918. The BEF was not the only body reliant upon the Channel ports, a factor that added a layer of inter-allied interest in the use of the space. Both the French and Belgian armies drew supplies from the northern French coast, and the demand for imports was exacerbated by the loss of much of France's industrial heartland to the Germans at the outset of the war. The territory relinquished by the retiring allied forces left the French increasingly dependent upon Britain for deliveries of coal, as the area responsible for approximately three-quarters of France's coal and coke production was directly affected by the German advance.[5] Vast quantities of the fuel were necessary for the heating of homes, the powering of factories and the operation of the railways between the coast and the front line. By November 1914 a 'coal famine' had begun to emerge in France.[6]

Deliberations over the use of Dunkirk illustrate the reality of the Franco-British relationship after the outbreak of war and the constraints placed upon the BEF by its hosts. As early as December 1914 the British requested access to Dunkirk, both to shorten the BEF's lines of communications and to assist with the supply of the troops expected to arrive in 1915. The IGC, Ronald Maxwell, projected – based upon the requirements of a British force that comprised thirty infantry divisions and six cavalry divisions – that the BEF needed sufficient port space to deal with the supplies necessary to feed and equip 706,200 men and 244,200 horses. He was confident that a total of 350,000 men and 120,000 horses could be fed through the southern ports of Le Havre and Rouen, designated as group 'A'. However, it was impossible for him to calculate the capacity of group 'B' – Boulogne, Calais and Dunkirk – until a detailed reconnaissance of Calais and Dunkirk had taken place. Maxwell had assembled a committee of British officers to undertake the investigation, but felt it expedient to request the company of a French officer. He wrote to the QMG Sir William Robertson on 1 December that

---

[5] J. Lawrence, 'The transition to war in 1914', in *Capital Cities at War: Paris, London, Berlin, 1914–1919*, ed. J. Winter and J. Robert (2 vols., Cambridge, 1997), i. 135–63, at p. 152.

[6] TNA, WO 95/3951, headquarters branches and services. Inspector general, Cowper to Marrable, 27 Nov. 1914.

'it would be of great assistance to the officers to have some idea of what may be expected to be the attitude of [*GQG*] in connection with the details of this subject'.[7] The response was unequivocal. Joffre refused to authorize the mission, telling Robertson that he preferred to await developments on other fronts before discussing the allocation of port space behind the western front.[8] Consequently, the BEF did not gain access to Calais until April 1915 while Dunkirk's principal contribution to the British war effort was as a seaplane rather than supply base.

Dunkirk was a 'sore point' in inter-allied relations throughout 1915 and beyond, which consistently reinforced the French authorities' primacy in the Franco-British coalition. In January 1915 Maxwell reiterated the importance of the port to the successful development of a mass army in France, but was offered facilities at Cherbourg instead. Joffre rebuffed Kitchener's request for ten British ships to be accommodated at Dunkirk two months later with a claim that the port was part of the French front line 'and any British installation would interfere with its defence'. Maxwell made a further demand for port space at Dunkirk prior to the battle of the Somme, warning the French that 'imports of ammunition for the battle could not be processed in sufficiently large amounts' unless the British were provided with more berths by their hosts. Once again Joffre held firm, and instructed the British to manage better their existing resources – a refrain commonly repeated in inter-allied discussions before November 1918.[9]

Alongside French reluctance to relinquish berths, the space available for the BEF's imports was further limited both by the volume of traffic leaving continental Europe and the extent of the German occupation of Belgium. As the storm clouds gathered and burst in early August 1914 thousands of British and American tourists, together with 'a certain number of well-to-do Belgians', had travelled across the Channel from Ostend and Antwerp. As the front line swept ever farther across Belgian soil, thousands of desperate refugees descended on the ports in search of passage to England on the Great Eastern Railway's ferries. The link between the Belgian ports and Britain did not last for long. The final allied craft, which included three vessels owned by the SECR, departed Ostend on 14 October – five days after the multinational force sent to defend Antwerp had evacuated the city. While a 'fairly constant stream' of Belgians travelled to Britain from the neutral Netherlands over the course of the war, the only route available to those driven west in the upheaval of 1914–15 was via Calais. Consequently,

7   TNA, WO 95/3951, IGC war diary, Maxwell to Robertson, 1 Dec. 1914.
8   TNA, WO 95/3951, IGC war diary, Robertson to Maxwell, 3 Dec. 1914.
9   E. Greenhalgh, *Victory through Coalition: Britain and France during the First World War* (Cambridge, 2005), pp. 34–5, 244–5.

that port was rendered inaccessible to BEF supply ships for the entirety of the war's first winter by the incessant stream of human traffic seeking passage to England.[10]

The quantity of refugees, the higher-than-anticipated number of casualties that required evacuation to hospitals in Britain and the pace of the German advance combined to dislocate the pre-war arrangements agreed by the French and British authorities. The BEF's original intention had been to abandon the port of Boulogne – alongside Rouen and Le Havre one of the three ports utilized for the disembarkation of troops and supplies during the concentration period – on the sixteenth day of mobilization. Yet orders for the evacuation of Boulogne were not issued until 10 p.m. on 25 August, and only then as a precaution 'owing to the rapid advance of the enemy'. By 2 a.m. on the 27th, despite a 'continuous downpour of rain', all the stores save for small quantities of hay and wood had been loaded onto vessels and removed from the port.[11] After the German advance had been halted and their troops thrown into a retreat of their own following the battle of the Marne, Boulogne was reinstated as a port of entry for allied supplies on 14 October. Sir Percy Girouard visited the port during his investigation of the BEF's transport organization the following week, and 'thoroughly [examined] its capacity as an army base'.[12] His report succinctly concluded that the port was 'in a somewhat disorganised condition'.[13] The director of supplies' post-war report was more vivid: it described Boulogne's north quay as comprising 'one chaotic heap of coal, manure, discarded engineering fittings, and material originally intended for the completion of the harbour' upon the BEF's re-entry to the port.[14]

Over the following weeks the situation at Boulogne and the BEF's other base ports deteriorated. The docks possessed insufficient cranes to cope with the task of unloading military supplies for the growing force, and lacked covered accommodation under which to shelter items such as hay and oats from the deteriorating weather.[15] Major Moore of the ASC complained to the base commandant at Rouen in late November that only one crane had been available for the unloading of two vessels, a situation that made it 'necessary

[10] E. A. Pratt, *British Railways and the Great War: Organisation, Efforts, Difficulties and Achievements* (2 vols., London, 1921), i. 228–33.

[11] TNA, WO 158/2, director of supplies: British armies in France and Flanders pt. I. 147–8.

[12] TNA, WO 32/5144 Girouard report, Girouard to Cowans, 24 Oct. 1914.

[13] TNA, WO 32/5144 Girouard report, p. 13.

[14] TNA, WO 158/2, director of supplies I, p. 147.

[15] TNA, WO 95/74, branches and services: director of supplies, diary entries, 9 and 13 Dec. 1914.

to man handle a lot of cargo, which in these times of pressure is an absolute waste of labour'.[16] Land for the expansion of sidings, the erection of storage accommodation and the construction of additional harbour space was available at Boulogne, and – acting under the assumption that the BEF's residence at the port was likely to be far longer than that envisaged before the war – Maxwell had a scheme for extension work at the Bassin Loubet (one of two docking basins at the port) prepared and submitted to GHQ. Attached to the plan was a letter, in which Maxwell described the proposed works as 'urgent' and 'vital' to the BEF's ability to develop Boulogne as an effective base.[17]

However urgent and vital, such projects were inevitably going to be time-consuming, expensive and reliant upon the skilled and unskilled labour of thousands of workers. As the French army had suffered colossal losses in the war's opening encounters – almost one million French soldiers became casualties before the end of 1914 – the coalition's senior partner was in no position to provide the manpower necessary to bring large-scale engineering projects into being. Yet while *GQG* granted the British permission to construct additional railway sidings at the Bassin Loubet to improve Boulogne's capacity as a port, the BEF also lacked sufficient spare manpower to carry out the work.[18] The onset of winter had begun to take its toll on the transport network behind the front line, and a few days earlier the BEF's senior engineer had written to Lord Kitchener to complain about the paucity of men available to complete the extensive repairs required on the French road network.[19] Both Kitchener and Sir John French were reluctant to 'despatch [a] gang of navvies' from Britain to repair the French roads, but the former was happy to allow the REC to identify a suitable authority to undertake the engineering works at Boulogne.[20]

Percy Tempest, the SECR's chief engineer, was particularly suited for the role. Tempest had been a major in the ERSC since March 1902, was well regarded within the railway industry for having upgraded the SECR's outdated network, and was thoroughly acquainted with the French railways. In the opening months of the conflict Tempest had acted as an agent for the Chemins de Fer du Nord and Belgian State railways. In this role he had overseen the purchase and inspection of railway materials destined for the reconstruction of lines destroyed by the Germans. The allies' failure to

[16] TNA, WO 95/3951, IGC war diary, Moore to Marrable, 25 Nov. 1914.
[17] TNA, WO 95/3951, IGC war diary, Maxwell to Robertson, 30 Nov. 1914.
[18] TNA, WO 95/3951, IGC war diary, Robertson to Kitchener, 28 Nov. 1914.
[19] TNA, WO 95/3951, IGC war diary, Kitchener to French, 27 Nov. 1914.
[20] TNA, WO 95/3951, IGC war diary, Kitchener to French, 27 Nov. 1914; French to Kitchener, 27 Nov. 1914.

liberate much of the conquered territory meant that such work was pursued in vain, but the emergence of the Bassin Loubet construction project provided Tempest with an opportunity to directly improve the transport infrastructure that supported the coalition's military efforts. He accepted the REC's offer to direct the work and 'started at once on the necessary plans and preparations'.[21]

Between December 1914 and September 1916 – when the work was finally completed – the SECR provided all the tools, materials, labour and supervisory staff required for the construction of sidings, loading platforms, roads and railways, storehouses and workshops at Boulogne. The scheme initially involved the removal, via a specially constructed light railway, of 34,000 cubic yards of soil to a dumping ground three-quarters of a mile along the coast. Then, before the sidings and roads could be laid, an extensive drainage system comprising almost three miles of pipe and 143 manholes, gullies and grids was installed. Even so, the nature of the soil caused extreme difficulties during the construction of the Bassin Loubet's 56,774 square yards of new storage space, 1,317 yards of roads and 17,644 yards of new and replacement sidings. A 700-foot-long retaining wall was also erected by the labourers, nine-tenths of whom were sent from England specifically to work on the project.[22]

Tempest's input considerably increased Boulogne's value to the BEF as a base. Between November 1914 and October 1916, the month before the work had begun and the month after it had been completed, the tonnage despatched by rail per month from Boulogne rose from 12,357 tons to 57,590 – an increase of 366 per cent (see Figure 4.1). The number of trucks used within the port underwent a correspondingly large increase during the same period, from 1,737 to 7,918. In the following spring the material handled through Boulogne reached a wartime peak of 70,506 tons, for which 9,202 trucks were required. From April 1917 onwards, Boulogne was responsible for issuing the rations to a monthly average of over half a million men per day, except in March 1918 when the number dipped to 483,000 in the wake of the German spring offensive. The 'temporary' port – expected to be a component of the British war effort for just sixteen days in August 1914 – remained a crucial link in the army's supply chain for the duration of the war. Between 14 October 1914 and December 1918 the port handled 2,366,919 tons.

---

[21] Pratt, *British Railways and the Great War*, ii. 634.

[22] The spoil from the initial excavations was used to form the foundations for a third group of sidings and an ammunition dump later in the war. See 'Special war services by the South-Eastern and Chatham Railway', *Railway Magazine*, May 1920, p. 347.

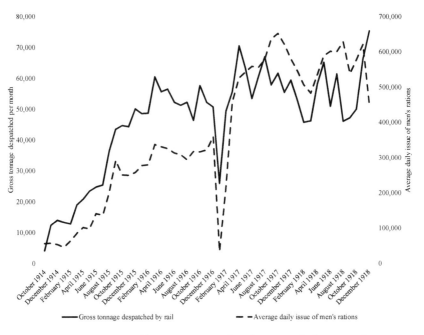

Figure 4.1. Monthly tonnage record and average daily issue of
men's rations from the port of Boulogne, 1914–18.

Note: The sharp decrease in Boulogne's output in January 1917 was caused by the accidental
stranding of the SS *Araby* in the mouth of the port on 23 December 1916. The port was closed
for twenty-seven days while the craft was salvaged, and only reopened in full on 18 January
1917.
Source: TNA, WO 158/2, director of supplies: British armies in France and Flanders pt. I, pp.
151–2.

Through its improvement of the facilities available for the storage and
movement of goods in the Bassin Loubet, the SECR played a significant
role in permitting the BEF's colossal expansion after 1914. However, the
company's contribution to events at Boulogne was not restricted to the
provision of labour, materials and construction expertise. In mid December
1914 Tempest was joined at the port by the SECR's general manager, Francis
Dent. Alongside submitting Tempest's estimates for the costs and duration
of the projected works in the Bassin Loubet, Dent observed to the director
of supplies that the cramped space and risks of exposure at Boulogne were
likely to cause heavy losses to the BEF's stocks of oats and forage in the near
future.[23] Rather than being dismissed out of hand by an intransigent and

[23]   TNA, WO 95/74, director of supplies war diary, diary entry, 11 Dec. 1914.

144

obstinate military command, the suggestion Dent went on to make on 11 December led to the conduct of a civil–military experiment at Boulogne that ran for much of the following year.

As with every other area of the French transport network, the pre-war agreement between the allied authorities made no provision for the British to acquire administrative responsibilities at the ports after the outbreak of war.[24] In his report of October 1914, Girouard emphasized that an 'arrangement with the French Government whereby they will allow us to organise bases, allocated to our use, in our own way and with our own men' was 'essential'.[25] The SECR's peacetime operation of the Folkestone–Boulogne ferry service meant that the company had offices and staff attached to the French port, and the company's employees at Boulogne had already been placed at the army's disposal before Tempest and Dent arrived in France. However, as 'full use' had not been made of the civilian manpower by the military authorities, the DRT suggested in November 1914 that the SECR might take over supervisory responsibilities within the Bassin Loubet with an 'adjoint' from the army as a liaison.[26]

The BEF was clearly receptive to the idea of securing civilian assistance, as Colonel Twiss's suggestion was by no means an isolated example within the force's administrative echelon. A month earlier the director of supplies, Major-General Frederick Clayton, had raised the possibility of deploying civilians on the BEF's lines of communications. He argued that the army could benefit from the experience of moving goods around Britain and across the world in a timely fashion that employees of the railway companies and large department stores possessed. Clayton had served on the advisory board of the LSE's 'Mackindergarten' course before the war, and was fully aware of the applicability of civilian methods to military requirements. He believed that suitably talented men could be used in 'essentially the same roles in France as they had filled with their civilian firms in Britain', thereby releasing trained soldiers for duties on the front line.[27] So when Francis Dent offered to 'study the situation on the spot' at Boulogne for a fortnight with a view to improving efficiency in the Bassin Loubet, Clayton gratefully accepted the proposal.

[24] M. G. Taylor, 'Land transportation in the late war', *Royal United Services Institution. Journal*, lxvi (1921), 699–722, at pp. 700–1.

[25] TNA, WO 32/5144, Girouard report, p. 13.

[26] TNA, WO 95/64, DRT war diary, Twiss to Murray, 12 Nov. 1914.

[27] I. M. Brown, *British Logistics on the Western Front, 1914–1919* (London, 1998), p. 87.

The application of civilian expertise: the Dent scheme

The so-called Dent scheme has garnered little attention in previous accounts of the BEF's logistical organization.[28] This is not merely a historiographical omission. Such was the perceived inconsequence of the scheme that the directorate of supplies, Dent's first point of contact within the army with regards to the Bassin Loubet, made no reference to either the individual or the experiment in its post-war report.[29] Ian M. Brown, despite noting that the experiment 'had [both] the potential to radically alter the way in which the BEF operated [the] port' and to rework the balance between civilian and military labour behind the lines, offered sparse coverage of events at the Bassin Loubet in 1915 in his published work.[30] The official history provided an even briefer assessment. Henniker's conclusion that 'it was [considered] inadvisable' to entrust the work of operating the port to 'civilian management and labour' has been unquestioningly reproduced in subsequent texts.[31] Yet the implementation of the Dent scheme at the Bassin Loubet merits reconsideration, as it underlines the character of the assistance that Britain's transport experts could provide to the BEF in the first half of the war and the constraints applied to that support by factors present on either side of the Franco-British divide.

Francis Dent's pre-war career, like that of his SECR colleague Percy Tempest, made him a suitable candidate for the task of solving the problems he identified at Boulogne. Born in 1866, Dent had begun his career in the railway industry at the age of seventeen. By August 1914 he had risen from a junior role in the general manager's office of the LNWR, through appointments in the goods traffic departments of the LNWR and SECR, to the position of general manager of the latter in 1911. The SECR's network covered the Channel ports at Folkestone and Dover as well as the prominent military sites of Woolwich and Chatham. Therefore, the company was an integral component of the British railways' preparations for war, and acted as the secretary railway to the army's eastern command in the development of the WF scheme. Unsurprisingly, given the line's importance to the mobilization process, Dent both obtained a commission into the ERSC

---

[28] C. Phillips, 'Early experiments in civil–military cooperation: the South-Eastern and Chatham Railway and the port of Boulogne, 1914–15', *War & Society*, xxxiv (2015), 90–104 represents the only extended piece of research on the SECR's operations at Boulogne.

[29] TNA, WO 158/2–3, Director of supplies: British armies in France and Flanders pts. I and II.

[30] Brown, *British Logistics*, pp. 88–9; I. M. Brown, 'Growing pains', pp. 46–7.

[31] A. M. Henniker, *History of the Great War: Transportation on the Western Front, 1914–1918* (London, 1937), pp. 91–2; J. Starling and I. Lee, *No Labour, No Battle: Military Labour during the First World War* (Stroud, 2009), p. 80.

upon his appointment as general manager of the SECR and was appointed a member of the REC upon its formation in November 1912.

In addition to his direct contact with the military, Dent was also able to bring previous experience to bear upon the task at Boulogne – where restricted space made the employment of more efficient working practices the only immediately available solution to the BEF's problems. In 1901 he had been appointed as goods traffic superintendent for the LNWR's Metropolitan district. At that time the company was faced by increasing congestion at Broad Street Station. Situated at the heart of the financial district, Broad Street was the third busiest station in turn-of-the-century London – the destination for thousands of commuters entering the capital each morning and a vital freight hub linking the Thames dockyards with industrial Birmingham. The LNWR's board feared that a significant expansion to the station was required in order to help it cope with the increased volume of traffic expected to pass through Broad Street in the future, a costly venture in such a heavily built-up urban environment. However, through a reorganization of working methods in the station, the establishment of a bonus payment system and his 'personal tact and influence', Dent accelerated the turnaround of goods within Broad Street to such an extent that 'the scheme for the enlargement of the station ... was abandoned'.[32] By August 1914 Francis Dent was a highly experienced professional manager with an established talent for the promotion of efficient working methods and a track record of success.

Furthermore, by the time he arrived at the Bassin Loubet in December, Dent had already made numerous contributions to the nascent British war effort. His service in the conflict's opening months encapsulates the uncoordinated nature of Britain's response to the multitude of challenges thrown up by its increasing involvement in the war, and indicates the transferability – from civilian to military applications – of the skills possessed by those who managed the nation's largest transport enterprises. In late July he was called to London to oversee the SECR's share of the armed forces' mobilization procedure alongside his colleagues on the REC, before he turned his attentions to the provision of ambulance trains for the wounded soldiers who required evacuation from the theatre of combat. The volume of casualties in the early weeks of the war, particularly among the troops of the French army engaged in the battle of the Frontiers, rapidly outstripped the capacity of the ambulance trains available for service on the western

---

[32] 'Retirement of Sir Francis Dent, general manager, South-Eastern and Chatham Railway', *Railway Magazine*, Apr. 1920, p. 253. That a review of Dent's professional career dedicated considerable space to the Broad Street reorganization demonstrates the significance attached to the project within the railway industry.

front. The War Office requested that the REC despatch an ambulance train to France for the BEF's use at the end of September 1914. Prior to its shipment Dent accompanied the LNWR's carriage superintendent and the Great Eastern Railway's chief mechanical engineer on a visit to France, where they discussed the technical requirements for ambulance trains with representatives of the Chemins de Fer du Nord. The group ascertained that no insurmountable obstructions existed between the rails and loading gauges used in France and Britain, and the LNWR immediately began work on an ambulance train for use on the French railways. Just three weeks after the REC had received the War Office's request vehicles were on their way to the western front.[33]

By December, when the casualties of First Ypres had been added to the lists of the wounded, further demands for bespoke vehicles to serve on the western front had been forwarded to the REC by the War Office. In response, and in acknowledgement that periodic requests for ambulance vehicles were likely to recur in the future, Dent took the chair of the REC's ambulance trains for the continent sub-committee. The sub-committee brought together representatives both of the railway industry and the army. From the former, Dent was accompanied by William Forbes, general manager of the LBSCR, and representatives of the engineering departments of twelve of Britain's largest railway companies. From the latter, to ensure the suitability of the trains from both technical and medical points of view, the sub-committee received the input of professional soldiers from the Royal Army Medical Corps (RAMC) and the military railway authorities from France and Britain. In fact, the ambulance trains for the continent sub-committee provided the reason for Dent's visit to Boulogne in December 1914. He attended a meeting at the port with Major Burke of the RAMC, at which they discussed the type of train best suited to wartime conditions and conferred with representatives of the Red Cross on financial and medical matters.[34] The result of this 'complete coordination' was the standardized ambulance train, which evolved under Dent's direction and combined the 'wisdom of experience' possessed by the various stakeholders invested in the production of effective and useful carriages over the course of the conflict.[35]

Further to his concern for the comfort of wounded soldiers, Dent participated in two other prominent REC sub-committees from October 1914 onwards. In the first he was responsible for the recruitment of officers

---

[33] Pratt, *British Railways and the Great War*, i. 201–3.

[34] TNA, WO 158/11, ambulance trains: sub-committee; minutes, Burke to Barefoot, 12 Dec. 1914.

[35] On the work of the ambulance trains for the continent sub-committee, see Pratt, *British Railways and the Great War*, i. 201–8.

and men for what became the ROD; in the second, he took charge of a sub-committee established to identify and organize the Belgian railwaymen who had found refuge in Britain since the start of the war. The Belgian railways refugee sub-committee comprised representatives of four other British railway companies, Major Leggett of the war refugees committee and a senior figure from the Belgian State Railways. As Edwin Pratt noted, 'it was a matter of military as well as of economic importance' that the Belgian railway experts 'should be readily available whenever the need for them arose, either during the course of the war or afterwards'. However, the Belgian railwaymen, 'in common with the other refugees [had] been allocated where suitable hospitality was available, so that while some found themselves in seaside resorts or country villages where no railway work could be provided for them, others were traced as far away as Tipperary or the North of Scotland'. The sub-committee dealt with 3,681 Belgian railwaymen during the war, 2,801 of whom lived in Great Britain and Ireland as refugees. They found employment in Britain for 1,755 of them, either on the British railways or – for those such as clerks and stationmasters who lacked suitable language skills – in the munitions factories.[36]

Dent set his existing commitments aside over Christmas 1914 to spend time at Boulogne observing operations. In a letter to Sir John Cowans he stated that:

> There is no doubt stores are suffering to a great extent through there being insufficient provision for stacking and storing under cover. Boulogne is a very good port for quick handling and, by using it properly, the transit of supplies to the front is much accelerated. In view of the increase in the army, it is desirable that we should get on as quickly as possible.[37]

To ensure that proper use was made of Boulogne, Dent proposed that the SECR should be given responsibility for the operation of all areas of the port reserved for the BEF's use – replacing the existing system whereby the naval staff discharged vessels onto the quayside and the army looked after the onward transport or storage of the goods.[38] His offer entailed the railway company taking over the 'work of discharging ships, stacking supplies and loading trains, [and] providing all the personnel' for these tasks in place of

---

[36] Pratt, *British Railways and the Great War*, i. 239–42.

[37] TNA, WO 95/3952, headquarters branches and services, inspector general, Dent to Cowans, 31 Dec. 1914.

[38] TNA, WO 95/3953, headquarters branches and services, inspector general, Proceedings of second meeting of committee on Mr Dent's scheme held at Boulogne, 15 Feb. 1915; Brown, *British Logistics*, p. 88.

the BEF's current reliance on an ever-dwindling supply of French labourers.[39] There was, Dent wrote, 'nothing in the way of checking or loading that would not be easy enough for a railway checker to perform' at the Bassin Loubet.[40] The object of the basin in peacetime was 'to ensure quick transit between steamer and train. The hangars were laid out with a view to easy checking and customs examination', and in many cases were supplied during the war by the same railway steamers that operated the routes in peacetime. Therefore, the task of discharging ships quickly, loading trains for the front and stacking supplies for later despatch was almost identical to the work undertaken at the railway ports under the SECR's control. In fact, Dent claimed, 'the military supplies business [was] simpler than ordinary trade practices' as most of the BEF's supplies arrived in bulk and did not require lengthy customs examinations upon arrival in France.[41]

Dent believed that a number of factors combined to retard the efficiency of operations at Boulogne in late 1914. Alongside the 'want of railway accommodation for internal movement' and the 'rough and unfinished state of parts of the yard' that Tempest's construction work was set to remedy, Dent considered the mixture of French, Belgian and British labour at the port, the import of huge reserves of forage before there was suitable covered accommodation available and the presence of *en-cas mobile* trains as the predominant causes of delay at Boulogne. His proposed solution to these issues involved on the one hand the separation of the various nationalities of labour at the port – the British assigned responsibility for the stacking of goods and other duties in the sheds, while the French and Belgians merely handled them – and on the other hand the establishment of a system whereby the majority of supplies were transferred direct from ship to rail upon arrival in France. Items required at the front urgently could be sent forward immediately, while those not considered to be priorities could be shifted to storage sites away from the docks. This would ensure that the quayside remained free of obstructions that complicated, and inevitably slowed down, the discharge of arriving ships. Dent claimed that if these problems could be rectified the Bassin Loubet was capable of handling over 5,000 tons per day.[42] (The true capacity of the Bassin Loubet,

---

[39] TNA, WO 95/3952, IGC war diary, Dent to Clayton, 28 Dec. 1914; Clayton to Dent, 30 Dec. 1914; diary entry, 13 Jan. 1915.

[40] TNA, WO 95/3952, IGC war diary, Dent to Clayton, 28 Dec. 1914.

[41] TNA, WO 95/3952, IGC war diary, Boulogne – memorandum by F. H. Dent, 28 Dec. 1914.

[42] TNA, WO 95/3952, IGC war diary, Boulogne – memorandum. The term *en-cas mobile* refers to a group of wagons 'kept permanently under load ready for immediate despatch' in the event of an emergency. See Henniker, *History of the Great War*, p. xxxi.

in Dent's opinion, was 7,000 tons per day. The lower estimate given in his memorandum reflected the nature of the labour available for work in the port, which consisted principally of 'boys and men not of military age'.)

With the BEF's projected demands for food alone set to reach 4,400 tons per day once the first troops of Kitchener's armies arrived on the western front, Dent's estimates were understandably appealing to the officers tasked with keeping the BEF fed and equipped.[43] However, Clayton was sceptical that Dent's projected figures could be achieved at Boulogne and had reservations over the practicability of the railwayman's proposed quick transit scheme. A central tenet of Dent's plan to maximize throughput at the Bassin Loubet involved loading cargo in Britain so that 'each ship should have approximately sufficient of everything to make the greater part of one or more supply trains'. Under such an arrangement, the trains for the front line could be made up directly from the quayside – with any surplus stocks on each ship or perishable items that had to be regularly turned over placed in systematized stores. The benefit of the proposal lay in the fact that it would involve comparatively less double-handling than the extant system in which ships were unloaded, the goods moved from the quayside for storage elsewhere in the harbour and then loaded onto rail at a later date.[44]

Dent's quick transit system was ideal in terms of operational efficiency, but unfeasible as a solution for the requirements of an industrial army with a multitude of demands. Clayton's response to Dent's memorandum illustrated the civilian's under appreciation of the difficulties faced by the military authorities between 1914 and 1918.[45] In response to Dent's criticism that forage had been imported into France before sufficient accommodation was available for it, the director of supplies pointed out sardonically that 'this I am afraid is one of the necessities of war; the Germans would not wait until we had everything in readiness'.[46] Yet the German army was not the only factor that had made Clayton's life difficult. The sustenance of the BEF's troops and machines depended upon a range of commodities – to prevent the risk of contamination, items such as petrol and lubricating oils were not transported on the same vessels as food. Nor was the food consumed by the BEF entirely despatched from Britain. Meat was taken from cold storage ships berthed at Boulogne, while bread was baked in

[43] TNA, WO 95/3952, IGC war diary, diary entry, 16 Jan. 1915.

[44] TNA, WO 95/3952, IGC war diary, Boulogne – memorandum.

[45] Dent was by no means alone in considering the suitability of direct loading. As Brown demonstrated, in Jan. 1915 correspondence on the subject passed between the IGC in France and the QMG at the War Office. See Brown, *British Logistics*, p. 82.

[46] TNA, WO 95/3952, IGC war diary, Clayton to Dent, 30 Dec. 1914. Unless otherwise stated, all quotations in this passage are taken from this source.

vast quantities at bakeries near the port and transported to the railway by lorry. The volumes of the other items in the soldiers' diet were 'trifling' in comparison to these two staple foodstuffs. Therefore, neither of the integral components of the troops' daily rations were going to be on board the ships that were central to Dent's concept. In addition, the type of food sent forward was frequently altered: 'We change from fresh meat to preserved and then to meat and vegetable rations; change bacon for butter and give the soldier as much variety as possible', Clayton noted. In short, the director of supplies summarized that it was impossible to 'pack a train for any formation straight from the ship except as regards hay and oats'.

Yet despite his detailed criticisms of Dent's proposals, Clayton was sufficiently amenable to the idea of greater civilian involvement in the war effort – at this stage of the conflict at least – to encourage further discussion of the scheme. As he recognized at the end of his reply to the SECR's general manager, 'after all, we are all out for the same object – the good of the country and to end the war as speedily as possible, so if you can help in this the Army will be grateful to you'.[47] Both Clayton and Robertson saw the potential benefits of transferring responsibility for the operation of Boulogne to the SECR. Consequently, a committee was formed to amend and improve Dent's outline – the membership of which illustrates the scale of the organization required to supply a modern army in the field. Officers from the directorate of supplies (Clayton took the chair), the directorate of works and the directorate of ordnance services joined the principal naval transport officer in the committee's deliberations.[48] Both the naval and military elements saw an 'advantage' in the centralization of responsibility at Boulogne, and indicated their willingness to accept the SECR's offer subject to approval from GHQ, the War Office and – as the BEF's hosts and partner – the French authorities. Even Dent's subsequent downward revision of the Bassin Loubet's capacity from 5,000 tons per day to 3,536 tons per day did not alter the committee's resolve.[49] Fred West, the goods superintendent of the SECR's London district, was instructed to 'ascertain the system of work of the various departments and to discuss various points with the officers in charge' following the committee's first meeting.[50]

---

[47] TNA, WO 95/3952, IGC war diary, Clayton to Dent, 30 Dec. 1914.

[48] TNA, WO 95/3952, IGC war diary, Robertson to Maxwell, 9 Jan. 1915.

[49] TNA, WO 95/3952, IGC war diary, diary entry 29 Jan. 1915; WO 95/64 DRT war diary, French to Kitchener, 23 Feb. 1915.

[50] TNA, WO 95/3952, IGC war diary, Commandant, Boulogne base to Clayton, 27 Jan. 1915. Clayton had taken over as IGC the day before this letter was written, in the administrative reshuffle that accompanied Sir William Robertson's appointment as CIGS. Lieutenant-General Ronald Maxwell vacated the role of IGC and became the new QMG in

West's report was a combination of observations on the existing situation at Boulogne and recommendations to help the BEF 'obtain the maximum amount of efficiency and economy' in the future.[51] His comments were circulated ahead of the second meeting of Clayton's committee, at which Dent played a key role. The civilian fielded questions from the military and naval officers, and elaborated upon his vision of the SECR's position within the new organization.[52] The members unanimously agreed that the navy, due to their ignorance of the landside procedures for the removal of goods from the quayside, should cede responsibility for the work of discharging ships to the port's 'single authority'.[53]

Multiple factors combined to recommend the SECR as a suitable organization to take on the duties of the 'single authority' at the Bassin Loubet. The company already had experience in the operation of railway ports, a working understanding of the port of Boulogne and a pre-existing working relationship with the Chemins de Fer du Nord. In addition, the company had a strong presence at the port – both in terms of the men placed at the army's disposal during the autumn and those connected with Tempest's engineering works – and Dent had demonstrated an evident willingness to participate in the experiment.[54] A request for permission to change the procedure at the port was duly despatched to the War Office in early February, but despite persistent appeals from Clayton ratification from London was inexorably slow to arrive. The SECR was finally authorized to take over operations within the Bassin Loubet on 17 March – a delay that effectively put the new system into stasis for six weeks, after which further time was required for Dent to 'collect his own staff' for work in the port, for those men to 'observe the routine working of a [military] port' and for arrangements to be finalized between the SECR and the French rail authorities.[55] Following discussions between Dent, the SECR and representatives of the *commission regulatrice*, the British railway company was eventually authorized to take over 'all the work of shunting, marshalling and the making up of trains in the Bassin Loubet' from 25 April 1915.[56]

France, while Clayton's post as director of supplies was handed to Colonel E. E. Carter – a graduate of the LSE's 'Mackindergarten'.

[51] TNA, WO 95/3953, IGC war diary, Bassin Loubet – Boulogne, Mr West's report, 13 Feb. 1915.

[52] TNA, WO 95/3953, IGC war diary, Proceedings of second meeting.

[53] TNA, WO 95/3953, IGC war diary, Clayton to Shortland, 16 Feb. 1915.

[54] TNA, WO 95/3953, IGC war diary, Clayton to Maxwell, 16 Feb. 1915; Clayton to Maxwell, 20 Feb. 1915; WO 95/75, Branches and services: Director of supplies, diary entries, 24 and 26 Feb. 1915.

[55] TNA, WO 95/3953, IGC war diary, diary entries, 5 and 27 Feb. 1915; WO 95/3954 Headquarters branches and services. Inspector general, diary entries, 8 and 17 March 1915.

[56] TNA, WO 95/3954, IGC war diary, Clayton to Maxwell, 21 March 1915; WO 95/27,

The delays that postponed the company's takeover of operations at Boulogne meant that the SECR inherited a port that had experienced increasing congestion during the spring. The demands for material to support the offensive at Neuve Chapelle in March overloaded the Channel ports' capacity, as the War Office had responded to the unprecedented scale of the fighting by despatching vessels in the direction of the battlefield as swiftly as possible. Ships were arriving in France without sufficient intervals to allow for one ship to be unloaded and cleared from the quay before the next arrived. As highlighted in a post-war article by the shipping expert Charles Ernest Fayle, the significant advantage in maritime carrying capacity available to the allies was of little value if that capacity was inefficiently utilized:

> The number of voyages a ship can make … depends not only on her speed at sea but on the rapidity with which she can be loaded and discharged, and this, in turn, depends not only on the actual equipment of the ports, but on the prompt arrival of cargo at the port of loading and on the rapid distribution of cargo from the port of discharge. The ports, docks, quays, and wharves; the railways, roads, and canals by which the ports are served; the offices in which arrangements for the voyage are made; the cables by which fixtures are effected and instructions given, are all as important as the ships themselves.[57]

Ships in port awaiting discharge, or delayed due to inefficiencies in the discharging process, could not quickly return to their port of origin to acquire their next cargo.

Poor communications between the ports of origin and those on the French coast hampered operations at the Bassin Loubet. The staff at Boulogne, both before and after the SECR took over, were frequently left with incomplete information as to the contents of incoming ships. On 13 March the SS *Juno* set out for the port behind an advanced notification from Britain that informed operators at Boulogne only that the ship contained 'general cargo'.[58] With limited crane facilities available it was imperative that the port authorities received detailed prior notice of the composition of each ship's contents, so that they could be directed to the berth best suited to their discharge immediately upon arrival. Without such information,

---

QMG war diary, diary entry, 12 Apr. 1915; WO 95/3955, Headquarters branches and services. Inspector general, diary entry, 20 Apr. 1915; WO 95/58, Branches and services: Director of ordnance services, diary entry, 19 Apr. 1915.

[57] C. Ernest Fayle, 'Carrying-power in war', *Royal United Services Institution Journal*, lxix (1924), 527–41, at p. 531.

[58] TNA, MT 23/353/1, naval transport officer, Boulogne. As to his advance notification of general nature of stores on transports, Hamilton to Shortland, 15 March 1915.

Clayton warned, the supply services could not guarantee that supplies required urgently at the front could be processed through the port in a timely fashion.[59] To alleviate the communications difficulties between the Bassin Loubet and Britain, Dent recommended that a bespoke telephone connection be installed to link Boulogne with the SECR's offices in London, Dover, Folkestone, Calais and Dunkirk. Precise information as to the contents of each ship could be received by telephone prior to the vessel's arrival, allowing staff to direct it to the most suitable berth and arrange for any specialist equipment to be provided to the stevedores responsible for its discharge.[60]

The responses to Dent's suggestion illustrate the limitations of the Franco-British coalition. The War Office in London raised no objections to a scheme with an obvious benefit to the BEF's supply operations behind the western front. However, although the BEF had been granted 'every latitude' for the improvement of local facilities within the zones populated by its fighting forces during the war, projects that included the installation of more permanent equipment also had to be signed off by the French authorities.[61] The provision of telephone facilities for the SECR's use was, somewhat unsurprisingly, clearly far from the top of *GQG*'s list of priorities. By the end of October 1915 Clayton had received no decision on the request. He had believed all along that the French were 'unlikely' to accede to Dent's wish but – following an appeal to 'badger' Joffre's staff – a further enquiry was made, which generated a firm refusal from the French in early November.[62] The potential benefits of the telephone line for the prosecution of the allied war effort were acknowledged by the French authorities. However, the French government realized that the system's installation would have conferred significant competitive advantages to the SECR after the war. Combined with a perception among French authorities at the port that a 'custom' of unauthorized telephone use had 'grown up' in the SECR's offices at Boulogne over the course of 1915, the BEF's hosts asserted that the existing telephone facilities – if used properly – were adequate for British requirements.[63]

The disagreement between the French authorities and Britain's civilian specialists may appear superficial, but the 'telephones incident' underscores

[59] TNA, WO 95/3954, IGC war diary, diary entry, 23 March 1915.
[60] TNA, WO 95/3954, IGC war diary, diary entries, 12 and 23 March 1915.
[61] TNA, WO 95/3953, IGC war diary, diary entry, 27 Feb. 1915.
[62] TNA, WO 95/3955, IGC war diary, diary entry, 7 Apr. 1915; WO 95/3961, Headquarters branches and services. Inspector general, diary entry, 31 Oct. 1915; WO 95/3962, Headquarters branches and services. Inspector general, diary entry, 8 Nov. 1915.
[63] TNA, WO 95/3962, IGC war diary, 8 Nov. 1915.

the instability of Franco-British relations during the First World War. Throughout the conflict French and British authorities were involved in a complex web of negotiations, to which were added the voices of Belgians, Italians, Americans and other allies as the war progressed. Political and military leaders discussed and sought conciliation in conference rooms across Europe, yet the post-war economic and strategic concerns of the coalition's individual components provided an underlying context that militated against absolute cooperation. Even the provision of a unified command in the person of General Ferdinand Foch in the latter months of the war did not eradicate the preponderance of national interests and underlying suspicions. A Franco-British disagreement over who should provide the manpower and materials required for the repair of Dunkirk continued until the armistice came into force. Both the British army and the Admiralty acknowledged privately that the port was a more suitable candidate for improvement than any of the others that served the BEF in October 1918, but the then QMG Travers Clarke was unable to ignore misgivings that the French wanted to see Dunkirk repaired for commercial reasons. Clarke stated baldly in the immediate aftermath of the conflict that 'unless absolutely demanded by the interests of victory, it was no part of our military or national duty to enlarge or modernize the equipment of foreign ports for after-the-war trade'.[64]

The war aims of France and Britain in western Europe – despite both ostensibly seeking the defeat of Germany – were in many respects profoundly different. These disparities, coupled with the changing nature of the two nations' comparative contributions, required French and British leaders to engage in constant discussion and compromise to preserve the delicate coalition. The fact that a formal contract between the two countries did not exist, and the absence of any organ for collective decision-making within the coalition, helped reinforce the primacy of national considerations over inter-allied requirements throughout 1915. The Dent scheme was implemented at Boulogne during a period in which the relative strength of the French in terms of land power – and the BEF's dependence upon the French transport network as a conduit for the output of the munitions factories across the Channel – acted as powerful bargaining chips in Franco-British negotiations. Within the tense atmosphere of an allied war effort that continued to achieve relatively little on the western front, the installation of a telephone system to improve throughput at the Bassin Loubet was not

---

[64] TNA, WO 95/40, branches and services: quarter-master general, minutes of conference held in the QMG's office on the subject of the use of the ports of Havre, Rouen and Dunkirk, 7 Oct. 1918; explanatory review, Nov. 1918, p. 14.

deemed sufficiently important to override French considerations of their post-war industrial strategy. Yet it was far from the sole reason why the Dent scheme was abandoned before the year of the Somme had even begun.

The growing complexity of Britain's war machine meant that Francis Dent became increasingly detached from events at Boulogne after April 1915. The proliferating demands for men of recognized managerial ability were such that Dent's commitments to the REC were already substantial by the time the SECR became responsible for operations at the Bassin Loubet. He continued to oversee the identification and deployment of displaced employees of the Belgian State Railways as part of the Belgian railways refugee sub-committee, maintained his central role in the ambulance trains for the continent sub-committee and became immersed in the recruitment of personnel for the ROD – Dent's portion of the work of the railway recruitment sub-committee – which took on fresh importance just as the Dent scheme finally got underway.

Dent's duties in relation to the new division in April 1915 were significant. He dealt 'with the multitudinous questions which arose in regard to the methods of enlistment, rates of pay, [and the] nature of duties' for recruits, and personally interviewed all applicants for commissions. He was, according to Pratt, 'accustomed to "father" the division's early recruits, and took great care to provide for their needs'.[65] Yet his most important contribution to the ROD's development lay in the selection of Cecil Paget as the division's commanding officer. In an atypical career that had taken in both engineering and traffic management positions at the Midland Railway before the war, Paget had gained a reputation as a 'brilliant organiser and administrator'.[66] Alongside his 'precise knowledge of the French language' – critical in a role that demanded constant liaison with the BEF's hosts – his rounded knowledge of locomotive engineering and traffic operations made him a perfect candidate to lead a division with mechanical and operational responsibilities. Colonel L. S. Simpson, the ROD's chief mechanical engineer, spent most of 1915 interviewing Belgian and French railway and military personnel alongside Paget, and observed that his chief

> had no difficulty in working in with the French or in carrying out the orders of higher authority, often involving complicated movements of men and materials, and it is entirely due to him that the Railway Operating Division took such a large and important part in contributing to the success of our arms. Both on the operating and the mechanical side we came to be regarded as a seventh

[65] Pratt, *British Railways and the Great War*, ii. 615, 617.
[66] E. G. Barnes, *The Midland Main Line, 1875–1922* (London, 1969), p. 224.

railway company, and our relations with the French staff and the officials of the State and five private companies were always most cordial.[67]

Paget retained command of the ROD throughout the war, a clear indication of his suitability for a role in which he 'acted as the equivalent of a superintendent of the line in conjunction with the French railway officers'. Sir Sam Fay wrote approvingly after the war that he 'could have been graded a general if he had so wished', such was the army's appreciation of Paget's contribution on the western front. However, 'he was more intent upon his duties than upon advance in military rank'.[68]

The ROD's first contribution to operations in France took place in the Bassin Loubet, where it became responsible for marshalling the supply trains made up at the port in June 1915. As the number of ROD units in France increased the division became responsible for more marshalling yards and depot sidings, before on 1 November the French authorities transferred operations on the Hazebrouck–Ypres line to the British.[69] Before the ROD took over the Hazebrouck–Ypres service, Simpson and his troops acquired responsibility for the repair of engines earmarked both for the division and the BEF's construction troops. A temporary workshop was established in a sugar factory near Calais for the overhaul of thirty-five Belgian locomotives, which were 'in a terrible state owing to their not having been touched since they were rescued from Belgium' the previous summer. However, once again the requirements of the French economy superseded the convenience of the British army in France. As early as mid August 1915 Simpson was 'obliged to find some other place where the repairing of engines could go on' as the premises they had occupied were required for sugar production. The British mechanics removed their tools and equipment to a temporary site in an ancient chalk pit between Calais and Boulogne, before they finally moved into a permanent locomotive repair shop at Audruicq in December.[70]

The provision of ambulance trains for use at home and abroad competed with the ROD for Dent's attentions in the spring and summer of 1915. Alongside his duties as chairman of the ambulance trains for the continent sub-committee of the REC, Dent acted as a conduit for communications between the predominantly civilian sub-committee in London and the ambulance train advisory committee in France – which comprised British

[67] L. S. Simpson, 'Railway operating in France', *Journal of the Institution of Locomotive Engineers*, xii (1922), 697–728, at p. 701.

[68] S. Fay, *The War Office at War* (London, 1937), p. 91.

[69] Henniker, *History of the Great War*, p. 168; *History of the Corps of Royal Engineers*, ed. H. L. Pritchard (11 vols., Chatham, 1952), v. 594.

[70] Simpson, 'Railway operating in France', pp. 699–700; Henniker, *History of the Great War*, pp. 168–9.

and French military railway authorities and RAMC officers. Dent's role was crucial as it facilitated the standardization of equipment on both sides of the Channel, a process that reduced both costs and production times for new vehicles.[71] By the end of the conflict the British railway companies had provided thirty ambulance trains for the British army's use, consisting of 518 carriages and all of the spare parts and materials necessary for their maintenance. As well as serving on the western front, British-built ambulance trains were sent to Salonika and Egypt and provided transport for the evacuation of casualties along the Mediterranean line of communications. In addition, following consultation between the REC and the American military authorities in the summer of 1917 – at which Dent was present – the British railway companies had constructed nineteen ambulance trains (comprising 304 vehicles) for the American Expeditionary Force (AEF) and had a further twenty-nine in varying states of readiness when the armistice came into effect.[72]

Within the maelstrom of work on the REC's behalf, Dent continued to be the general manager of a critically important railway line in the south-east of England – one that experienced its own significant challenges as the war continued. As noted above, the SECR connected London to the two shortest ferry routes between Britain and Europe, and in peacetime it principally operated as a passenger line for commuters into the capital and tourists to the coast and continent. However, the character of the SECR's traffic changed radically once Britain's productive capacity was geared towards the war effort. By the summer of 1915 passenger traffic – excluding troop movements – no longer predominated, and the SECR's system had to be adapted as quickly as possible to carry a heavy goods traffic 'for which they had never been designed and were, at first, not fully prepared to meet'.[73] In July 1915 the SECR's London district handled 56 per cent more wagons than it had twelve months earlier (see Table 4.1), even though the company's workforce had been diminished by the enlistment

---

[71] See, e.g., the agendas for meetings of the ambulance train advisory committee in February, June and August 1915. Each meeting discussed agenda items supplied to the committee by Dent on behalf of the ambulance trains for the continent sub-committee in London. TNA, WO 158/9, ambulance trains: advisory committee meeting; general correspondence, Agenda for ambulance train committee on Saturday 13 Feb. 1915 at 5:30 p.m.; Agenda for meeting of advisory committee on ambulance trains to be held at Boulogne at 10 a.m., on Sunday 6 June 1915; Agenda for meeting of advisory committee on ambulance trains to be held at Abbeville on 18 Aug. at 5:30 p.m., 1915.

[72] Pratt, *British Railways and the Great War*, i. 206–7; TNA, WO 158/11, Ambulance trains, Minutes of meeting held at 35 Parliament Street, Westminster, at 2:45 p.m. on Monday 18 Nov. 1915, pp. 1–2.

[73] Pratt, *British Railways and the Great War*, ii. 1075.

Table 4.1. Wagon turnover for the London district of the
South-Eastern and Chatham Railway, 1914–18.

Date	Number of wagons	Increase over 1914
July 1914	223,798	———
July 1915	340,193	56 per cent
July 1916	432,896	93 per cent
July 1917	464,121	107 per cent
July 1918	447,000	100 per cent

Source: E. A. Pratt, *British Railways and the Great War; Organisation, Efforts, Difficulties and Achievements* (2 vols., 1921), ii. 1078.

of 2,689 employees to the colours. The patriotic sacrifice of the railway's servants left Dent and his management team facing a significantly increased workload with a workforce some 12 per cent smaller than it had been when the war began.[74]

The army's decision to exploit Folkestone more thoroughly presented the SECR with further traffic to deal with from April 1915 onwards. As a locally produced chronicle of the town's experience of the war recorded:

> The port of Folkestone was opened for [the] transport of troops about the end of March 1915, when the Authorities discovered that it was very much the quicker route. After that date a steady flow of troops to and from France was maintained. On an average six ships, not including cargo ships and lighters, sailed daily all through the war with reinforcements and leave men.[75]

The Folkestone–Boulogne service became the principal route for British personnel both on their way home from the front on leave and on their return journey. There is some disagreement over the accuracy of Yelverton's and Carlile's figures for the number of men and women who passed through Folkestone during the war, which provides a caveat to the authors' claim that the SECR handled '3,416 motor cars; 192,468 tons of the Company's traffic; nearly 91,000 tons of Government stores; 11,641 tons of material for Red Cross societies; 383,098 mails and parcel post; and 63,985 tons for Expeditionary Force Canteens; making a total tonnage, outwards and inwards, of 742,188' tons between August 1914 and February 1919.[76]

[74] TNA, ZPER 7/103, records of railway interests, pp. 80–1.
[75] B. J. D. Yelverton and J. C. Carlile, 'The cross-Channel service', in *Folkestone during the War: a Record of the Town's Life and Work*, ed. J. C. Carlile (Folkestone, 1920), pp. 186–98, at p. 195.
[76] Yelverton and Carlile, 'The cross-Channel service', p. 197. On the variety of figures

Regardless of the exact figures, the quantity of traffic that passed over the SECR's lines was far beyond anything carried before 1914. Under such arduous circumstances Dent felt unable to oversee the day-to-day operations of the Bassin Loubet, and he handed over control of the dock to the superintendent of the SECR's northern district, Francis Flood-Page. Flood-Page was clearly a capable official – he received the Military Cross in 1916 – but he lacked both the experience and the authority of the company's general manager. Flood-Page's name, unlike Dent's in late 1914 and early 1915 – is conspicuous by its absence from the war diaries of the BEF's senior supply officers in the second half of 1915. His influence did not reach beyond the confines of the Bassin Loubet, and his presence did not command the same degree of attention among the military authorities as that of Dent – a pre-war senior executive of a large company, a lieutenant-colonel in the ERSC and a member of the REC from its establishment.

The effects of the war's evolution into a battle of material were rapidly felt at the Bassin Loubet. By 3 May – a week after the SECR had taken over as the single authority at the port – the director of supplies, Colonel Carter, recorded that 'considerable progress' had been made in the arrangement of storage accommodation within the area.[77] However, within a fortnight the congestion at Boulogne reached the point at which Carter felt compelled to authorize the stacking of stores 'in the open'.[78] Sustained calls for ammunition from the front the following month forced GHQ to shift additional labour to Boulogne, to ensure that the shells required by the artillery were discharged and sent forward as a priority.[79] The ASC were called upon to transfer men from Calais to deal with the potentially hazardous and specialist task of handling explosives, while officers stationed at Boulogne for training purposes ahead of assignments elsewhere found themselves pressed into temporary action to help clear backlogs within the port. Eric de Normann, destined for Salonika, was one such officer. He wrote to his mother that he was relieved to be involved in the 'very interesting' work of unloading munitions, and saw it as a diversion from the parades and drills that had hitherto dominated his wartime experience.[80]

recorded for passenger traffic through Folkestone during the war, see C. Fair, 'The Folkestone harbour station canteen and the visitors' books', *Step Short: Remembering the Soldiers of the Great War*, 2008 <http://www.kentfallen.com/PDF%20REPORTS/FOLKESTONE%20HARBOUR%20STATION.pdf> [accessed 6 Sept. 2016].

[77] TNA, WO 95/75, director of supplies war diary, diary entry, 3 May 1915.

[78] TNA, WO 95/75, director of supplies war diary, diary entry, 14 May 1915.

[79] TNA, WO 95/3957, headquarters branches and services. Inspector general, diary entry, 15 June 1915.

[80] IWM, Private papers of Sir Eric de Normann, 72/72/1, de Normann to his mother, 3, 5 and 10 Sept. 1915.

De Normann had a grandstand seat over the summer, during which he observed a 'very heavy traffic in the port'. 'Everything', he wrote, was 'being accumulated for *der Tag*' at Boulogne.[81]

The additional support failed to remedy the mounting congestion, and by the end of August Carter had lost patience with what he dubbed the 'so-called Dent scheme's' inability to meet the standards promised by the transport expert the previous winter.[82] Following an inspection of the port and discussions with Clayton about the difficulties that had been experienced since the adoption of the Dent scheme at Boulogne, Carter ordered the 'old method' of working at the Bassin Loubet to be adopted for a fortnight's trial on 1 September.[83] The ASC regained responsibility for the removal of stores from the quayside and the personnel of the SECR were retained purely for the discharge of arriving ships – to act as civilian labour under military direction. After the two weeks had elapsed, officers from the departments that had initially authorized the Dent scheme's implementation adjudged the trial to have been 'an unqualified success'. They reported that ships had been offloaded and dealt with more quickly than had been the case under the SECR's management, even though the average daily tonnage handled through the port remained far below the targets set by the IGC.[84] However, while the army representatives wished to make the organization trialled during September a permanent fixture at Boulogne, the navy demurred. Instead, a report that proposed a reversion to the system in place before April 1915 was forwarded to the committee for its consideration.[85] Clayton requested that nothing be done to 'disturb the existing arrangement', but the War Office was forced to concede that it was illogical to resist the navy's demand to regain authority over the labour employed to discharge ships now that the army once again controlled the onshore workforce.[86]

When placed within the wider context of the BEF's supply operations in 1915, the navy's argument was particularly compelling. Only at Boulogne had the 'single authority' experiment deviated from the procedures to which the navy sought a return. By reverting to the working practices familiar to the soldiers, sailors and labourers at each of the other ports that contributed

---

[81] IWM, De Normann papers, 72/72/1, de Normann to his mother, 5 Sept. 1915.

[82] TNA, WO 95/75, director of supplies war diary, diary entry, 25 Aug. 1915.

[83] TNA, WO 95/75, director of supplies war diary, diary entry, 1 Sept. 1915.

[84] TNA, WO 95/3960, headquarters branches and services. Inspector general, diary entry, 18 Sept. 1915; WO 95/3961, IGC war diary, diary entry, 12 Oct. 1915.

[85] The report itself does not appear to have survived. However, its contents can be deduced from TNA, MT 23/443/4, naval transport work overseas. Report of proceedings of principal naval transport officer, 3 Oct. 1915.

[86] TNA, WO 95/3961, IGC war diary, diary entries, 12 and 26 Oct. 1915.

to the BEF's subsistence – which happened on 24 October 1915, when the SECR surrendered responsibility for the unloading of ships in the Bassin Loubet – operations were standardized across the Channel ports for the remainder of the war.[87] Just six months after it had begun, the civil–military experiment set in motion by Francis Dent the previous December was modified to restore the managerial authority of the navy and army at the port.

The reassertion of military control at Boulogne, and the official history's later emphasis on the 'inadvisability' of entrusting such work to 'civilian management and labour', implies that the civilian operators at Boulogne had proven uniquely incapable of dealing with the growing strain of servicing the needs of Britain's expanding continental commitment. However, Boulogne was not alone in experiencing difficulties during 1915. Congestion at ports both in France and Britain became pronounced in the first full year of the conflict, as the allies attempted to respond to the traffic changes engendered by the war's outbreak. The provision of adequate labour to deal with the BEF's increasing traffic requirements – whether from civilian or military sources – was a constant source of correspondence between London and GHQ during 1915, as the latter received an incessant stream of departmental demands for men to offload ships, shift road stone, build defences and myriad other unspectacular but necessary duties.[88] The virtual closure to international traffic of the ports on the east coast of England, and the continued use of Southampton as a military port, shifted a colossal amount of traffic onto the port of Liverpool. By 11 January 1915 there were forty-four steamers waiting to berth, and by the start of March several vessels containing perishable goods had been unable to discharge their cargoes for over a month. The formation of the Liverpool Dock Battalion by the War Office in March 1915 – a military unit under military law – did not alleviate the problem. As Starling and Lee recorded, by June an average of sixty vessels per day were recorded as awaiting berths on the Mersey.[89]

The SECR's tenure as the Bassin Loubet's 'single authority' was comparatively brief, but its withdrawal did not spell the end of civil–military cooperation at the port; the ROD's civilians in uniform retained command over the railway operations within the 'small and inconvenient marshalling yard' at the Bassin Loubet following the termination of the experiment.[90]

---

[87] TNA, WO 95/3961, IGC war diary, Thomson to Macgregor, 24 Oct. 1915; Clayton to Macgregor, 25 Oct. 1915.

[88] Starling and Lee, *No Labour, No Battle*, pp. 77–86.

[89] Starling and Lee, *No Labour, No Battle*, pp. 34–6.

[90] TNA, WO 95/3963, headquarters branches and services. Inspector general, British lines

Nor did it diminish the opportunities for Francis Dent, the architect of the scheme, to make direct contributions to the supply operations of the wartime British army. Following his appointment as DGT, Sir Eric Geddes chose Dent to lead an investigation into the organization of railways in Egypt and Salonika, which provided the former with a thorough understanding of the EEF's and BSF's resource requirements for 1917. Dent left Marseille for Cairo on 31 October 1916 where, after discussions with the EEF's senior command, he visited the Alexandria docks and drew upon his experience at Boulogne to recommend improved wagon-loading methods at the port.[91] He then spent much of November at Salonika before returning to Cairo to produce a comprehensive report on the railway situation in Egypt with Sir George Macauley.[92]

Dent ensured that the eastern theatres were placed on a solid transport footing ahead of 1917, and that precious railway equipment was not distributed inefficiently. Unsurprisingly, given Macauley's knowledge of the country and expertise as a railway engineer, Dent was able to report to London that the line east towards Palestine from Qantara had been well constructed and was capable of supplying a force twice the size of that being readied for the advance. In addition, 'the rolling stock position was not acute, and, provided greater use was made of water rather than rail communication in Egypt itself, the State Railway rolling stock might be considered sufficient for the time being. It would, however, be necessary later to supply additional stock' if the EEF's advance crept further into Ottoman territory.[93]

The absence of an equivalent to Macauley within the BSF made the Salonika portion of Dent's investigation more complicated. Following his arrival at Salonika he inspected operations at that port and then viewed the construction of a light railway between Stavros and Tasli.[94] Ahead of his arrival at Stavros Dent had forewarned Sir Guy Granet, the director-general of military railways (DGMR) at the War Office, to expect an order for '5 locomotives and 140 wagons, 20 miles of flat-bottomed rails not less than 80 lbs and new sleepers'. 'It is of greatest importance', Dent continued, 'for [the] Army here to know if you can supply standard gauge material

---

of communication in 1915, p. 2; Simpson, 'Railway operating in France', p. 700.

[91] TNA, WO 95/4389, branches and services: Director of railway transport, diary entries, 7–9 and 12 Nov., 2 Dec. 1916.

[92] TNA, PRO 95/4389, DRT war diary, diary entries, 16–27 Dec. 1916.

[93] 'The Palestine campaign', *Railway Gazette: Special War Transportation Number*, 21 Sept. 1920, pp. 119–28, at p. 119.

[94] TNA, WO 95/4784, branches and services: director of railways, diary entries, 17–18 and 28 Nov. 1916.

as suggested above instead of the 40 miles of narrow gauge and 18 locos ordered already and probable dates of shipment in either alternative. Latter information is key to situation'.[95] The standard gauge material was never despatched. Dent's confirmation that the port at Stavros was unsuitable for the unloading of standard gauge equipment – combined with the lack of material immediately available from British or Egyptian sources; the fact that necessity for the railway to be completed before the summer months made construction impracticable; General Milne's preference for the 'rapidity' of narrow-gauge laying; and the eventual movement of the British forces westward to Doiran – meant that the locomotives, wagons, rails, sleepers and shipping capacity required to transport them to the Mediterranean could be deployed more effectively elsewhere.[96]

Conclusion

Geddes's decision to entrust Dent with the leadership of the transportation mission to Egypt and Salonika, alongside his multiple ongoing commitments to the REC's work, underlines the high regard within which the SECR's general manager's technical skills were held by civilian and military authorities during the First World War. Dent's wartime service was not curtailed by the perceived failure of the civil–military experiment at Boulogne in 1915. Instead, he was awarded a knighthood in January 1916 in recognition of his ongoing service to the nation at war, and was unanimously elected by his peers to the chairmanship of the Railway Clearing House general managers' conference in July 1917.[97] Nor were the acknowledgements of Dent's expertise restricted to Britain. In 1917 he was despatched to the United States to provide the American military authorities with advice on the use of ambulance trains ahead of their own troops' introduction to the fighting on the western front. Yet it was from the French that Dent received the most fulsome praise. The French government awarded the British railway manager a *Légion d'Honneur* in the summer of 1915, and he was issued with a replica of the award at a special gathering of French and British dignitaries that August.[98] However, his wartime contribution was eclipsed in the railway industry's post-war records of the conflict. Dent – unlike his

[95] TNA, WO 95/4764, branches and services: Adjutant and quarter-master general, Dent to Granet, 27 Nov. 1916.

[96] LHCMA, Milne papers, BSF: summary of information, 15 Nov. to 8 Dec. 1916; TNA, WO 95/4389, DRT war diary, diary entry, 13 Dec. 1916; 'Railways and the Salonica campaign', *Railway Gazette: Special War Transportation Number*, 21 Sept. 1920, pp. 110–18, at p. 114.

[97] 'Ministerial changes', *The Times*, 26 July 1917, p. 8.

[98] 'General news section', *Railway Gazette*, 3 Sept. 1915, p. 233.

contemporaries Sir Sam Fay, Sir Guy Granet, Sir Guy Calthrop and Sir Eric Geddes among others – was not highlighted for praise within the pages of the *Railway Gazette's* 1920 special issue on wartime transportation or Edwin Pratt's exhaustive account of British railways in the First World War.[99]

He and his colleagues at the SECR may have proven unable to match the ambitious projections he had made in the winter of 1914, but they embodied the approach of Britain's transport experts to the challenges of the First World War. Dent and Percy Tempest made themselves available to the service of the nation from the very outset of the conflict – having already been active participants in Britain's pre-war preparations – and made demonstrably important contributions to the establishment of a supply chain capable of sustaining the expanding BEF.

However, Dent's exertions at Boulogne also exemplify the weaknesses of the coalition's approach to the unprecedented growth of the British contribution to the war's principal theatre of operations. The single port experiment at Boulogne was essentially nothing more than 'tinkering' with a thread in a large and complex web, one with a multitude of potential weaknesses that lay dormant until the colossal demands of 1916 exposed the structural frailties in the BEF's transport infrastructure. The relative paucity of the demands made upon it in 1915 – before the French and British war economies had achieved maximum output – meant that transportation facilities had not yet replaced production shortages as the constraining factor on allied operations on the western front. Under such circumstances, the SECR's failure to reach the throughput levels estimated to be possible at the Bassin Loubet – which were partly a consequence of French protectionism in addition to Dent's overambitious projections – overshadowed the long-term improvements to the dock's capacity that were overseen by Percy Tempest and his team. Through the construction of new facilities within the port, the SECR played a vital role in facilitating the thorough exploitation of the Bassin Loubet in the second half of the war. Elsewhere in the BEF, another civil–military collaboration initiated in the war's opening winter produced similar results. However, in the same month that the SECR was removed from managerial responsibilities at the Bassin Loubet, those in charge of the department of IWT experienced a very different fate.

---

[99] 'British railways and the war', *Railway Gazette: Special War Transportation Number*, 21 Sept. 1920, p. 7; Pratt, *British Railways and the Great War*, i, pp. ix–xiii.

5. Commitment and constraint II: Commander Gerald Holland and the role of inland water transport

The nomenclature that identifies the actions of the First World War in the English language is peppered with references to individual towns – Ypres, Verdun and Amiens among others – and by association with local waterways, such as the Marne, the Somme and the Lys. Their presence indicates that the patch of Europe that became the western front was served by a communications network comprising more than merely roads and railways. The canal and river systems of northern France and Belgium in 1914 were, from a transportation point of view, 'undoubtedly among the finest in the world'.[1] Across the two nations ran almost ten thousand miles of navigable waterways that – unlike in Britain, where the spread of the railways had severely curtailed the use of canals for the bulk carriage of goods – remained an integral component of the local and regional freight traffic infrastructure. In 1905 the total quantity of freight carried by water in Belgium amounted to 53,345,000 tons, approximately half of the nation's entire goods and merchandise traffic.[2] The outbreak of war in August 1914 brought this traffic almost entirely to a standstill. However, the 'permanent way' of the canal and river network remained relatively undamaged by the opening campaigns and – in many areas – within the hands of the allies.

This chapter examines the manner in which the BEF exploited waterborne transport during the First World War. As at the port of Boulogne in 1914–15, the development of an inland water transport (IWT) service on the western front demonstrates the British army's open minded approach towards the input of suitably qualified experts before Sir Eric Geddes's transportation mission in August 1916. Gerald Holland, the man responsible for establishing an IWT service on the western front, was embedded within the BEF's command hierarchy from the outset – unlike the SECR's Francis Dent and Percy Tempest – and provided specialist technical advice to the British army

[1] TNA, WO 95/56, branches and services. Director of inland water transport, memorandum number 1, 19 Sept. 1915, p. 1.
[2] TNA, WO 158/851, director general of transport: history of inland water transport, p. 2; WO 95/56, director of IWT war diary, memorandum 1, p. 2.

until his death in 1917. Rather than offering guidance at arms-length or overseeing the delivery of a solitary endeavour, Holland was incorporated into the army and given the freedom to devote his energies and expertise to the provision of canal and river freight services in France. His talents were recognized and respected by the army's most senior administrative officers, and the British commenced upon a multitude of expensive engineering projects (in terms of time, resources and costs) upon the strength of Holland's advocacy.

However, in addition to illustrating the BEF's willingness to engage with Britain's transport experts, the development of Holland's department in the first half of the war demonstrates the inadequacies of the allies' response to the conflict's evolution. Holland's experiences at GHQ provide evidence both of the limitations of the Franco-British coalition and the deficiencies within the BEF's extant transportation organization before Geddes's arrival. The British army's compartmentalized approach to transport meant that IWT remained an under-exploited resource on the western front in 1915–16, which had significant implications for the road and rail networks behind the BEF before and during the battle of the Somme. Holland, although part of the army's command structure, possessed responsibility for solving issues related to only one link in the transport chain. Throughout 1915, and until the great battles of 1916 had devastated the existing infrastructure, there was neither the political will to broaden the scope of civil–military cooperation nor the military imperative to reassess the coalition's existing administrative structures. Consequently, the allies' approach during this period was characterized by a fixation on the resolution of short-term, localized, specific transport challenges rather than a consideration of the long-term organizational and infrastructural improvements that underpinned the war effort in France and Belgium.

The establishment of the department of inland water transport

The British army was thoroughly aware of the presence and importance of the French and Belgian waterways prior to the First World War. In a lecture delivered to the Aldershot command in 1908, the then Brigadier-General William Robertson directed his audience's attention to the fact that – with the 'aid of canals here and there' – the Scheldt, Sambre and Meuse rivers were all navigable 'in their course through Belgium'.[3] Five years later the War Office produced a report on the available communications in Belgium, which dedicated almost eighty pages to reconnaissance of the Sambre, the

---

[3] LHCMA, Robertson papers, 1/3/2, text of a lecture given by Robertson at Aldershot, Hampshire, on the military geography of western Europe, 1908, p. 14.

Meuse and the Blaton–Ath Canal.[4] Yet despite the acknowledged existence of a network of waterways in a potential theatre of operations, the War Office's studies were not buttressed by the creation of a procedure for the operation of waterborne transport in the event of war.[5] The possible exploitation of IWT was almost entirely absent from the instructions issued to the IGC upon mobilization.[6] Whereas the French and British pre-war discussions had created distinct – if in the event unworkable – demarcations of responsibility around the operations of the French railway network, the canal and river systems were overlooked in the arrangements under which the Franco-British coalition entered the war.

There were three reasons for the almost complete omission of waterborne traffic from the BEF's considerations at the outset of the war. First, the army had not utilized IWT during its most recent large-scale conflict, the South African War of 1899–1902. Coupled with the comparatively insignificant use of canals within peacetime British industry by 1914, the army had consequently become blinded to the potential advantages of an efficiently operated fleet of IWT vessels. Waterborne carriage was briefly touched upon in the lectures delivered to students on the LSE's course for administrative staff officers, but as John Cowans stated in December 1914: 'from a transport point of view, and my knowledge is fairly wide in this respect, I am unaware of any [Royal] Engineer officer who has much idea of working inland water transport'.[7]

Second, the limitations of IWT were stark when compared to railway and road transport. Waterborne traffic routes were fixed by nature and the process of altering the flow of rivers or canals took far longer than the equivalent task for a railway line or road. Furthermore, repairs to waterways damaged during operations required a far greater commitment of manpower and resources than were necessary to reconstruct a similar length of railway. The sedate rate of progress of rivercraft also made them unsuitable for supplying an army engaged in a war of manoeuvre – the type of conflict most pre-war strategists predicted would characterize the conflict. Barges were restricted to travel during daylight hours only as they possessed no lights, and had to deal with the negotiation of lock gates, currents, adverse winds and ice that – except for the latter – did not unduly affect rail or

---

[4] TNA, WO 33/615, report on roads, rivers and billeting in Belgium, volume ii, 1913.

[5] A. M. Henniker, *History of the Great War: Transportation on the Western Front, 1914–1918* (London, 1937), p. 174.

[6] TNA, WO 33/686, instructions for the IGC, Part V, sub-section 5.12.

[7] TNA, WO 158/851, history of IWT, p. 126; C. W. Gwynn, 'The administrative course at the London School of Economics', *Royal Engineers Journal*, vi (1907), 229–35, at p. 232; LHCMA, Robertson papers, 2/2/43, Cowans to Robertson, 14 Dec. 1914.

road services.[8] Millicent Peterkin, a nurse who spent much of 1918 aboard an ambulance barge, experienced the impact that poor weather could have upon the operations of IWT during the war. Following a period of leave at the start of the year, Peterkin arrived at No. 10 Stationary Hospital in St Omer on 1 February to await the arrival of her barge from Calais. However, ten days later she was still at St Omer as there had been 'a great deal of wind for several days'. The barges 'should have been back three or four days ago … I wish my barge would hurry up, for I badly need a change of clothes!'[9] Throughout the war's final year Peterkin's letters home frequently described the wind as a 'beastly nuisance', and catalogued the disruption caused by the weather upon her barge's ability to complete its journeys to and from Calais.[10]

In addition to their susceptibility to inclement weather, canal and river traffic was governed by strict speed limits well below those permitted for trains and mechanical transport. Barges were restricted to a top speed of six kilometres per hour for single vessels and just four-and-a-half kilometres per hour for convoys. In the same way that speed limits for motor transport were carefully managed to protect the condition of both vehicles and the road surface, the restrictions placed on IWT were introduced to ensure that the wash that emanated from the craft did not damage the banks. Consequently, even under perfect conditions a self-propelled barge required two days to traverse the fifty-two miles between Dunkirk and Béthune; a tug with four barges required a further twelve hours to make the same journey.[11] Such a stately rate of progress made IWT an unsatisfactory medium for the conveyance of supplies demanded urgently at the front line.

Finally, the use of IWT was largely excluded from the thoughts of British commanders as the opening exchanges of the conflict left a large stretch of the Belgian system and key connections to the French waterways (such as the St Quentin Canal) either within German-held territory or unsafe for navigation until the location of the front line settled.[12] These impediments did not deter Commander Gerald Holland from approaching the War Office in the opening weeks of the war with a conviction that the 'splendid' waterways of France could provide a useful supplement to

---

[8]   Henniker, *History of the Great War*, p. 174.
[9]   BLSC, Bamji collection/PET, hospital barges in France: correspondence from a nursing sister, with the British Expeditionary Force, during World War I, p. 8.
[10]   BLSC, hospital barges in France, pp. 9, 10, 12, 16.
[11]   TNA, WO 158/851, history of IWT, p. 34; A. C. Fewtrell, 'The organisation of the transportation services of the British armies on the western front', *Minutes of Proceedings of the Engineering Association of New South Wales*, xxxiv (1919), 153–72, at p. 171.
[12]   TNA, WO 158/851, history of IWT, p. 4; Henniker, *History of the Great War*, p. 173.

the existing road and rail facilities.[13] Like Eric Geddes – whose offer to recruit a battalion of skilled railwaymen from among the employees of the North-Eastern Railway was made to the War Office around the same time – Holland's approach fell victim to the nature of the French and British staffs' pre-war arrangements, which located responsibility for the provision of all the BEF's transport requirements with the French. The comparatively low strain placed on the French railways by the 'contemptibly' small BEF, combined with the fluidity of the front line and the dearth of navigable waterways in the section of front initially held by the BEF, meant that the formation of an IWT service was not a priority for the army in August 1914. Consequently, Holland's proposal was declined by the military 'as it was at that time considered that rail transport, supplemented by adequate road transport, would fully meet the [BEF's] requirements' for logistical support.[14]

The nature of Holland's rebuff, unlike that issued to Geddes, did not engender a longstanding animosity between the former and the soldiers at the War Office. As the rank suggested, Commander Holland possessed previous experience with the armed forces. He had served with the Royal Indian Marine between 1880 and 1905, and saw service in Burma, India, and during the South African War. Following his departure from the navy he had entered the employment of the LNWR and, after a brief stint at Fleetwood, in 1907 Holland became the marine superintendent at Holyhead (a role formerly held by Francis Dent's father, Admiral Charles Bayley Calmady Dent). According to Cowans, who knew Holland personally, he was 'a most able officer'.[15] Therefore, when the circumstances in France changed, Holland's proposal was revisited.

Two factors compelled the army to reassess the potential role of IWT on the western front. First, the decision to raise and deploy a large force in France brought with it the requirement to create and maintain lines of communications capable of feeding and equipping that force. Second, the BEF's relocation to Flanders in October 1914 placed it within proximity of the northern network of waterways (see Figure 5.1). Therefore, by the winter of 1914–15 the exploitation of IWT was far more practicable than it had been when Holland approached the War Office in the summer. On 10 December the loading of barges began at Bergues, with the loaded vessels placed under the command of non-commissioned officers from

[13] TNA, CAB 45/205, Lieutenant-Colonel G. E. Holland, information dictated by Major Bradbury.

[14] TNA, CAB 45/205, Holland, information dictated by Major Bradbury.

[15] LHCMA, Robertson papers, 2/2/43, Cowans to Robertson, 14 Dec. 1914.

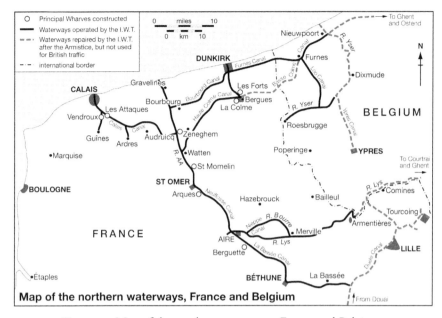

Figure 5.1. Map of the northern waterways, France and Belgium.

Source: A. M. Henniker, History of the Great War: Transportation on the Western Front, 1914–1918 (London, 1937), p. 173. Map drawn by Cath D'Alton.

the ASC for their journeys to Estaires and Béthune.[16] Two days later the DOM at the War Office, Brigadier-General Richard Montagu Stuart-Wortley – correctly identifying that the provision of IWT in France was likely to become a large enough job to require the establishment of a separate department – wrote to Sir William Robertson to recommend Commander Holland, 'a most energetic and useful officer', to head the department.[17]

In the same month that Geddes experienced his uncomfortable meeting with Lord Kitchener and Cowans at the War Office, the army was open minded enough to recommend an 'outsider' to a position of seniority within the BEF's command structure. Before Christmas, Holland had called upon Stuart-Wortley to discuss matters in France, and on 28 December he was offered a temporary commission in the Royal Engineers. Within forty-eight hours the newly commissioned Lieutenant-Colonel Holland had crossed the Channel to 'report as to the steps which

---

[16] TNA, WO 95/27, QMG war diary, the use of canals for supply purposes, 15 Dec. 1914.
[17] TNA, WO 32/5162, formation and organisation of inland water transport in France and Belgium; establishment and appointments, Stuart-Wortley to Robertson, 12 Dec. 1914.

should be taken to enable the waterways to be utilized for transport work for the British Army'.[18]

Holland's private diary from this period survives, and illustrates the unpromising foundations upon which he had to construct an IWT department within the BEF. On 30 December 1914 he reported for duty at GHQ and was placed under the authority of the director of railways in the QMG's office.[19] A day after his arrival he interviewed a local tug captain and ascertained that the French custom was for a barge to be operated and lived upon by an entire family.[20] Rather than work to orders and provide transport on routes identified and requested by the British, the barge owners preferred to choose their own routes and only carry cargoes between docks they themselves had selected.[21] A meeting with the French army's canal expert on 1 January 1915 quickly revealed that this obstinacy was not based on any kind of xenophobic prejudice or nationalist intransigence; the French boating community happened to be just as truculent in the face of French military authority. Holland met with the Belgian canal representative the next day, which meant that within four days of his arrival on the western front he had established what appear to have been friendly and progressive relations with his counterparts in both the French and Belgian armies. However, his conversations with the BEF's coalition partners had not engendered an agreement for them to provide the British with either the manpower or materials required to create an IWT service behind Sir John's expanding army. As Holland reflected later, in January 1915 the IWT department in France comprised 'two officers, no men, one hired tug and 34 barges'.[22]

Holland's diary also documents both the complexity of the task that lay ahead of him and the assistance he received from the army and the British transport industry. Without a pool of reliable local barge operators to call upon, the only alternative open to Holland was the enlistment of personnel from Britain to crew the vessels and provide the technical and administrative support required to maintain an efficient fleet. His diary recorded the names and occupational backgrounds of those chosen to

---

[18] TNA, WO 32/5162, formation and organisation of IWT, note 6, 19 Dec. 1914; WO 158/851, History of IWT, p. 5.

[19] TNA, CAB 45/205, Holland, diary entry, 30 Dec. 1914.

[20] See IWM, private papers of Brigadier A. E. Hodgkin, Documents.12337, diary entry, 2 May 1916 for a vivid account of the extraction of a civilian barge from difficulties. Hodgkin's narrative is accompanied by two sketches of the methods by which young children aboard the civilian barges were tethered to the vessel to prevent them from falling into the water. My thanks to Edward Spiers for alerting me to Brigadier Hodgkin's diary.

[21] TNA, CAB 45/205, Holland, diary entry, 31 Dec. 1914.

[22] TNA, WO 95/56, director of IWT war diary, memorandum 1, pp. 3–4.

populate the new department. The entries also emphasize the breadth of skills necessary to manage an industrial army. The majority of recruits, such as Horace Pitman, were selected because of their previous associations with waterborne transport. Pitman possessed ten years' experience as a yachtsman, while Corporal William McKinlay was transferred into the department thanks to his having trained as a surveyor with Lloyd's before the war. George Tagg, despite being fifty-two years of age, was appointed both for his knowledge of the French and Belgian canal systems and his familial links to the boat-building industry.[23] Others were chosen for less obvious but equally important qualities. E. G. Weston, for example, brought his experience as assistant secretary in the colonial civil service to the provision of clerical support to the newly established department, while the War Office also contributed a cadre of officers. Stuart-Wortley agreed to release Lieutenant Baugh from the directorate of movements and to Colonel Albert Collard's attachment to Holland's fledgling outfit.[24]

Yet like the BEF as a whole, the force's IWT department was heavily dependent for personnel upon those from outside the pre-war British army. Holland's own pre-war career, both at sea and on the railways, provided the nucleus around which the BEF's IWT organization was constructed. His three most senior subordinate officers in France were former Royal Indian Marine officers, while the LNWR contributed numerous labourers and administrative staff keen to serve under their former manager.[25] On 13 January a list of fifty men who were willing to enlist from the marine department at Holyhead was compiled; they were medically examined soon after and sent to the Royal Engineers' training camp at Longmoor for instruction under Major Cyril Luck – one of Holland's former naval colleagues who had been appointed officer commanding IWT troops.[26] Elsewhere, Private R. H. Williams transferred into the department from the 16th (Service) Battalion (Public Schools), Middlesex Regiment – and received a promotion to lieutenant – following an interview with Holland on 20 January. Williams, Holland recorded, possessed previous LNWR locomotive experience and was 'intelligent looking'.[27]

[23] TNA, CAB 45/205, Holland, diary entries, 20 and 23 Jan. 1915.

[24] TNA, CAB 45/205, Holland, diary entries, 9 and 10 Jan. 1915.

[25] TNA, WO 158/851, history of IWT, pp. 8–9; IWM, Hodgkin papers, diary entry, 19 June 1916.

[26] TNA, CAB 45/205, Holland, diary entry, 13 Jan. 1915; WO 158/851, history of IWT, p. 9.

[27] TNA, WO 32/5162, formation and organisation of IWT, particulars of officers; CAB 45/205, Holland, diary entry, 20 Jan. 1915.

Whereas Williams came to Holland's attention as a result of a direct recommendation from the former's commanding officer, an 'active campaign of enlistment' at various ports across the country accounted for the lightermen, watermen, seamen, engineers and other assorted trades required to ensure the department's ability to fulfil its duties.[28] Brigadier Adrian Hodgkin, who was attached to IWT between December 1915 and August 1916, recorded the assortment of technical skills that were collected upon each barge during the war. A chemical glass manufacturer prior to the war, Hodgkin served on the water purification barge A.174 after being wounded at Ypres in July 1915. Nobody on A.174 matched the maritime experience of Corporal Mapplebeck, the commander of A.412, whom Hodgkin described as 'a delightful old man, one of the dirtiest looking villains I have ever seen; he talks in the broadest Yorkshire, comes from Hull, and has lived "all us lives" on a boat similar to A.412'.[29] Yet most of Hodgkin's crewmates on A.174 were employed in shipping-related occupations. Corporal Fernandez had worked on the Mersey ferries in peacetime; Sapper A. Arnold was a second mate sailor; Sapper D. Applegate had been a motorboat driver; Sapper Humphreys was a Thames waterman; and Sapper Marsh had been both a Thames waterman and a former stoker in the Royal Navy. In fact, only Hodgkin and Sergeant J. W. McCririck – a 'general engineer in peacetime' – possessed no previous experience in the operation of waterborne transport.

The directorate's headquarters, to which Hodgkin transferred in June 1916, was less reliant upon waterborne experience but still contained a higher proportion of Britain's transport experts than it did professional soldiers. In addition to the former Lloyd's surveyor William McKinlay, who by that point had attained the rank of major, two of the ten deputy assistant directors of IWT in mid 1916 had backgrounds in the merchant navy. Another, Captain Daniels, had worked before the war in ship repairs. Three of the remainder were servants of Holland's immediate pre-war employer, the LNWR.[30]

Holland's technical expertise, alongside the skills of the officers who helped him to build the IWT department, were clearly valued highly by those charged with ensuring that the expanding BEF continued to receive ample logistical support. The paper strength of the IWT department when Holland was appointed stood at thirty-six officers and 654 other ranks – although by the end of January 1915 just five of each were actually at work

---

[28] TNA, WO 158/851, history of IWT, p. 19.
[29] IWM, Hodgkin papers, diary entry, 24 Apr. 1916.
[30] IWM, Hodgkin papers, diary entry, 19 June 1916.

in France – and throughout the army there existed an expectation that IWT would be called upon to undertake 'an immense amount' of work 'sooner or later'.[31] In acknowledgement that the British army lacked the knowledge base to make the most effective use of the canal and river systems in France, senior British officers on both sides of the Channel favoured the employment of men with comparatively little military experience to positions of significant responsibility over the use of inadequately prepared professional soldiers.

Concurrent with the appointment of officers and the enlistment of other ranks, work in France began in earnest. On the morning of 5 January 1915 road stone from Guernsey was discharged direct to a barge drawn alongside a ship berthed at Calais, and a second vessel was loaded in the same way that afternoon. The following day Holland agreed a price of ninety centimes per ton with a local contractor in Armentières for the vessels' discharge.[32] As the units raised in Britain passed through the training camp at Longmoor and crossed to France, the local labour withdrew and the IWT department gradually came to more closely resemble a recognizable provider of military logistics. By the end of June, Holland's department had swollen from just two officers and no men to comprise twenty-five officers and 423 other ranks, with plans in place for further expansion. In the same period the BEF's fleet of craft had provided transportation for 15,926 tons of supplies, 27,241 tons of road metal, and 3,216 tons of miscellaneous supplies (which included bridging materials and coal), while 628 officers and men had been evacuated from the battle zone by ambulance barge – reducing the loads carried by rail and providing the wounded men with a comfortable, if sedate, journey back to the base hospitals.[33] In September, when Holland wrote his first memorandum on the progress of IWT in France, he wrote with evident pride that barges were carrying 1,200 tons every day across the northern waterways and that requisitions for over 156,000 tons to be moved by IWT before the end of the year had been received.[34]

To fulfil such obligations, a more appropriate fleet of vessels had to be introduced. The pre-war traffic on the northern waterways was predominantly towed by horses, which Holland identified 'would not cope effectively with the heavy demands which he foresaw would be made for the transport of war material by water'. He recommended the use of tugs capable

[31] LHCMA, Robertson papers, 2/2/43, Cowans to Robertson, 14 Dec. 1914; 2/2/44, Robertson to Cowans, 16 Dec. 1914; TNA, WO 158/851, history of IWT, p. 8.

[32] TNA, CAB 45/205, Holland, diary entries, 5 and 6 Jan. 1915.

[33] TNA, CAB 45/205, Holland, IWT Corps, British army France 1915. Summary of organization and development, pp. 6–7.

[34] TNA, WO 95/56, director of IWT war diary, memorandum 1, p. 9.

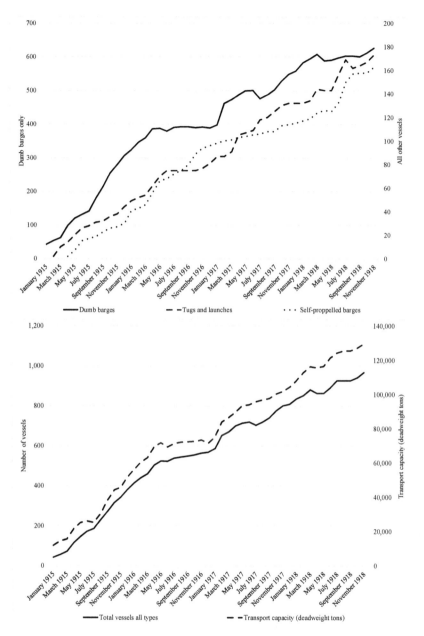

Figure 5.2. The development of inland water transport
resources on the western front, 1915–18.

Source: TNA, WO 158/851, director general of transport: history of inland water transport,
Appendix B2: Schedule showing development of inland water transport resources in France
month by month, 1915–18.

of hauling conveys of 'dumb' barges – which relied upon either a horse or tug for propulsion – and the provision of self-propelled barges capable of operating independently and responding more swiftly to urgent demands from GHQ.[35] Orders for the construction of vessels were placed with British firms immediately, and suitable craft already in use in Britain, Belgium and the Netherlands were purchased by the War Office and despatched to the front.[36] As the department grew in size the craft originally hired from French sources – which in the first instance came complete with their original crews – were purchased outright and manned by military personnel, 'as it was found that the civilian crews were very unsatisfactory'.[37] In just nine months the department expanded from 'one tug and 34 hired barges' to control a fleet of over two hundred vessels, with a total carrying capacity in excess of 38,000 tons (see Figure 5.2). The standard barge was capable of carrying about 280 deadweight tons of material.[38] Therefore, as Hodgkin noted in his diary, the IWT service quickly acquired the ability to remove a significant volume of traffic from the railways: three barges held the same quantity of material as two trains comprised of fifty ten-ton wagons.[39]

In addition to his duties with regard to the provision of adequate personnel and the equipment to maintain an effective delivery service, Holland's department was also made responsible for the repair of vessels and waterways, the operation of inland quays and docks, the regulation of traffic on the water and the establishment and upkeep of a communications network across the entire IWT system.[40] Telephone communications played a vital role in the IWT department's operational practices, which drew heavily upon the organizational structures employed by the railways in peace and war. In the same way that the railway transport establishment was formed to oversee the BEF's use of railways – and to provide a conduit for British requests to be presented to the French rail authorities – responsibility for the management of IWT on the western front was divided into districts.[41] District officers supervised the loading and unloading of vessels within their jurisdiction, maintained contact with the British and French military authorities in the area, ensured the safe passage

---

[35] TNA, WO 158/851, history of IWT, p. 6.
[36] TNA, WO 158/851, history of IWT, pp. 10, 32.
[37] TNA, WO 158/851, history of IWT, p. 32.
[38] TNA, WO 158/851, history of IWT, p. 3.
[39] IWM, Hodgkin papers, diary entry, 28 June 1916.
[40] The complete list of duties devolved upon the IWT department upon its formation are given in TNA, WO 95/56, director of IWT, war diary, memorandum 1, p. 4.
[41] TNA, WO 32/5162, formation and organisation of IWT, memorandum – IWT, 4 Jan. 1915, p. 2.

of craft through their zone and circulated information about the traffic in their district to their colleagues in neighbouring areas. In essence, the IWT department established a system of decentralized responsibility, which devolved the detail of everyday work to the district officers. The existence of these 'men on the spot' freed Holland and his senior subordinates at GHQ to concentrate upon the establishment of the principles and procedures required to obtain the highest degree of efficiency from the fleet.

The effective deployment of its fleet was crucial to the department's success, as the construction of new waterways was practically impossible. To ensure smooth operations, Holland's department had to coordinate the movements of the BEF's traffic, the smaller number of craft being used to supply the French and Belgian armies, and the relatively insignificant volume of civilian traffic that continued to ply the waterways.[42] To keep track of the whereabouts of these craft – particularly once the British IWT presence in France spread beyond the northern waterways – Holland's department implemented a system of control that had been pioneered in the previous decade by the Midland Railway, which relied heavily upon the presence of a reliable telecommunications network.[43] On 2 February 1915, just over a month after he had arrived in France, Holland was issued with twenty-five expert telephone linesmen who undertook all the communications work required to make the IWT department a self-sufficient unit of the BEF. Telephone lines ran across the northern waterways and were later installed along the River Somme as the department's sphere of operations expanded.[44] The telephone system provided the link required for district officers to pass vessels from district to district and to update their colleagues on their impending traffic commitments. The detailed information gathered through the system gave district officers advanced warning of likely busy periods, affording them opportunities to source extra labour for deployment at lock gates to reduce transit times.[45] The locations of all craft were relayed back to GHQ each night and recorded on a diagram board – a principal component of the Midland's train control system – which gave Holland's staff a regularly updated, graphic illustration of the fleet's distribution. Such innovations aided decision making within the department in relation to the redistribution of craft and personnel as circumstances dictated during the conflict.[46]

[42] TNA, WO 158/851, history of IWT, p. 46.

[43] TNA, ZLIB 6/88, Midland Railway train control issued by Midland Rly, 1914, p. 62.

[44] TNA, CAB 45/205, Holland, diary entry, 2 Feb. 1915; WO 95/27, QMG war diary, Maxwell to Kitchener, 12 March 1915; WO 158/851, history of IWT, p. 61.

[45] TNA, WO 158/851, history of IWT, pp. 45–46.

[46] TNA, WO 158/851, history of IWT, p. 22; ZLIB 6/88, Midland Railway train control,

The growth of the directorate of inland water transport

The IWT department began its work in 1915 on just a small section of the northern waterways, which connected the ports of Dunkirk and Calais with the towns of Armentières and Béthune. However, the limited scope of the department's initial zone of operations did not limit Holland's focus. From his arrival on the western front, he adopted a 'policy of looking well ahead and forecasting the probable requirements of the future ... both as regards new demands for transport and new constructional and repair work' on waterways acquired by the allies in the event of an advance.[47] As we have seen, such an outlook was not unique within the British military effort. A core component of Sir Percy Girouard's report into the BEF's transport organization in October 1914 surrounded the question of ensuring the British possessed a 'voice' in the administration of captured transport infrastructure, while a significant proportion of the IGC's voluminous correspondence over the winter of 1914–15 covered the preparations required to maintain Kitchener's armies once they arrived in France.[48] In the summer of 1915 an IWT service began to operate on the River Somme in support of the British divisions stationed in the area, and by the end of the year over 215 miles of navigable waterways behind the western front lay open to British traffic.[49] By the autumn of the same year, the number of men whose work was coordinated from Holland's office under the DRT had grown to 830.

The continued growth of the IWT department, and the ambitious plans for further expansion made by its head, were retarded by its subordination to Colonel Twiss. Holland had been placed under Twiss's authority in December 1914 to mirror the structure of the French army, which regarded canals and railways as 'one question' and administered them within the same department.[50] Therefore, Holland had no direct access to the QMG. He could not personally represent his department's needs and advocate for a more thorough exploitation of the French and Belgian waterways. Instead, he was forced to 'express all his views' through Twiss who – like all of his colleagues within the pre-war British army – 'had not technical knowledge [or] experience' of IWT and 'whose time moreover was fully occupied with matters appertaining to the railway transport problem'.[51] Twiss's ignorance of the peculiarities of waterborne traffic led to the detachment of Holland's

---

pp. 17–20.
[47] TNA, WO 158/851, history of IWT, p. 15.
[48] See, e.g., TNA, WO 95/27, QMG war diary, Maxwell to Robertson, 1 Nov. 1914.
[49] TNA, WO 158/851, history of IWT, pp. 11–12, 14.
[50] LHCMA, Robertson papers, 2/2/44, Robertson to Cowans, 16 Dec. 1914.
[51] TNA, WO 158/851, history of IWT, p. 17.

department from the DRT's oversight in October 1915. Two weeks later, and just four days after the Dent scheme at the Bassin Loubet had been finally terminated, Holland was gazetted as a temporary colonel at the head of an independent IWT directorate. Instead of working through Colonel Twiss's office, from 28 October onwards Holland answered directly to the QMG in France, Sir Ronald Maxwell.[52]

The re-establishment of military command at the port of Boulogne in October 1915 should not overshadow the successful formation and development of an IWT organization heavily influenced by non-military personnel. Far from being gripped by anti-civilian phobia at this point of the war, the continued expansion of IWT in 1915 demonstrates that the BEF's senior commanders were far more open to the application of civilian expertise than has previously been asserted. Within such an environment, Holland's contacts, knowledge and attitude were highly sought-after attributes in the expanding British war effort. However, the period following the separation of the command link between IWT and the railways was not one of steady, unbroken expansion. Holland's experience of directorship in France over the following year exposed two weaknesses: the threshold of the BEF's freedom of action on the territory of a powerful, sovereign nation; and the perceived limitations of a slow means of transport, which operated outside of the pre-existing, land-based supply hierarchy behind the western front.

There were a number of reasons why Holland pressed for the acquisition and deployment of 'double, even treble, and possibly still a greater number of vessels' than the 330 accounted for by his department in September 1915.[53] First, the pre-war transport arrangements between the French and British had already begun to unravel, and Holland was quick to identify that responsibility for the maintenance and operation of the waterways could not be divorced from control over the network if the BEF were to gain the maximum possible benefit from IWT. Inter-allied discussions under the umbrella of the railways and canal commission had taken place that summer, and decreed that the British were to undertake the repair, maintenance, and operation of any waterways behind the BEF in the event of an advance.[54] Colonel Henniker, who represented the British on the commission, did not record the specific outcomes of the discussions on waterborne transport in his account of the commission's deliberations. However, it is clear from

[52] TNA, WO 158/851, history of IWT, pp. 17–18; WO 95/56, director of IWT war diary, memorandum number 2, 5 May 1916, p. 2.

[53] TNA, WO 95/56, director of IWT war diary, memorandum 1, p. 9.

[54] TNA, WO 158/851, history of IWT, p. 64; Henniker, *History of the Great War*, pp. 94–101.

Holland's diary that he had emphasized the importance of securing British control over the Belgian waterways – should an advance take place – to Henniker long before the commission met.[55]

A second reason provided by Holland for his promotion of an expanded IWT provision was financial. The British incurred a charge for all freight carried by the French railways on the BEF's behalf, which generated a vast correspondence within the QMG's offices. In contrast, the French did not request payment for cargo moved in British vessels along the rivers and canals behind the front. Furthermore, Holland observed, the vessels constructed for the BEF's use during the war represented an investment. At an initial estimate of £138,140 for the construction of barges, cranes, tugs and the provision of stores, the IWT department could not have been developed without a significant material and monetary commitment.[56] However, the devastation wrought by the conflict upon the European boat-building industry had created what Holland anticipated would be significant post-war demand for the BEF's fleet. The port of Antwerp alone required six million tons of lighterage per year, while the engagement of French workshops on war-related activities – and the ongoing military recruitment of men from the French labour force – meant that the stock of IWT craft in France was likely to be severely depleted when the fighting ended. Consequently, Holland argued, 'any vessels we may have will be of great value to replace losses, and will assuredly be bought by those, who then turn their attention to the restoration of commercial business, at prices which will, I confidently expect, recoup a large proportion of our outlay'.[57]

The possibility of recouping some of the costs incurred in the provision of a substantial IWT fleet on the western front was persuasive, but by far the most compelling justification for the sustained expansion of Holland's directorate lay in conjunction with the difficulties experienced at the docks under the BEF's control throughout 1915. As demonstrated from the very outset of their employment in France, IWT vessels drawn up alongside ships berthed at the ports eliminated the need for supplies to be landed on the quayside. This arrangement saved the labour needed to move goods from the shore into the storage depots, reduced the demand for space within the confined accommodation immediately surrounding the harbours and did not require locomotives and rolling stock to be deployed to remove supplies from the port area by rail. Goods that were transported several miles inland

---

[55] TNA, CAB 45/205, Holland, diary entry, 25 Feb. 1915.

[56] TNA, WO 32/5162, formation and organization of IWT, Estimate of cost of craft now required for IWT, BEF, 13 Jan. 1915.

[57] TNA, WO 32/5162, formation and organization of IWT, conference on canal transport, 12 Jan. 1915, pp. 1–2; WO 95/56, director of IWT war diary, memorandum 1, p. 3.

by water allowed the wagons in circulation in northern France to be worked over shorter distances – an action that ensured individual wagons returned to the depots more frequently and consequently increased the number of journeys each could make to and from the front line. Furthermore, the extra capacity offered by the hulls of IWT craft provided the BEF's senior supply officers with the option to remove from the railways altogether stores whose demand was stable and predictable. The availability of a waterborne alternative freed up rolling stock to respond to more volatile and unpredictable requests for stocks such as ammunition, whose expenditure could not be accurately predicted in advance – particularly in the event of an enemy attack.[58] In March 1916 Maxwell noted that approximately 2,000 tons of road 'stone metal, engineering stores and material, [and] hay and oats' were among the stores delivered daily by IWT – the majority of it achieved 'without [the cargo] touching a road or railway' between the coast and its destination.[59]

As the pressures on the available rolling stock became acute over the winter of 1915–16, and congestion at Calais and Dunkirk threatened both the despatch of trains and the turnaround of ships, GHQ decided to pursue the construction of an IWT depot capable of handling stores removed by barge from the two ports. The project's principal goal was merely to reduce the BEF's reliance upon the limited railway facilities around Calais and Dunkirk, and upon the communications that linked the ports to the wider French rail network (see Figure 5.3). A suitable location for the depot, known as Zeneghem, was found at the junction of the Calais Canal and the River Aa. The site was within 'a summer day's journey by barge' of both Calais and Dunkirk, and had the added advantage of offering a separate return route for traffic from the latter. The operation of different inbound and outbound routes meant that congestion at lock gates was minimized, which made fluidity in the network easier to sustain.[60]

However, Holland's ambitions for the depot grew as the BEF's expansion continued and the strain on the Channel ports – and British shipping capacity – became consequently greater. Rather than simply alleviate congestion at the docks by permitting the discharge of ships direct to barge, the director of IWT began to envisage the site near St Pierre Brouck as the French hub of a cross-Channel service that could bypass the congested ports entirely. He laid down his views on the subject on 29 April 1916 in

[58] Henniker, *History of the Great War*, pp. 175–76.
[59] TNA, WO 107/15, inspector-general of communications, Maxwell to Clayton, 25 March 1916.
[60] TNA, WO 158/851, history of IWT, pp. 47–8; WO 95/56, director of IWT war diary, memorandum no. 3, Dec. 1916, p. 2.

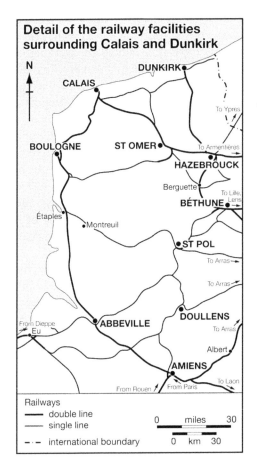

Detail of the railway facilities surrounding Calais and Dunkirk

N

DUNKIRK

CALAIS

To Ypres

BOULOGNE   ST OMER   To Armentières

HAZEBROUCK

Berguette

To Lille,

BÉTHUNE   Lens

Étaples

Montreuil

ST POL

To Arras

To Arras

From Dieppe

Eu   ABBEVILLE   DOULLENS

To Arras

Albert

AMIENS

From Rouen   From Paris   To Laon

Railways
— double line
— single line
– · – international boundary

0   miles   30

0   km   30

Figure 5.3. Detail of the railway facilities surrounding Calais and Dunkirk.

Source: J. H. F. Le Hénaff and H. Bornecque, *Les chemins de fer français et la guerre* (Paris, 1922); A. M. Henniker, *History of the Great War: Transportation on the Western Front, 1914–1918* (London, 1937). Map drawn by Cath D'Alton.

response to an enquiry from Sir John Cowans into the practicability of 'bringing material from England to France … with a view to economising on shipping'. Holland argued that goods could be loaded onto barges in Britain rather than ships, despatched across the Channel, and unloaded at an inland site away from the cramped conditions at the coast. Such a service would help relieve some of the pressure on the limited dock space available to the BEF, free up shipping and reduce the travelling distance for the locomotives and rolling stock that connected the docks and the front line. 'Provided it was understood that the service could not be looked upon as a daily one, as the barges could only cross as and when weather conditions permitted', Holland believed the proposal for a cross-Channel service that terminated at Zeneghem was 'a practical one'.[61]

[61] TNA, WO 158/851, history of IWT, p. 56.

It was also expensive. In addition to the development of Zeneghem as a depot capable of receiving the goods transported by cross-Channel barge, the service required the construction of a suitable departure point for the vessels in southern England. Richborough on the River Stour was chosen to host the British terminus of the cross-Channel service because of 'its geographical position, relative to the Channel ports and the Continental canal system, the existence of large deposits of sand and gravel, ease of railway access, and the extensive areas available for camps and store yards'. As John Kerr Robertson's exhaustively detailed paper in the *ICE Proceedings* demonstrated, the creation of a 'great military depot ... consisting of camps, workshops, power houses, shipyards, wharves with extensive basins, warehouses, store yards, [and] salvage depots' at Richborough required significant investments of materials, manpower and money.[62] The construction costs incurred at Richborough amounted to some £433,476, while over one-and-a-half million concrete blocks were manufactured on site for the construction of camps, offices and workshops.[63] Despite the existence of competing demands for the finite resources available to the leaders of Britain's war effort, Holland's opinion was held in sufficiently high regard both at the War Office and GHQ for work to be commenced at Richborough even before the battle of the Somme had highlighted the deficiencies in the BEF's transport infrastructure. By early May 1916 Colonel Collard was immersed in the 'very extensive' work of placing orders for the construction of craft capable of operating in the English Channel and on the northern waterways, and by September enough progress had been made at Richborough to permit the first barge to be loaded up. The cross-Channel barge service commenced regular operations in December of the same year.[64]

While senior military figures in Britain were quick to respond to the expanding transport requirements of the war, on the other side of the Channel the French authorities took a different approach. In February 1916,

[62] J. K. Robertson, 'Richborough military transportation depot', *Minutes of the Proceedings of the Institution of Civil Engineers*, ccx (1920), 156–207, at p. 157.

[63] Using 1916 as a base year, the Bank of England's inflation calculator suggests a total of £35,845,130.77 (at 2017 prices) was spent upon construction at Richborough. See 'Inflation calculator', *Bank of England* <http://www.bankofengland.co.uk/education/Pages/resources/inflationtools/calculator/default.aspx> [accessed 23 Oct. 2018]; Robertson, 'Richborough military transportation depot', pp. 174–5, 185.

[64] University of Warwick Modern Records Centre (UWMRC), papers of Sir William Guy Granet, MSS. 191/3/3/51, memorandum to Sir Guy Granet, 19 Oct. 1916, p. 3; Robertson, 'Richborough military transportation depot', p. 188; Henniker, *History of the Great War*, p. 238. The specifications of the cross-Channel barges, designated as '&c. barges', are given in TNA, WO 158/851, history of IWT, pp. 56–7.

the French had made an initial request to GHQ for the British to provide rolling stock to relieve the pressure on the French railways' own reserves. Yet the advantage for the railway network of Zeneghem's development as an alternative to rail transport in France did not engender automatic approval from *GQG* for the British to proceed with construction. Work on the depot did not begin until 25 July 1916, almost a month into the Somme offensive and nearly three months after Collard had begun to source vessels to ply the cross-Channel route. The location of a suitable site and the accumulation of building materials for the depot contributed to the delay, but the chief cause lay in the fractious relationship between Britain and its host. The site near St Pierre Brouck was only, in Holland's words, 'eventually agreed upon' after 'several proposals' and numerous meetings between representatives of the French and British armies.[65] Construction on the first quay, which measured 1,575 feet in length and ultimately contained fifteen berths, was not completed until 14 October 1916.[66] Only four barges were able to make the crossing during the following month.[67] Consequently, the cross-Channel barge service was unable to provide any meaningful support to the allies' major offensives in 1916.

GQG's insistence that they retained overall control of the decision-making process acted as a significant retardant on the growth of Holland's directorate. Even before the discussions surrounding the development of Zeneghem had begun, French bureaucracy had served to frustrate the former railwayman's ambitions for the IWT service. As early as October 1915 Holland had suggested that barges could be loaded direct from ships at Le Havre, to facilitate the discharge of vessels and reduce the growing levels of congestion at the port. The loaded barges could then be sent inland to Rouen via a combination of the River Seine and the Tancarville Canal. Both Maxwell and Clayton approved of Holland's plan. However, following 'protracted negotiations', Holland recorded that the French authorities 'would not hear of the proposal although it would undoubtedly have done much to relieve the congestion on the railways'.[68] The relatively dispassionate language of the directorate's post-war report – written in the glow of victory – claimed that the French authorities had 'at all times, given courteous, prompt, and ungrudging aid' to the BEF.[69] Holland's

---

[65] TNA, PRO 95/56, director of IWT war diary, memorandum 3, p. 1.

[66] TNA, WO 158/851, history of IWT, Appendix C 1A – Particulars of quays constructed and equipped by the IWT, p. 1.

[67] TNA, WO 158/851, history of IWT, p. 57.

[68] TNA, CAB 45/205, Holland, summary of organization and development, p. 11; WO 158/851, history of IWT, p. 52.

[69] TNA, WO 158/851, history of IWT, p. 16.

contemporaneous remarks provide a stark contrast to the diplomacy of the official document. In a memorandum written on 5 May 1916 Holland noted that 'all sorts of reasons were put forward by the French against [the Tancarville Canal] project, none of them convincing'.[70] In his private diary he went even further, defacing the page that recorded the chrysalis of the idea with a note scrawled in red pencil and denoting obvious frustration: 'Finally French refused permission for any British service'.[71]

As with the delays to the cross-Channel service between Richborough and Zeneghem, what British officers perceived to be French obstinacy meant that the military operations on the western front in 1916 did not benefit from the foresight displayed by Holland during the previous eighteen months. Rather than commence in or around October 1915, when Holland first received his superiors' support for the Tancarville Canal scheme, the French only withdrew their objections to the BEF's use of the canal on 4 August 1916. The colossal demands of the fighting around Verdun and the Somme had created severe congestion and rolling stock shortages at Le Havre, which finally persuaded the French to sanction a 'limited inland water transport service' to receive cargo from ships berthed at the port for onward transport.[72] The IWT directorate was unable to respond immediately. The barges required to operate the Tancarville Canal route had to be transferred from their locations on the northern waterways and the River Somme via the English Channel, a journey that took thirty-three days to complete. Consequently, IWT did not begin to load goods at Le Havre direct from ship to barge until 22 September 1916 and construction on an inland depot at Soquence (near Rouen) was only completed a month later – almost exactly one year after Holland had made the proposal to utilize the canal.[73]

Holland was attempting to be proactive, and was planning for the continued expansion of the BEF's logistical capabilities. At the same time the French authorities appeared to want the British to take on a larger share of the responsibility for the sustenance of their troops, while simultaneously acting to constrain their ability to do so – until the necessities of the military campaign intervened. However, this is a highly Anglo-centric perspective on events, which does not take into account the wider considerations of the allied war effort. Alongside the BEF's demands, the French authorities

[70] TNA, WO 95/56, director of IWT war diary, memorandum 2, p. 1. Sadly, Holland did not elaborate upon the reasons offered to justify the French authorities' decision.

[71] TNA, CAB 45/205, Holland, summary of organization and development, p. 11.

[72] TNA, WO 95/56, director of IWT war diary, memorandum 3, p. 2.

[73] TNA, WO 158/851, history of IWT, Appendix C 1A – Particulars of quays constructed, p. 1.

in the winter of 1915–16 had to balance the requirements of a much larger French army and the needs of the civilian population upon whose shoulders the French war effort rested. The BEF was not the only institution seeking to make use of IWT as the pressures on the French railway network intensified. The French military authorities began to appreciate the value of their waterways, and attempted to reintroduce pre-war traffic in heavy goods to the rivers and canals as the war intensified. The volume of coal transported on French vessels along the River Seine was almost doubled – increasing from 350,000 to 600,000 tonnes per month – and improvements were made to IWT depots on the Marne, the Moselle and other navigable waterways across the country. With the capacity of the canals limited in the same way as that of the railways and roads, the vessels deployed in support of the BEF had to be integrated with the local traffic rather than imposed upon it.[74]

Furthermore, to lay the blame for the lethargic expansion of IWT in 1915–16 purely at the feet of Britain's principal ally is unwarranted and creates a deceptive impression of the extent to which transportation's complexities were appreciated within the BEF prior to the battle of the Somme. There was a clear willingness to engage with and support the development of IWT among senior administrative soldiers such as Maxwell and Cowans. However, such commitment was by no means universal in the British army. In this respect, the decision to sever the command relationship between waterborne transport and the DRT in October 1915 may have reduced the influence Holland and his directorate were able to have over decisions made at army and corps level. As in the pre-war British economy, where the bulk carriage of goods over long distances was dominated by the railways (although coastal shipping also made a significant contribution where available),[75] the independent IWT directorate was unable to attract a substantial demand for its services.

Individual formations, each desirous of obtaining the resources they believed were necessary to ensure the continued efficiency and security of their own troops, were reluctant to utilize the canals. In the absence of a centralizing authority to collate and coordinate the BEF's transport demands – and until the sheer volume of goods entering France made the identification of priorities a fundamental requirement for the sustenance of operations on the western front – there was little Holland could do to persuade commanders to embrace the canals and reduce their dependency

[74] J. H. F. Le Hénaff and H. Bornecque, *Les chemins de fer français et la guerre* (Paris, 1922), pp. 176–7.

[75] J. Armstrong, 'The role of coastal shipping in UK transport: an estimate of comparative traffic movements in 1910', *Journal of Transport History*, viii (1987), 164–78, at p. 176.

on the overburdened rail network. Holland's assistant directors of IWT, along with the various district officers stationed across the waterway system, were responsible for 'keeping in close touch' with the units in their area and ensuring that local transport requirements were met.[76] However, there appears to have been little desire among army, corps and divisional officers to reduce their demands upon the fastest method of transport available – even when the railway network became incapable of answering all of the requests that emanated from the front. Few followed the lead of Major-General Sir John Moore, the director of veterinary services, who approached Holland directly in June 1916 to arrange for the evacuation of sick horses by barge. Following Moore's enquiry, small open barges with a capacity of approximately seventy tons were drawn together and equipped with gangways for the loading and unloading of animals almost immediately. The first barges for the veterinary service were put into operation on 5 July 1916 – a reaction that highlighted both the IWT directorate's responsiveness to the army's needs and the existence of spare capacity within the fleet. By the end of the year a fleet of ten specialist craft had been sourced from the Yorkshire canals to operate the service, and 4,675 horses and mules had been transported away from the front by barge.[77]

The supply of road stone provided a ludicrous counterpoint to the success story of equine evacuation. The carriage of stone by train was suspended at the outset of the battle of the Somme to free up railway capacity for the movement of munitions and other supplies required by the fighting forces.[78] Further back along the line of communications, shipments of 1,000 tons of stone per day into Dunkirk were maintained throughout July but were 'suddenly' reduced to 400 tons per day – without GHQ's knowledge – the following month.[79] In the middle of September Major-General Charles Dawkins wrote to Lloyd George to stress that the 'operations which are now taking place have resulted in greatly increased use of all roads in the Fourth and Reserve Army areas, and the roads on territory taken from the enemy require entirely remaking. The requirements for road metal are increasing daily, and still further demands on a larger scale may be expected at any time'. The latest fortnightly returns illustrated that only 76 per cent of the BEF's 193,000-ton requirement for road stone had been received.

---

[76] TNA, WO 158/851, history of IWT, pp. 23–4, 42.

[77] TNA, WO 95/56, director of IWT war diary, memorandum 3, pp. 3–4; WO 107/296, report of British armies, p. 27; WO 158/851, history of IWT, p. 37.

[78] R. U. H. Buckland, 'Experiences at the Fourth Army headquarters: organization and work of the R.E.', *Royal Engineers Journal*, xli (1927), 385–413, at p. 389.

[79] TNA, WO 95/3970, headquarters branches and services. Inspector general, diary entry, 23 Aug. 1916.

'These figures', Dawkins complained, were 'both actually and relatively, considerable [sic] worse than any previous figures with which I have been furnished'.[80] However, despite the BEF's shortages of a material described by Haig as 'of vital importance for the conduct of operations' on the western front, no demands for the carriage of road stone by barge had been made by British commanders. During the same period IWT vessels within Holland's fleet *were* being utilized for the conveyance of road stone along the River Somme, but at the request – and for the use – of the French army.[81]

The BEF's failure to thoroughly exploit IWT as part of an integrated transport solution had deleterious effects both on the efficiency of the railways and the capacity of the road network directly behind the front-line troops. The 'deplorable state of the roads' in the BEF's zone of operations soon became the 'chief source of anxiety' both for the Fourth Army's chief engineer and for Colonel Woodroffe, the deputy QMG, who catalogued his concerns in a series of notes written to his superior.[82] A lack of materials in France was not to blame for the state of the roads behind the BEF. As Woodroffe noted on a visit to the railhead at Belle-Église on 12 August, '[t]he stone dump is assuming a really ridiculous size. It is now so high in places that it requires a big lift to get the stone out of the truck onto the dump'. So much stone had accumulated at the railhead that quantities of the material fell back off the dump onto the tracks after they had been unloaded, an occurrence that caused the removal of empty wagons from the railhead to be delayed while the line was cleared – reducing fluidity on the railway network.[83] In early August Brigadier Hodgkin passed through the area of the Somme occupied by French troops. On his journey to Froissy to investigate the disappearance of chemicals from one of the filtration units attached to the IWT directorate he travelled through Longueau, Villers-Bretonneux, Lamotte-en-Santerre, Proyart and Chuignolles, and remarked that the roads behind the French troops were 'kept as smooth as glass … Compare these with the roads in Flanders, in our charge … where there are more shell holes than roads'.[84]

The British approach to road repairs was only one component of the war effort that Hodgkin criticized within his wartime diary. Unlike many of his

---

[80] TNA, WO 95/3970, IGC war diary, Dawkins to Lloyd George, 16 Sept. 1916.

[81] TNA, WO 95/3970, IGC war diary, 23 Aug. 1916; WO 95/56, director of IWT war diary, memorandum 3, p. 3.

[82] Buckland, 'Experiences at the Fourth Army headquarters', pp. 391–2; IWM, papers of Brigadier-General C. R. Woodroffe, 3/38/1/2, notes and reports (forwarded to QMG), June to Nov. 1916.

[83] IWM, Woodroffe papers, diary entry, 19 Aug. 1916.

[84] IWM, Hodgkin papers, diary entries, 8 and 9 Aug. 1916.

colleagues within the higher echelons of the IWT directorate, as a pre-war territorial officer Hodgkin was thoroughly conversant with the particular requirements of a military organization. The discipline within the IWT service on the western front did not meet Hodgkin's expectations, and upon his departure from the directorate in mid August 1916 he reflected that:

> I shall be sorry to go in one way, because I have found most of the officers here very jolly; but the organization from a military point of view is non-existent, and neither the other officers nor myself are able to feel any pride in the Corps, a state of things which does not conduce to efficiency. And the CO [Holland] won't allow the organization to be improved, which is annoying.[85]

Hodgkin singled out the 'most miscellaneous collection of personages' that comprised the IWT directorate's senior team as the chief culprits behind its inefficiency, and claimed that Holland's organization contained 'no chain of command and no discipline worth speaking of'. He perceived that the training the men had received at Longmoor prior to their despatch to France, and the minimal supervision that could be applied to men who were constantly circulating around the waterways, had resulted in the IWT directorate suffering an 'undue proportion of courts martial and other nuisances'.[86]

However, Hodgkin's observations should be approached with caution. Throughout the portion of his diary dedicated to his time as an IWT officer he recorded his frustrations at not making what he believed to be a valuable contribution to the British war effort. While on board barge A.174 Hodgkin peppered his diary with entries that claimed he had experienced a 'week of absolute idleness', that he was 'still hard at work doing nothing', and that he had 'wasted, absolutely – and-without-any-extenuating-circumstances-wasted, five months, and I am sick and tired of it'.[87] He desired a return to combat duties and had requested a transfer to the Special (Gas) Brigade within two months of having received his assignment with the IWT directorate. His comments demonstrate a common attitude among British officers that 'combat remained the measure of the soldier'.[88] Through his contribution to the provision of safe drinking water, Hodgkin possessed a role within the IWT directorate where his scientific background was of direct benefit to the BEF. Yet throughout what he dubbed his 'six months of unbroken idleness' aboard A.174 and at IWT headquarters, Hodgkin failed

---

[85] IWM, Hodgkin papers, diary entry, 19 Aug. 1916.
[86] IWM, Hodgkin papers, diary entry, 28 June 1916.
[87] IWM, Hodgkin papers, diary entries, 6–10 March and 1 May 1916.
[88] I. M. Brown, *British Logistics on the Western Front, 1914–1919* (London, 1998), p. 86.

Table 5.1. Approximate tonnages of materials carried by
inland water transport on the western front, 1915–18.

Year	Tons	Percentage increase over previous year
1915	205,047	—
1916	839,519	309.43
1917	2,378,342	183.30
1918	2,843,793	19.57

Source: TNA, WO 158/851, Director general of transport: History of inland water transport, p. 15.

to fully appreciate the work he and his colleagues had undertaken.[89]

The same lack of appreciation for IWT's role affected the BEF more widely. For all the unprecedented scale of the demands generated by the fighting on the Somme in 1916, Holland was forced to return barges requisitioned from French civilian sources due to a lack of military material for them to convey.[90] Regardless of the 309 per cent increase in the tonnage conveyed by IWT in 1916 over the figures from the previous year (see Table 5.1), significant spare capacity existed across the fleet. A total of 73,500 deadweight tons carrying capacity was available to the BEF in October 1916, but Sir Eric Geddes recorded that the maximum quantity conveyed by IWT in a single month was just 69,000 tons. 'Each deadweight ton', Geddes observed, 'was not fully occupied once in the month … A great carrying capacity has been provided and no adequate use found for it'.[91] The man who had been more responsible than anyone else for the provision of that great carrying capacity was Gerald Holland, marine superintendent of the LNWR. However, the task of making the best use of it ultimately fell elsewhere; to Geddes and his successors as DGT, and to Cyril Luck, who replaced Holland as director of IWT in the summer of 1917.

Luck's promotion did not come about as a result of any deficiency in Holland's execution of the role of director. Rather, in 1917 the latter became 'the highest ranking and most decorated railwayman to die in the Great War'.[92] Holland's commitment to the war effort was a significant factor in his death. Following the German retirement on the Somme in early 1917, Holland chose to personally survey the devastation wrought on the canal

[89] IWM, Hodgkin papers, diary entry, 19 June 1916.
[90] UWMRC, Granet papers, MSS. 191/3/3/4, Geddes to Lloyd George, 15 Sept. 1916, p. 2.
[91] UWMRC, Granet papers, MSS. 191/3/3/102, memorandum by Sir Eric Geddes, 26 Nov. 1916, p. 23.
[92] J. Higgins, *Great War Railwaymen: Britain's Railway Company Workers at War 1914–1918* (London, 2014), p. 237.

network east of Péronne. He recommended that the section between Frise and Péronne be repaired for traffic immediately, but beyond that point he reported on 31 March that 'the destruction was so complete as to make the expenditure of labour, material, etc. necessary for its rehabilitation, inadvisable, having regard to the limited use to which this waterway could be put owing to its relation to the Army on its new line'.[93] The exertions of a thorough survey in the adverse weather conditions of a notoriously cold spring took a heavy toll on the fifty-six-year-old.[94] He contracted an illness and returned to England on sick leave the following month, but did not recover. He died at St Leonards-on-Sea in Sussex on 26 June 1917 and was buried near his pre-war home at Holyhead. The post-war history of the directorate he created recorded its appreciation of Holland's influence on IWT in France in glowing terms:

> [Holland was] an officer of great foresight and powers of initiative with wide experience in connection with the services, civil, marine and mechanical engineering problems, a born administrator with a particularly strong capacity for the mastering of details, he had worked whole-heartedly to make the IWT service in France efficient and capable of meeting any demands upon its resources.[95]

Yet Holland's influence upon the use of waterborne transport in the British war effort stretched far beyond France, as the sustained growth of Richborough as a home base for the IWT directorate in the second half of the war illustrated.

Every item that did not pass through the Channel ports freed up space on the French coast for cargo ships, and the development of Richborough allowed it to play a vital role in the BEF's supply operations. By the end of 1916 the 'mystery port' in Kent had evolved from a concept into a working dock, and the cross-Channel ferry service championed by Holland had despatched 1,969 tons of goods to France. In the two years that followed 'a fleet of sixty tugs and 160 craft delivered ... 1,400,000 tons, of which 1 million were delivered at inland depots', from a sprawling complex that employed almost 16,000 people.[96] As the BEF drove towards victory on

---

[93] TNA, WO 95/56, director of IWT war diary, memorandum number 4, 29 Apr. 1917, p. 4.

[94] A series of recollections from those who experienced the winter and spring of 1916–17 on the western front can be heard at IWM, 'Winter 1916–17', *IWM Voices of the First World War* <http://www.iwm.org.uk/history/podcasts/voices-of-the-first-world-war/podcast-25-winter-1916-17> [accessed 31 Aug. 2017].

[95] TNA, WO 158/851, history of IWT, p. 21.

[96] *History of the Corps of Royal Engineers*, ed. H. L. Pritchard (11 vols., Chatham, 1952), v. 621; H. Best, *'The Mystery Port', Richborough* (Blackpool, 1929), p. 9.

the western front in October 1918 the cross-Channel barge service moved 25,000 tons per week.[97] Stores from Richborough were also sent to the other theatres of war in which British troops were employed, supplied to waterways and docks depots throughout Britain and even used for the construction of aerodromes across southern England.[98] Furthermore, the port was connected to the SECR's main line and, from 10 February 1918, to the French rail network thanks to the installation of a train ferry service.

The establishment of the train ferry owed a great deal to civilian expertise. The idea for a cross-Channel train ferry service between Britain and France had been raised prior to the war, but it was given fresh impetus in late 1916 by Follett Holt, the former chief engineer and general manager of the Entre Ríos Railway in Argentina; Brodie Henderson, whose pre-war career in railway, dock and bridge construction spanned Africa and South America; and Alexander Gibb, the Scottish civil engineer responsible for the extension of Alexandra Docks at Newport and the construction of Rosyth naval base in Scotland. Holt had opened a train ferry service across the Paraná River in 1907, and drew upon that experience when he 'made a strong appeal to the British government to install a system of train ferries' across the Channel. The ferries allowed a train made up in any part of Britain to be transferred to a railhead in France without the need for transhipment at any point of its journey. 'In this way, still further relief would be given to the shipping situation; there would be a greater saving of labour at the ports, while the distribution on the other side of the Channel would be much more effective than was possible under the barge system' due to the relative abundance of railheads when compared to 'canalheads'.[99] Instructions to proceed with the construction of train ferry termini at Richborough and Southampton were issued by the War Office on 17 January 1917, and sanction was received by the French authorities for the provision of French termini at Dunkirk, Calais and Dieppe within a month. Gibb was appointed chief engineer of ports construction and oversaw the building work required to prepare the landing facilities in France, while Henderson designed the 'berths and accessory works' associated with the termini.[100] The first train

[97] S. D'A. Crookshank, 'Transportation with the B.E.F.', *Royal Engineers Journal*, xxxii (1920), 193–208, at p. 197.

[98] E. A. Pratt, *British Railways and the Great War: Organisation, Efforts, Difficulties and Achievements* (2 vols., London, 1921), ii. 1107.

[99] Pratt, *British Railways and the Great War*, ii. 1109–10.

[100] A thorough account of the engineering challenges overcome in the development of the train ferry termini is given in F. O. Stanford, 'The War Department cross-Channel train ferry', *Minutes of the Proceedings of the Institution of Civil Engineers*, ccx (1920), 208–38.

ferry arrived in France before the end of 1917, less than twelve months after the scheme had been authorized.[101]

The train ferry service handled a comparatively modest volume of traffic during the war. Between February and December 1918 just over 200,000 tons were conveyed to France aboard the three vessels constructed to ply the route – a pale shadow of the weekly average of 224,000 tons discharged from vessels at the French ports in May 1917 alone.[102] However, the real value of the train ferry lay in the conveyance of railway materials and large, bulky items such as tanks and siege guns. A total of 164 locomotives and tenders, seventy narrow-gauge locomotives, 7,142 railway wagons, and 658 tanks were despatched to the western front aboard the train ferry.[103] 'Originally', Henniker noted in the official history, 'tanks were sent across [the Channel] in special vessels, but after the ferry service started they were sent by rail, loaded on the specially-built tank wagons, direct from the testing centres at home to the tank depots in France' without the need to be loaded and unloaded several times on their journey.[104] Similar advantages accrued in the transportation of the BEF's heaviest guns, such as the 14-inch railway guns 'Scene Shifter' and 'Bosche Buster' that weighed almost 300 tons and measured eighty-seven feet each.[105] The latter of these guns made a particularly lethal contribution to the fighting in France. Under the watchful eye of King George V, a shell fired from 'Bosche Buster' on 8 August 1918 from a location near Marœuil scored a direct hit on the railway station at Douai some nineteen miles away. A German troop train was destroyed by the shell at a cost of 400 casualties, and the Germans were prevented from making full use of the railway for the remainder of the conflict.[106] Under war conditions such vast weapons could only have been transported overseas by train ferry. Therefore, the transport experts who conceived of, constructed, and operated the service between Britain and France had a deadly impact upon the conduct and character of industrial warfare.

[101] Best, 'The Mystery Port', pp. 29–30.
[102] Pratt, British Railways and the Great War, ii. 1112; Henniker, History of the Great War, p. 238.
[103] 'The directorate of inland waterways and docks', Railway Gazette: Special War Transportation Number, 21 Sept. 1920, pp. 141–51, at p. 147; D. Chapman-Huston and O. Rutter, General Sir John Cowans, G.C.B., G.C.M.G.: the Quartermaster-General of the Great War (2 vols., London, 1924), ii. 219–20.
[104] Henniker, History of the Great War, p. 241.
[105] Best, 'The Mystery Port', p. 56; Crookshank, 'Transportation with the B.E.F.', pp. 197–8.
[106] C. Hooper, Railways of the Great War (London, 2014), p. 104.

Conclusion

The colossal scale of demands placed upon the BEF's transportation services as the force's presence in France and Flanders grew were such that IWT could only ever play a subsidiary role in their fulfilment. The relatively diminutive quantity and capacity of the IWT fleet in France, coupled with the reluctance of the French authorities to permit the establishment of new traffic routes and the position of Holland's directorate as a scion of the recognized supply chain, constricted IWT's growth in the year that followed its separation from the directorate of railway transport. The minor role afforded to the development of IWT on the western front in Henniker's official history has combined with these factors to overshadow the evolution of a small, under-exploited, but effectively managed civil–military partnership at GHQ. He may not have been able to generate demands within the BEF for canal transport that kept pace with the IWT fleet's growth, but Gerald Holland's talents for organization and his technical expertise were recognized and respected by the British army's senior administrative officers. The leaders of Britain's war effort commenced upon engineering projects that involved considerable commitments of time, money and resources – both human and material – on the strength of Holland's advocacy.

However, Holland's proactive approach was insufficient to counteract the constraints caused by deficiencies in the command structure within the BEF's transport hierarchy. Prior to Sir Eric Geddes's arrival on the western front the abilities of Britain's transport experts were, as a consequence of the compartmentalized approach to transportation within the QMG's department, only applied to the solution of individual problems in solitary links of the supply chain. Throughout 1915 and 1916 there was neither the political desire to further embed civilian specialists in the military hierarchy, nor the operational necessity to conduct a wholesale replacement of the BEF's administrative foundations. The result was a tendency towards a focus on tinkering with individual transport challenges rather than reconsideration of the infrastructure and systems that underpinned the British and allied war efforts.

The concentration on such localized responsibilities rendered individuals like Holland and Francis Dent unable to negotiate successfully with their French allies, who were attempting to do two things simultaneously: first, to balance demands for further assistance from the British with a desire to retain superiority within the coalition; and second, to prioritize the short-term possibilities of bringing the war to a swift conclusion rather than divert attention towards the development of a coherent long-term strategy for the maintenance of the allied forces on French soil. The effects of the continued lack of a formal alliance structure to guide and govern the expansion of Britain's contribution to the land war in northern France – carefully omitted from the diplomatic post-war reports submitted by the officers of a victorious BEF – were evident in Holland's exasperated diary entries during 1916, as his vision for the IWT directorate was subjected to the pressures of coalition warfare. The absence of a coordinated Franco-British consideration of the logistical requirements of a greatly expanded BEF, combined with a continued trend towards decentralization and self-sufficiency within the British force, impaired the development of a fully integrated, centrally directed transportation system on the western front.[107] This pattern endured until the 'strain imposed by active operations' highlighted the 'non-appreciation of the real meaning of the service of transportation' within the British army.[108] It did so astride the Somme in the summer of 1916, and it precipitated the reorganization of military transportation in France and beyond. The man chosen for this task, which dwarfed that taken on by Gerald Holland in December 1914, was another of Britain's transport experts. Between August 1916 and May 1917 Sir Eric Geddes, the highest paid railway official in pre-war Britain, made a pivotal contribution to the British empire's conduct of the First World War.

[107] On the encouragement of a decentralized administration within the BEF, see TNA, WO 95/74, director of supplies war diary, diary entry, 20 Dec. 1914; WO 95/27, QMG war diary, French to Kitchener, 18 Jan. 1915; WO 107/15, IGC, general correspondence, Maxwell to Cowans, 18 July 1915; LHCMA, Robertson papers, 2/2/63, Robertson to Cowans, 8 Jan. 1915.

[108] M. G. Taylor, 'Land transportation in the late war', *Royal United Services Institution. Journal*, lxvi (1921), 699–722, at p. 704.

6. The civilians take over? Sir Eric Geddes and the crisis of 1916

In his post-war memoirs David Lloyd George predicted that when 'the whole story of British achievement in the sphere of transport during the war' was charted by historians it 'would reflect very high credit on those who were responsible for its development, most of all on Sir Eric Geddes'.[1] Sir Douglas Haig had already paid a glowing tribute to Geddes's contribution, when he reflected in 1919 that:

> The Directorate-General of Transportation's Branch was formed under the brilliant direction of Major-General Sir Eric Geddes in the autumn of 1916 ... To the large number of skilled and experienced civilians included by him on his Staff, drawn from the railway companies of Great Britain and the Dominions, the Army is greatly indebted for the general excellence of our transportation services.[2]

These commendations from the principal political and military figures in Britain's war effort, whose 'opinions diverged more often than they coalesced' during and after the First World War,[3] echoed earlier praise for Geddes's role in the reorganization of Britain's military transportation services recorded in the railway press. 'There is nothing like a war to make or break the reputations of a nation's leaders', wrote the *Railway Magazine* in 1917, 'and so far as the British Empire is concerned there is no subject of His Majesty the King who has made more remarkable progress in national service and public recognition than Sir Eric Geddes'.[4]

Few civilians could claim to have had a larger, more profound impact on the BEF's command structure during the conflict than Geddes. In August and September 1916 he investigated and reported upon the existing

---

[1] D. Lloyd George, *War Memoirs of David Lloyd George* (6 vols., London, 1933; 2 vols., London, 1938), i. 479.

[2] D. Haig, *Sir Douglas Haig's Despatches (December 1915–April 1919)*, ed. J. H. Boraston (London, 1919), p. 351.

[3] K. Grieves, 'Haig and the government, 1916–1918', in *Haig: a Reappraisal 80 Years On*, ed. B. Bond and N. Cave (Barnsley, 2009), pp. 107–27, at p. 121.

[4] TNA, ZPER 39/41 *The Railway Magazine*, vol. xli, 'British railway service and the Great War' [xvli], 1917, p. 186.

transport network in France and Flanders. From then until May 1917 he created, populated and directed entirely new transport management hierarchies on the western front and at the War Office. He then bequeathed fully functioning directorates to civilian successors drawn from among the ranks of Britain's transport experts.

This chapter examines the precarious nature of civil–military relations within the British war effort in the summer of 1916, the conclusions of Geddes's mission to GHQ and the establishments both of the directorate-general of transportation in France and the directorate-general of military railways in London. Geddes's unprecedented appointment to the directorships of both organizations – located *within* the military machine – exemplified Lloyd George's desire to more thoroughly exploit the skills possessed by Britain's civilian specialists, and it precipitated an overhaul of the personnel and procedures involved in the supply of British forces dispersed around the world. Yet Geddes's twin role could not have been conceived, let alone his duties discharged successfully, without the personal and professional support of Britain's political and military leadership. Lloyd George in London and Haig in France provided Geddes with the institutional assistance required to bring the concept of a centralized transportation service into being. Both protected Geddes from the criticisms and petty jealousies of those within and outside the 'military trade union', both understood the weaknesses in the BEF's supply foundations that were exposed by the battle of the Somme and both worked to ensure that transport requirements received a priority hitherto denied them in a war effort that had been predominantly focused upon the creation of a mass army and the manufacture of ever-larger volumes of firepower. Together they created the platform from which the British army unleashed its ultimately successful war of material.

Military attitudes to civilian 'interference' in 1916

On 19 November 1917, Lloyd George, by now the prime minister, delivered a statement in the House of Commons. He claimed during his address that he had acted against the advice of the military high command on just two occasions during the war. In the first instance he had ordered 'extravagant' quantities of guns and shells when he was minister of munitions. 'I was told that I was mad', he said. 'The second case where I pressed my advice on soldiers against their will', he continued, 'was in the appointment of a civilian to re-organise the railways behind the lines – my Right Honourable Friend (Sir E. Geddes) – and I am proud to have done it'.[5] Lloyd George reiterated his position on the latter incident in his *War Memoirs*, taking aim

---

[5]   Hansard, *Parliamentary Debates*, 5th ser., xcix (19 Nov. 1917), col. 904.

at a War Office that had, he claimed, 'held the opinion that [transport issues] were purely military matters, into the sanctity of which no profane civilian must be allowed to intrude'.[6] As the previous chapters have demonstrated, Lloyd George's enduring image of the British army as a narrow-minded and obstinate institution is not borne out by its multiple interactions with transport experts in the first half of the conflict.

Yet the notion of a British military clique, disengaged from the wider world and reluctant to engage with civilians, was not forged in the aftermath of a controversial battle or in the pages of post-war memoirs. In 1913 the French military attaché Colonel Huguet described the British army as 'insular and therefore mistrustful of whatever came from outside'.[7] Politicians were especially likely to raise the suspicions of the army authorities, and an atmosphere of apprehension about Lloyd George's motives was evident from the moment he became secretary of state for war in July 1916.[8] Asquith, sensitive that the 'fluttering of military dovecotes' could accompany Lloyd George's appointment, urged the latter to 'work intimately with the soldiers' rather than seek confrontation.[9] Lord Esher, himself no stranger to the inner workings of the military mind, also counselled Lloyd George to exercise 'care' in his use of civilian specialists within the army.[10]

The new secretary of state for war was not alone in being cautioned to tread carefully. In an 'unofficial' chat at the War Office, Auckland Geddes was informed that 'you can't do a war-dance on senior officers' pet corns and expect them not to kick'. Consequently, 'brother Eric' was implored 'not to start a row' or present himself at GHQ as Lloyd George's 'dogsbody'. Instead, Auckland advised his brother to 'talk the language' of the army, emphasize his education at the Oxford Military Academy and his experience of working on the railways, and stress to the officers in France that his purpose was to be an expert assistant rather than a civilian usurper.[11] Allied to his fraternal pep talk, Geddes's visit to GHQ was preceded by a letter from Lloyd George to Haig that set out the transport problem in plain terms:

---

[6] Lloyd George, *War Memoirs*, i. 471.

[7] E. Greenhalgh, *Victory through Coalition: Britain and France during the First World War* (Cambridge, 2005), p. 7.

[8] IWM, Wilson papers, HHW/2/83/65, Hutchinson to Wilson, 7 July 1916.

[9] PA, Lloyd George papers, LG/E/3/14/26, Le-Roy Lewis to Lloyd George, 8 Nov. 1916; LG/E/2/23/2, Asquith to Lloyd George, 6 July 1916.

[10] PA, Lloyd George papers, LG/E/2/11/2, Esher to Lloyd George, 13 Aug. 1916.

[11] A. C. Geddes, *The Forging of a Family: a Family Story Studied in Its Genetical, Cultural and Spiritual Aspects and a Testament of Personal Belief Founded Thereon* (London, 1952), pp. 233–5.

The output at home of munitions has now so greatly increased that we can meet with comparative ease the higher demands which you quite properly make on us, but I doubt whether, without careful preparation, the powers of absorption of the ports and lines of communication can expand to a commensurate degree. What I have specifically in mind is the desirability of ensuring such an expansion as will next year, and the year after if necessary, enable us to cope with the ever increasing volume of munitions and stores which will be needed for the services of your force.[12]

Put simply, Lloyd George felt confident that the colossal firepower demanded by the BEF's commanders could be produced. However, he could not guarantee that the munitions could be delivered to where they were required in a timely fashion – with obvious implications for the BEF's effectiveness as a fighting force.

The BEF's experience at the battle of the Somme reinforced Lloyd George's concern. From an artillery perspective alone the requirements of the Somme were prodigious. Until mid June 1916 some five to twelve ammunition trains per week were sufficient to meet the BEF's demand for shells. Yet in the weeks that preceded the offensive the number rose rapidly to between forty-five and ninety trains per week.[13] The equivalent of thirty-six miles of motor lorries per division were required to shift the forty-nine ammunition trains per week that arrived in the Fourth Army's area in the period after 5 June.[14] The Ministry of Munitions' success in raising the output of shells engendered hitherto unprecedented pressure upon the lines of communications behind the western front.

The demands of the Somme offensive exposed the inadequacy of the transport infrastructure in the battle zone, and underlined the subordinate position to which considerations of logistics had been relegated in the first half of the war. According to the official historian 'the railways were inadequate, [and] the roads in the area behind the front [in Picardy] ... were few and indifferent'. In 1916 'almost any part of the Arras–Ypres front was better furnished with villages, railways and roads'.[15] Two single-lines to Arras and the double-line between Amiens and Albert, which was within

[12] TNA, WO 32/5163, appointment of Sir E. Geddes and others to investigate transport arrangements in connection with the British Expeditionary Force at home and overseas, Lloyd George to Haig, 1 Aug. 1916.

[13] I. M. Brown, *British Logistics on the Western Front, 1914–1919* (London, 1998), p. 120.

[14] NLS, Haig papers, Acc. 3155/106, diary entry, 5 June 1916.

[15] J. E. Edmonds, *History of the Great War. Military Operations, France and Belgium, 1916* (2 vols., 1932), i. 271. Even in the more understated words of Colonel Henniker, 'the railways serving [the Somme] ... were not good'. See A. M. Henniker, *History of the Great War: Transportation on the Western Front, 1914–1918* (London, 1937), p. 120.

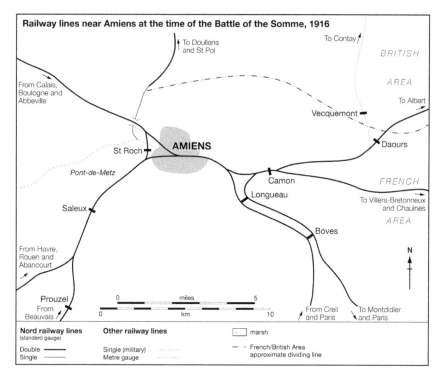

Figure 6.1. The railway lines near Amiens at the
time of the battle of the Somme, 1916.

Source: A. M. Henniker, *History of the Great War: Transportation on the Western Front, 1914–1918* (London, 1937). Map drawn by Cath D'Alton.

range of the German artillery, represented the only pre-war main line rail communications available in the twenty-three miles between Arras and the River Somme. Furthermore, the undulating countryside made the provision of reliable new railways impractical. Railway construction in the area had been underway since October 1914, and a seventeen-mile line that linked Fienvillers, Candas and Acheux had been handed over to the ROD in April 1916.[16] However, a considerable bottleneck around the key railway junction of Amiens could not be eliminated before the battle's scheduled start date (see Figure 6.1). The one-mile section heading east from St Roch comprised: the principal rail connection between the zone of operations and the BEF's southern line of communications (which ran inland from the ports of Dieppe, Le Havre and Rouen); the only inland north to south line

---

[16] Edmonds, *France and Belgium, 1916*, i. 273.

between the Béthune coal field and Paris; a heavily worked civilian traffic route; and the vital junction through which any strategic troop movements required during the battle would have to pass. At the Camon–Longueau interchange east of Amiens all the traffic heading to and from the BEF's Fourth Army crossed the route used by most of the traffic required to supply the French Sixth Army, which operated on the BEF's right flank during the offensive. To ensure the continued supply of the forces in the area the Camon–Longueau junction had to handle 240 trains each day, expected to intersect each other's routes at a rate of one train every six minutes.[17]

However, the BEF's supply trains did not keep to a regular schedule. Thanks to the ongoing problem of congestion at the ports, trains were despatched when they were ready to depart rather than according to a set timetable. Therefore, trains arrived at the Amiens bottleneck from three different directions at largely unpredictable intervals. Coupled with the need to attach extra engines to help heavy supply trains deal with the steep gradients on the so-called Plateau line between Albert and Amiens, delays were inevitable. One eyewitness recalled that 'eighteen miles of trains under load stood end-to-end waiting to get to railheads' outside Amiens within a few weeks after zero hour.[18] With so many trains held up on their journey to the front, too few locomotives and wagons were available to clear the docks of the voluminous quantities of material that poured into France every day. With the railways unable to clear the imported stock from the wharves, the quaysides became overcrowded and subsequently reduced the speed with which incoming ships were offloaded and put back to sea.[19]

Conditions in front of the railheads were no better. The roads in rural France had originally been constructed upon a chalk foundation to service a traffic that largely comprised farmers' carts and bicycles. Under the 'heavy pounding of the army's mechanical transport' the foundation of the road broke up, a 'chalky ooze' appeared on the surface, the granite setts worked loose, and the entire road began to disintegrate.[20] The traffic required to service the battle was incessant. In a twenty-four-hour period in late July – described by the provost marshal as 'one of the quietest we have had' – the traffic that passed Fricourt Cemetery was recorded as 26,516 troops, 568 cars,

---

[17] Brown, *British Logistics*, p. 184; Henniker, *History of the Great War*, pp. 136–7.

[18] J. C. Harding-Newman, *Modern Military Administration, Organization and Transportation* (Aldershot, 1933), p. 16.

[19] For a broader account of the events discussed in this passage, see C. Phillips, 'The changing nature of supply: transportation in the BEF during the battle of the Somme', in *At All Costs: the British Army on the Western Front 1916*, ed. S. Jones (Warwick, 2018), pp. 117–38.

[20] *History of the Corps of Royal Engineers*, ed. H. L. Pritchard (11 vols., Chatham, 1952), v. 293.

1,244 lorries and ambulances, 3,832 horse-drawn vehicles, 1,660 motorcycles and cycles and 5,404 horses. In just six hours over 2,500 vehicles passed along the Amiens–Albert road the following day.[21] When the weather conditions deteriorated, the problems caused by heavy traffic were exacerbated. Horse transport, which ordinarily travelled on open ground next to the roads, was forced to share the limited road space with the army's mechanical transport. The intensification of the traffic reduced the speed of all vehicles on the roads and made their repair and maintenance increasingly difficult. In a post-war article, Major-General Sir Reginald Buckland recalled having witnessed 'a man stooping down to spread stone between the feet of a team of horses while traffic was at a standstill, but as a rule congestion of traffic meant cessation of work'.[22]

Lloyd George had anticipated such a problem as early as September 1915, when he wrote to Kitchener to enquire whether the French transport network would be able to handle the enormous mass of stores projected to be available by mid 1916.[23] He received assurances at the time that it could.[24] The eighteen-mile-long queue of trains outside Amiens proved unequivocally that it could not. The demands of the Somme offensive, particularly the unprecedented scale of artillery expenditure and huge requirements for casualty evacuation, strained the BEF's transport infrastructure to breaking point.[25] However, Haig's cool response to Lloyd George's proposal that Geddes visit France to examine the transport situation gave little cause for the secretary of state to be optimistic. Haig replied to Lloyd George that 'you will, I am sure, realise that everyone behind the army, no less than at the front, is working at such high pressure at present that they will not be able to devote as much time to [Geddes] as we should like'.[26]

If Haig's reaction was cool, the attitude of his QMG, Sir Ronald Maxwell, was positively icy thanks in part to the existence of a document that had arrived at GHQ three weeks earlier. Acting on Lloyd George's behalf, Lord Derby handed Haig a memorandum on the transport situation on 11 July. The report outlined its anonymous author's concept for a new directorate that 'would be charged with the general supervision of the dock, rail and canal transport in France ... charged with making adequate provision in

[21] TNA, WO 95/441, the Fourth Army. Deputy adjutant and QMG war diary, Census of traffic at Fricourt Cemetery, 24 July 1916; Amiens–Albert road, census of traffic, 24 July 1916.
[22] R. U. H. Buckland, 'Experiences at Fourth Army headquarters: organization and work of the R.E.', *Royal Engineers Journal*, xli (1927), 385–413, at p. 389.
[23] TNA, WO 107/15, IGC, general correspondence, Cowans to Maxwell, 10 Sept. 1915.
[24] TNA, WO 107/15, IGC, general correspondence, Maxwell to Cowans, 12 Sept. 1915.
[25] Brown, *British Logistics*, p. 109.
[26] TNA, WO 32/5163, appointment of Sir E. Geddes, Haig to Lloyd George, 4 Aug. 1916.

rail, dock and canal facilities, and with arranging for the necessary material and personnel in connection therewith for the present and future needs of the armies in the field … responsible for the transport of all classes of military traffic' both to and from the front, and 'would report directly to the Secretary of State for War' in London rather than to Haig.[27] A note on the file, written by Haig, recorded that the BEF's commander-in-chief believed the memorandum to have been the work of one of 'Lloyd George's men' – most likely the one who would seek to run the new directorate. No evidence has been unearthed to confirm or allay Haig's suspicions. Yet as the memorandum drew upon the managerial structures of a large railway company to conceptualize the proposed organization it appears likely that Geddes had a significant influence over the document, which Maxwell dismissed as 'quite impracticable'.[28] Furthermore, in a demonstration of the QMG's mind set towards external interference in the BEF's forward planning activities, he noted that: 'It is not stated [in the memorandum] why the time has arrived to strengthen the transport arrangements of the BEF. So far as the work in France is concerned these arrangements have worked *perfectly smoothly and efficiently*: 1. in the ports; 2. on the railways and canals; 3. on the roads'.[29] As will be demonstrated further below, Maxwell's attitude was not unique among both the officers in France and those at the War Office.

Maxwell's reluctance to acknowledge the supply problems that were already manifest by the third week of the Somme was not the catalyst for Haig's response to Lloyd George. Nor should Haig's reply be taken as further evidence of an entrenched, insular military elite unwilling to embrace outsiders' criticisms of their procedures. Instead, the commander-in-chief's comments were a reflection of the fact that the BEF was engaged in the largest battle in British military history and, as a result, Haig felt unable to guarantee that an investigation into administrative processes and organizational responsibilities could receive priority at GHQ over the unfolding offensive. The emergence of severe logistical challenges during the first month of the battle meant that Haig was 'anxious to afford Sir Eric Geddes every possible facility for conducting his enquiry', and was 'glad

---

[27] NLS, Haig papers, Acc. 3155/215Q, memorandum (received from Lord Derby), 11 July 1916, pp. 1–2.

[28] The memorandum proposed that the directors of railways and IWT 'should occupy very much the same position to the [director-general of the new directorate] as superintendents of British railways occupy to the general managers'. See NLS, Haig papers, Acc. 3155/215Q, memorandum, 11 July 1916, p. 1.

[29] NLS, Haig papers, Acc. 3155/215Q, memorandum by Maxwell, 17 July 1916, p. 1. Emphasis added.

to make arrangements for his visit'.[30] As Ian M. Brown has demonstrated, Haig's interest in transport issues was apparent from the moment he replaced Sir John French as commander-in-chief on 19 December 1915.[31] On his first day at GHQ he met with his adjutant general and QMG before touring the various branches of the general staff, and within his first week he had discussed the railways and roads behind the Third Army with Lieutenant-General Edmund Allenby. Haig followed up his meeting with Allenby by urging 'the value of improved railway facilities' behind the Third Army to the French, and ordered Maxwell 'to hasten, and begin at once to build the railway projected into the Third Army area ... I also went into the reasons why the narrow gauge lines in the Second Army area (near Poperinghe) had not been turned to military use'. By 27 December he could write with evident satisfaction that: 'QMG recorded that work had begun on the railway in the Third Army area. A B[road] G[auge] line in the direction of Contay ... It is just four days since I begged Joffre to help us in this matter'.[32] Haig's understanding of the proper place of transportation in the army's activities – and the continued degradation of the BEF's lines of communications as the demands of the Somme took their toll on the roads, railways and docks – were sufficient grounds for him to authorize Geddes's visit to GHQ in late August.[33]

Where Haig's attitude encouraged dialogue between the military and one of Britain's leading transport experts, the War Office took a far less cordial stance towards Lloyd George's proposal. Brigadier-General Richard Montagu Stuart-Wortley, the DOM in London, was the chief protagonist behind the War Office's position. Lord Derby noted that Stuart-Wortley's 'intense dislike for Geddes' had not subsided in the aftermath of their frosty encounter at the start of the war, and his antipathy had been further fuelled by what Stuart-Wortley perceived to be increasing civilian encroachment into the military realm as the war progressed.[34]

Buttressed by the support of his commanding officer, Sir John Cowans, Stuart-Wortley's disinclination to support Lloyd George's proposition threatened to derail the transportation mission before it began. Mindful

---

[30] TNA, WO 32/5163, appointment of Sir E. Geddes, Haig to Lloyd George, 4 Aug. 1916.

[31] Brown, *British Logistics*, p. 104.

[32] NLS, Haig papers, Acc. 3155/104, diary entries, 22–27 Dec. 1915.

[33] TNA, WO 32/5164, facilities and arrangements for Sir E. Geddes in conducting his investigation on transport arrangements in connection with the British Expeditionary Force at home and overseas, Haig to Lloyd George, 22 Aug. 1916.

[34] PA, Lloyd George papers, LG/E/1/1/6, Derby to Lloyd George, 15 Sept. 1916; K. Grieves, 'The transportation mission to GHQ, 1916', in *'Look to Your Front!' Studies in the First World War by the British Commission for Military History*, ed. B. Bond et al. (Staplehurst, 1999), pp. 63–78, at p. 71.

of the fragility of civil–military relations, and the requirement that his investigations be conducted swiftly, Geddes wished to be accompanied in France by soldiers who could both explain the existing procedures and minimize the inconvenience to GHQ's staff officers.[35] Geddes identified the by now Lieutenant-Colonel Henry Mance, Stuart-Wortley's deputy DOM, as an ideal companion. Geddes and Mance had worked closely on REC business in the fifteen months that preceded the war, and the civilian wished to exploit the soldier's past experience of military railway operations – obtained within the railway directorate during the South African War. A letter was despatched from Lloyd George to Cowans, requesting that Mance be temporarily released from the War Office to join Geddes's team in France. Stuart-Wortley's response to the letter claimed that he 'could not possibly spare [Mance] for so long a time as three or four weeks', as to do so would 'seriously prejudice the work of my directorate and I do not consider that I can be held responsible for what may occur during his absence'. He described Mance as his 'head railway advisor', his technical assistant on 'all questions which involve dealings with the Railway Executive Committee or with the French and Belgian railways' and the man responsible for 'all questions connected with Mesopotamian, Egyptian and Salonika railways'. Stuart-Wortley argued that his deputy had an expertise that nobody within the directorate of movements could match, that he was the 'designated Acting Director of Movements in the event of an invasion … and [that] he has a knowledge of all home defence schemes which is unique'.[36] However, Stuart-Wortley's pleas and the ongoing threat of invasion, although enough to ensure that an enormous permanent garrison of 1.5 million men was retained in Britain during the war, were not enough to prevent Mance from crossing the Channel with Geddes in late August.[37]

GHQ in France proved far less obstructive. Haig made no attempt to dissuade Geddes from utilizing the services of Colonel Henry Freeland during his visit, despite the sustained stress on the BEF's staff as the battle of Somme continued. Like Mance, Freeland was handpicked by Geddes as the civilian had prior knowledge of his abilities. Geddes and Freeland had first met in India when they worked for adjoining railways at the same station – the former for the Rohilkund and Kumaon and the latter as deputy traffic manager with the North-Western. Alongside Geddes's knowledge of

[35] TNA, WO 32/5164, facilities and arrangements for Sir E. Geddes, Geddes to Lloyd George, 10 Aug. 1916.

[36] TNA, WO 32/5164, facilities and arrangements for Sir E. Geddes, Stuart-Wortley to Cowans, 7 Aug. 1916.

[37] D. Stevenson, *With Our Backs to the Wall: Victory and Defeat in 1918* (London, 2011), p. 260.

Freeland's work in India, the latter was deemed valuable as he had studied the methods employed by the French army.[38] He had observed the systems in use for the packing of French supply trains early in 1916, and his feedback had helped shape the composition of the BEF's daily supply trains in the second half of the war.[39]

The reasons behind Geddes's identification of Mance and Freeland emphasize the depth of interactions between the army and the railways in the period before the First World War. Both officers were chosen to participate in the transportation mission because of their demonstrable military experience and their pre-existing relationships with the civilian specialist. The railways of India, Africa, the United States and Britain had provided these men with the knowledge of railway operations in peace and war. They now came together on French soil to scrutinise the BEF's existing transport infrastructure and operating procedures alongside two civilians with whom Geddes was also highly familiar.

The first, Philip Nash, was another example of the diaspora of British expertise throughout the pre-war empire. He had been in India since 1899 and, after a series of promotions, had attained the position of joint secretary of the East Indian Railway (roughly equivalent to the role of assistant general manager) by 1911. It is unclear whether Nash and Geddes's paths had crossed while the latter was resident on the sub-continent, but they were definitely brought together when Nash was recruited to the Ministry of Munitions in 1915. Nash, as head of the national filling factories, was one of Geddes's assistants faced with the task of increasing the supply of gun ammunition to the British army. One of Geddes's other assistants, the North-Eastern Railway's statistics expert J. George Beharrell, comprised the final component of the civil–military transportation mission.[40]

The civilian members of the mission did not record any difficulties with regards to military attitudes prior to their departure from Britain, but Geddes observed that Mance joined the mission with some hesitancy. Stuart-Wortley's hostility towards Geddes may partially explain Mance's trepidation. However, his reluctance was also undoubtedly linked to the difficult position into which the civilian expert thrust his military companions. As Keith Grieves has explained, Mance and Freeland 'found

[38] TNA, WO 32/5164, facilities and arrangements for Sir E. Geddes, Geddes to Lloyd George, 10 Aug. 1916.
[39] TNA, WO 95/76, branches and services: Director of supplies, diary entry, 11 Jan. 1916.
[40] 'Philip Arthur Manley Nash', *Grace's guide to British industrial history*, 2016 <http://www.gracesguide.co.uk/Philip_Arthur_Manley_Nash> [accessed 7 Dec. 2016]; K. Grieves, *Sir Eric Geddes: Business and Government in War and Peace* (Manchester, 1989), pp. 19–24; Grieves, 'The transportation mission', p. 65.

themselves in the unenviable position of assessing the failure of GHQ to organise the free flow of supplies behind the line'.[41] The organization they were tasked to examine was in large part the creation of officers who outranked them, most notably the IGC, Major-General Sir Frederick Clayton. In 1915 Clayton had proven amenable to the idea of exploiting the SECR's civilian expertise to improve the BEF's supply operations at the port of Boulogne, but by the summer of 1916 his attitude towards civilian involvement on the lines of communications had become far less welcoming.

Clayton's correspondence with Cowans in the twelve months prior to the transportation mission charts both his frustrations at the BEF's approach to the challenges of trench warfare and his perception that his work as IGC had been underappreciated. He claimed that the 'combing out' of men suitable for front line duties in the summer of 1915 had robbed him of 'all the important trained men' within his department, and that 'to send out men and expect them to pick up in a month what others have taken ten months to learn, is asking, in my opinion, a little too much'.[42] By November 1915 the continued removal of his subordinates for deployments elsewhere inspired Clayton to pen a spirited lament towards the approach taken by his superiors in London:

> The War Office robbed me of my former M[ilitary] L[anding] O[fficer] (Watson), who went to the Dardanelles, and I understand they are now going to take away Blencowe, who manages all the movements of troops at Boulogne. Also I believe I am to lose Humphreys who does all the embarkation work at Havre, and in addition I have just had a man called Solomon, who was one of my Deputy Assistant Quartermaster-Generals here, taken away suddenly, after having trained him for three months, just as he was getting into the work.
>
> I shall have to fill all these places by retired officers or Territorials, in fact anyone I can get. Of course this is not conducive to good staff work, and makes things extremely hard for me; however I suppose we shall have to carry on somehow.[43]

Clayton's opinions on another letter within the same file, written by Cowans to Maxwell at the end of May 1916, were either not committed to paper or have not survived. However, as the letter requested the views of the BEF's senior supply officers on the 'possibility of transferring men from the Railway Transport Sections for more active duties in the field' just as the

---

[41] Grieves, *Sir Eric Geddes*, p. 30.

[42] TNA, WO 107/15, IGC, general correspondence, Clayton to Cowans, 8 July 1915.

[43] TNA, WO 107/15, IGC, general correspondence, Clayton to Cowans, 23 Nov. 1915. For further missives on the same theme, see Clayton to Cowans, 4 Dec. 1915 and 7 Jan. 1916.

supply services were gearing up for largest battle in British military history, one can surmise that the timing of the request alone confounded Clayton's frustrations.[44]

In addition to exasperation at the removal of 'his' trained men, Clayton perceived himself to be an isolated figure within the BEF's hierarchy. As his headquarters was located at Abbeville rather than GHQ, he felt cut off from the cluster of officers that surrounded the commander-in-chief. He bemoaned the lack of recognition for his endeavours in a letter to Cowans in February 1916, writing that:

> I was not mentioned in the previous dispatch [*sic*] and as I have told you have never had a mention since I have been IGC over twelve months now. [Sir Frederick] Robb who was not a brilliant success as IGC got a KCB. Maxwell who was IGC for three months and only had 250,000 men to deal with got a KCB. I have had over one million to deal with and have not even had a mention.[45]

Clayton's belief that his efforts had been insufficiently acknowledged was exacerbated by the number of investigations into logistical and administrative procedures that took place on the lines of communications during his tenure as IGC. Alongside pointing out that answering queries from soldier or civilian-led parties took up a 'great deal' of his and his staff's time, Clayton asked whether 'some steps' could be taken to 'stop these *constant attacks* and investigations being made on the lines of communication'. His reaction demonstrated that the aims of such examinations were not adequately understood by all the BEF's senior officers in the summer of 1916. Clayton's solitary concern was whether the work of his department had 'been done to the satisfaction of the C-in-C'.[46]

The IGC's growing antagonism towards outside interference threatened to jeopardize the efficacy of Geddes's transportation mission, and his mindset endangered the future of the BEF's supply operations. His argument, which was summarized in a response to the findings of a commission into the ongoing congestion at the Channel ports led by the shipping magnate Sir Thomas Royden, was that despite the BEF's colossal expansion over the preceding eighteen months it had 'been supplied with everything it requires with clockwork regularity; nothing had failed, all demands have been met and nothing but praise has been given to those who have done the work'. In addition:

[44] TNA, WO 107/15, IGC, general correspondence, Cowans to Maxwell, 31 May 1916.

[45] TNA, WO 107/15, IGC, general correspondence, Clayton to Cowans, 10 Feb. 1916.

[46] TNA, WO 95/3969, headquarters branches and services. Inspector general, Clayton to Maxwell, 14 June 1916. Emphasis added.

> The only conclusion one can come to after reading [the Royden report] is, that it is impossible for the ordinary business civilian to understand what are the conditions under which we have to work and that it is a mistake to allow them to interfere with an army business that most of us have studied all our lives … *when* we fail in any way to keep the army supplied it will be time for criticism.[47]

Clayton was not alone in besmirching the conclusions produced by the myriad examinations of the BEF's working practices. In April 1916 the director of supplies branded a report into the use of labour at Rouen by Major Ronald Williams of the Dockers Battalion, which had been commissioned by Haig rather than the government, as 'simply valueless and useless'.[48] Even Sir William Robertson, whose appreciation of supply questions at the start of the war had helped to sustain the BEF as a fighting force, believed that criticisms from London about congestion at the ports, poor storage practices, neglect of the canal network and the failure to develop railway traffic prior to the Somme were 'misinformed'.[49]

Until the supply link that reached back from the front line to Britain (and the world beyond) had *actually* broken down, Clayton believed that it was unfair of the War Office to continue bombarding his department with civilians hell bent on 'interfering' with his operations. At the very least his response to the Royden report illustrates that he was unwilling to countenance the potential problems that awaited the BEF should the transport network in France collapse under the weight of goods shipped to the western front. Nothing within Clayton's correspondence implied that he appreciated how examinations such as Royden's and Williams's were undertaken to ensure that catastrophic failure did not occur as the British war effort expanded. Inquiries that only took place *after* the system broke down – which in Clayton's view was the correct time for them – would theoretically occur too late to rectify the situation should the BEF wish to remain an effective fighting force.

By August 1916, despite the emergence of fruitful working partnerships between civilian specialists and army officers both prior to and during the initial stages of the conflict, there existed a clear and palpable sense of distrust among those responsible for managing the war effort. The growing stresses of an unfamiliar form of warfare, for which no clear blueprint

---

[47] UWMRC, Granet papers, MSS. 191/3/3/14, remarks on the report of the commission sent out by the shipping control committee, 30 July 1916. Emphasis added. The conclusions of Royden's report are summarized in Pritchard, *History of the Royal Engineers*, v. 610.

[48] TNA, WO 95/76, director of supplies war diary, diary entry, 24 Apr. 1916. Haig considered Williams to be 'broad minded and sensible'. See NLS, Haig papers, Acc. 3155/105, diary entry, 28 March 1916.

[49] LHCMA, Robertson papers, 7/6/60, Robertson to Haig, 28 July 1916.

for success had emerged, ratcheted up tensions between the political and military actors charged with the delivery of victory. Suspicions and reservations within the army over the motives of 'outsiders' – particularly those as closely connected to Lloyd George as Geddes – to do anything more than meddle with pre-existing structures and erode the army's jurisdiction, were mirrored by wariness and doubts over the competence of the soldiers responsible for overseeing the BEF's umbilical cord. Lord Derby, who was 'much impressed' by Geddes, described Clayton as 'very stupid, conceited and narrow-minded'.[50] Maxwell, Robertson acknowledged, was not 'the sort of man who would favourably impress Lloyd George' due to his 'hidebound manner'.[51] As IGC and QMG respectively, Clayton and Maxwell were the BEF's two senior supply officers and both viewed the transportation mission with naked hostility. However, their superior did not replicate their attitude. Even though Haig had adjudged Clayton's 'methodical system' to be 'very remarkable' in December 1915,[52] the BEF's commander-in-chief was thoroughly aware that the continued expansion of the British war effort necessitated frequent reassessments of the BEF's logistical foundations. Consequently, Geddes and his colleagues were received at GHQ on 24 August 1916 and began work the following day.

The transportation mission and the genesis of the directorate-general of transportation

The terms of reference issued to Geddes before his departure for France emphasize the enlarged scope of his mission in comparison to the localized, small-scale investigations of the previous eighteen months. His team were instructed by Lloyd George to: review the existing capacity of the BEF's transport infrastructure and ascertain whether it was capable of conveying the 'very considerably increased quantity of ammunition and other stores' to be despatched from Britain in preparation for the offensives of 1917; identify the repairs, extensions and operational improvements that were required at the docks, on the railways and on both the canal and road networks to render them capable of sustaining an advance towards the German border; and learn 'all that [was] possible from the very excellent transport arrangements of the French Army', so that the British could appropriate efficient practices for implementation within the BEF.[53] After

---

[50] PA, Lloyd George papers, LG/E/1/1/3, Derby to Lloyd George, 30 Aug. 1916.
[51] LHCMA, Robertson papers, 7/6/60, Robertson to Haig, 28 July 1916.
[52] NLS, Haig papers, Acc. 3155/104, diary entry, 30 Dec. 1915.
[53] TNA, WO 32/5164, facilities and arrangements for Sir E. Geddes, Lloyd George to Haig, 16 Aug. 1916; Lloyd George to Roques, 23 Aug. 1916.

the investigation had taken place Geddes was to produce both a series of statistical breakdowns, which detailed the quantities of various materials that the BEF were likely to require for future operations, and a number of reports that catalogued the range of variables involved in the transport network's maintenance and improvement.[54] In short, Geddes was directed to assess the past and present of the BEF's supply system before offering recommendations for the future of British transportation on the western front.

The mission began with a series of observational visits. Accompanied by the deputy QMG, Colonel Woodroffe, Geddes's team undertook a two-day tour of the BEF's rear areas. Over the course of forty-eight hours they visited ammunition railheads and newly constructed stations and sidings, and Geddes was afforded the opportunity to discuss the supply situation with officers in command of the artillery batteries deployed along the Mametz–Carnoy valley.[55] In his biography of Geddes, Keith Grieves stated that the tour was 'largely uninformative' due to the 'model' nature of the sites the civilian specialist was shown.[56] However, Woodroffe's account of the trip demonstrates that it provided the inspiration behind many of the improvements that were subsequently made to the transport infrastructure in France. The expedition impressed upon Geddes the immediate need for action to be taken to alleviate congestion and increase economy in the BEF's 'tail'. Furthermore, the brief overview of conditions behind the fighting troops provided him with the lines of enquiry upon which his follow-up investigation rested. As Woodroffe recorded:

> The points which appeared to impress themselves on [Geddes] most were: a) the enormous quantity of labour and material required to keep the roads in order and for the construction of the various station yards; b) the urgent necessity of some form of light railway to take the traffic off the roads and thus reduce the demand for road metal; c) the wastage of manpower particularly as regards the labour employed in transshipping stone either broad gauge to metre gauge, from rail to lorry or from rail to dump; d) the huge quantity of empty ammo boxes etc., the efforts which are being made to deal with the problem, and the large amount of labour employed for this purpose; e) the large quantities of brass 18-pdr. cartridge cases which are still lying about the British areas.[57]

Following his 'model' tour – but before he returned to London – Geddes

---

[54] A complete list of the statistics, projections, and reports originally demanded from the transportation mission is given in Appendix I.

[55] IWM, Woodroffe papers, 3/38/1/2, notes and reports, 25 Aug. 1916.

[56] Grieves, *Sir Eric Geddes*, p. 29.

[57] IWM, Woodroffe papers, 3/38/1/2, notes and reports, 25 Aug. 1916.

met with Haig and was asked by the commander-in-chief for his opinion on what he had seen. 'His reply was guarded', Auckland Geddes wrote later, 'to the effect that he had seen plenty to think about but as yet did not know what to think'.[58] Rather than risk sounding like he had arrived in France with pre-existing judgments, Geddes requested the opportunity to have a 'free run' of the BEF's lines of communications and full access to the statistics and information required to complete his mission. Increasingly concerned by the blockage of supplies around Amiens, Haig acquiesced and notified Maxwell of Geddes's requirements. Mindful of the QMG's antipathy towards examinations on the lines of communications, Haig also issued a circular to all armies and administrative departments, which ordered that 'all necessary information and any statistics required will be placed at the disposal of Sir Eric Geddes ... and the C-in-C desires that every facility will be afforded [Geddes] in the conduct of [his] enquiries'.[59]

Geddes's aspirations for the thoroughness of the 'free run' were emphasized by the composition of the party that returned to France in early September. The original team was augmented by the addition of John Blades, a 'very highly skilled dock superintendent' employed by the North-Eastern and Hull and Barnsley railway companies.[60] Blades joined Nash and Freeland at the French Channel ports, where they were tasked to analyse existing conditions and discover the capacities of the docks based on the nature of the goods anticipated to be despatched in support of the 1917 offensives. Meanwhile, Geddes and the remainder of the party surveyed the rest of the transport network with the aim of building up a 'complete statement' of the weight of traffic required to supply the BEF.[61] Within a fortnight, he felt sufficiently informed to offer a preliminary view of the situation on the western front to Lloyd George. It is clear from this letter that Geddes remained sensitive to the fragility of relations between his mission and many officers within the BEF. He implored the secretary of state for war not to reveal its contents to anyone in the War Office or at GHQ, as he feared that the criticisms contained within the document would severely jeopardize the remainder of the investigation if circulated.

The letter's conclusions – formed even before the bulk of the necessary data had been collected, let alone analysed – were an unequivocal condemnation of the BEF's logistical foundations and the innate reactivity of the force's administrative echelon. 'This is a war of Armies backed by machinery and

[58] Geddes, *The Forging of a Family*, p. 232.
[59] TNA, WO 95/31, branches and services: quarter-master general, circular to all armies, IGC and engineer-in-chief, 3 Sept. 1916.
[60] UWMRC, Granet papers, MSS. 191/3/3/4, Geddes to Lloyd George, 15 Sept. 1916, p. 2.
[61] UWMRC, Granet papers, MSS. 191/3/3/102. memorandum by Geddes, pp. 2–3.

"movement"', Geddes wrote, 'and I do not think that "movement" has received sufficient attention in anticipation of the advance. I judge this by the total absence of light railway or road organization, or policy for the use of waterways'.[62] The fact that canal barges had been returned to civil work even as the French railways continued to be clogged by ever-increasing quantities of road stone exemplified the issue. Rather than operating as an integral component of the BEF's transport mix, canals were considered as a carrier of last resort – to be requested only when rail transport was not available. The consequence of the BEF's decision to go to war without an integrated IWT directorate was that, despite Gerald Holland's efforts to develop capacity and promote waterborne transport, no guiding principles existed for the exploitation of the theatre's abundant canals and rivers. Holland believed that IWT was capable of transporting far more than had hitherto been requested of it. However, 'neither [in Britain] nor in France' could Geddes 'ascertain what the policy of canal user is. I doubt if one exists'.[63]

The BEF's problems were the consequence of insufficient forward planning and coordination, a result of the move towards decentralization instigated when the force began to expand in early 1915. At that time Robertson had acknowledged that the BEF had assumed 'too great a strength to admit of matters being centralized at GHQ to the extent they are now'.[64] However, the redistribution of authority over the various components of the transport infrastructure had resulted in the emergence of heavily compartmentalized departments. Officers were only able to adjust working practices in their own sections, and no oversight was in place to ensure that seemingly minor modifications in one area did not adversely affect the operations of other departments whose work was necessarily interconnected. As Colonel Henniker noted, 'the various transport agencies were a chain, the whole chain being no stronger than its weakest link'. In 1916 the links were not sufficiently connected 'so as to ensure a smooth uninterrupted flow of traffic' along the lines of communications.[65] The geographical barriers between Maxwell's offices at GHQ in Montreuil-sur-Mer, Clayton's at Abbeville and Holland's at St Omer (where IWT had remained following GHQ's transfer to Montreuil-sur-Mer) were a physical manifestation of an organizational deficiency. In a text on military transportation published after the war,

[62] UWMRC, Granet papers, MSS. 191/3/3/4, Geddes to Lloyd George, 15 Sept. 1916, pp. 7–8.
[63] UWMRC, Granet papers, MSS. 191/3/3/4, Geddes to Lloyd George, 15 Sept. 1916, pp. 2–3.
[64] LHCMA, Robertson papers, 2/2/63, Robertson to Cowans, 8 Jan. 1915.
[65] Henniker, *History of the Great War,* p. 192.

Major-General J. C. Harding-Newman was particularly scathing of the situation. He wrote that Clayton 'seldom, if ever, saw the QMG', and claimed evocatively that 'if ever there was more convincing proof of the dangers of separating the sub-divisions of a Staff, only the memorial to 77,000 unknown officers and men at Thiepval can provide it'.[66]

The BEF possessed no internal structures through which it could regularly review its procedures and consider the future of its transport organization. Facilities had been improved 'here and there' when experience proved they were incapable of handling the amount of work required, but no authority had been established to prioritize the distribution of materials and labour so as to ensure the most efficient use of the limited resources in France. The system was a 'hand-to-mouth' one, which had not kept pace with the growing demands on it or conducted accurate forward planning activities.[67] While railway construction in the event of an advance on the Somme had been planned between the DRT and French authorities, the extra quantity of rolling stock required to bridge the extended gap between the depots and the front line had not. Instead, the question had been subjected to 'rule-of-thumb' estimates generated within the railways directorate that illustrate the inadequacy of the BEF's planning mechanisms in 1916. The DRT tasked two officers to identify the number of wagons required to service the BEF's railway requirements to the border between Belgium and Germany. Lieutenant-Colonel Henniker predicted that 22,501 wagons would be required to work the BEF's daily traffic under such circumstances, whereas Lieutenant-Colonel Paget believed a mere 11,240 wagons would suffice. The wide discrepancy between the two figures was partly explained by the different parameters the officers had set themselves – Henniker, for example, added a 25 per cent margin to his estimate to take account of traffic dislocation and the unauthorized use of wagons as storage vehicles at railheads and in construction areas – yet neither soldier had based their calculations on a realistic prediction of the composition of the BEF's likely traffic in 1917. The Somme demonstrated the artillery-intensive nature of the industrial battle, and its failure to dislodge the German army from French soil highlighted that even larger exertions would be required if future operations were to be successful. Geddes's 'scrutiny' of Henniker's estimate – the larger of the two – revealed that the latter had considerably underestimated both the BEF's projected strength and the 'tonnage of certain commodities' that individual directorates had told the former would

[66] Harding-Newman, *Modern Military Administration*, p. 16.
[67] Henniker, *History of the Great War*, p. 184.

be necessary to fulfil their 'ultimate requirements' on the western front.[68]

The absence of a comprehensive statement of the BEF's needs meant that estimates like those produced by Henniker and Paget were at best misguided and at worst essentially worthless. Furthermore, the dearth of accurate forecasts concealed the scale of the challenge that the extant transport infrastructure was going to face in 1917. Nowhere was this more apparent than at the ports responsible for receiving all the BEF's imported supplies. Prior to Geddes's departure for France, Stuart-Wortley had provided him with a statement written by Clayton in July 1916, which outlined the 'tonnage which he was prepared to discharge at the French ports'. The figure of 138,000 tons per week, which the IGC referred to as 'the ultimate requirement' for sixty divisions, almost exactly matched the maximum weekly tonnage discharged at the ports during August 1916.[69] Following a discussion with Geddes on what the civilian referred to as 'general matters', Clayton revised his estimate of the ports' maximum possible discharge upwards to 160,916 tons. For reasons Geddes chose not to speculate upon, between July and September 1916 Clayton's calculation for the volume of work he believed the BEF's ports to be capable of increased by over 16 per cent.

Through the production of a comprehensive statement of the BEF's requirements, acquired from the force's individual directorates and departments, Geddes discovered that even Clayton's higher estimate was woefully inadequate for the war the British hoped to fight in 1917. With the provision of a 'margin for irregular arrivals and for contingencies', Geddes established a maximum discharge to be provided for at the BEF's ports of 248,327 tons per week – 40,225 tons every day (see Table 6.1). The provision of an accurate forecast meant that, for the first time, the discrepancy between the force's demands and its ability to fulfil them was made tangible and clear. To meet the projected requirements for 1917 the capacity of the ports under the BEF's control would have to increase by over 54 per cent.[70]

Geddes's almost immediate exposure of the inadequacy of the BEF's forecasting capabilities convinced him that the transport mission as originally conceived could not continue. The time for investigations, formal enquiries and interviews with overworked officers had passed. 'Executive action' was called for on both sides of the Channel – to install a comprehensive,

[68] UWMRC, Granet papers, MSS. 191/3/3/155, railway arrangements for advance through Belgium, 28 Oct. 1916, pp. 1–2.

[69] UWMRC, Granet papers, MSS. 191/3/3/102, memorandum by Geddes, p. 2. In the week ending 20 Aug. 1916 a total of 138,897 tons were discharged at the ports allocated to the BEF. The weekly average over the four weeks ending 27 Aug. was 129,024 tons.

[70] Henniker, *History of the Great War*, p. 185.

Table 6.1. Estimate of probable daily requirements
for the British Expeditionary Force, 1917.

Item	Daily tonnage required
R.E. stores and material	
– General	1,129
– Timber, bricks and gravel	3,782
– Stone	12,000
– Railway	1,500
Total	18,411
Supplies	10,425
Ammunition	8,600
Ordnance	1,513
Miscellaneous	1,276
Total	40,225

Source: *History of the Corps of Royal Engineers*, ed. H. L. Pritchard (11 vols., Chatham, 1952), v. 561–2.

centrally directed policy for transportation that took account of the myriad questions of coordination, resourcing, staffing and expansion which arose in the management of a modern army's supply arrangements.[71] As Geddes concluded in his private letter to Lloyd George on 15 September 1916:

> It is beyond argument that there is today no one who controls the continuous transit from this country to the front. There is no one who can tell you throughout where his weak places are, or coordinate the policy and resources, present and future, of the various means of transit. It is not possible for the C-in-C or QMG in France to do it; it is alone a big job for the best man you can find. If the C-in-C is not satisfied with his transport arrangements and desires someone to go into them in anticipation of the spring, he must, I think, appoint a man for the job, put him in charge of it, and back him strongly.[72]

The BEF's existing organizational structure was incapable of producing, analysing and interpreting the data streams required to maintain the efficiency – and increase the capacity – of the transport network upon which a vast force was dependent. As far as Geddes was concerned, the mass of special reports and memoranda originally requested by Lloyd George were no longer the priority were the BEF to be capable of successful offensive

[71] Grieves, 'The transportation mission', p. 65.
[72] UWMRC, Granet papers, MSS. 191/3/3/4, Geddes to Lloyd George, 15 Sept. 1916, p. 9.

operations in France in 1917. The secretary of state for war agreed. Crucially, so did Haig.

The common ground between the army's political head and its senior field commander became a platform both for the restructuring of the BEF's administrative organization and for the appointment of some of Britain's leading transport experts into positions of seniority within the military hierarchy. In London, Lloyd George acted quickly upon Geddes's plea for executive action. On 18 September, just three days after the latter had penned his 'preliminary opinion' on matters in France, the secretary of state for war established the directorate-general of military railways at the War Office. The new directorate was initially created 'with a view to improving transport facilities at present existing in this country and France', and the position of director-general conceived to act as a deputy to the QMG of the forces. However, the constitution of the new directorate also explicitly stated that the DGMR was to have 'direct access' to the secretary of state for war and would attend meetings of the Army Council at which matters of military transportation were under discussion.[73]

Reasons of both practicality and personality governed this decision. In terms of the former, access to the Army Council and the secretary of state permitted the DGMR to attend conferences with policy makers and argue the case for resources to be made available for military transport rather than other components of the war effort.[74] Raw materials such as steel were vital to the production of military necessities as diverse as helmets and tanks, and also to the construction of the locomotives, ships and railway tracks necessary to transport supplies and maintain Britain's connection with the world's markets. As the DGMR was to 'assume responsibility for the purchase of material for the construction, equipment, maintenance, repair and working of railways, light railways, canals, docks and roads',[75] his success in the role was dependent upon his ability to acquire sufficient money, materials and manpower to fulfil these duties. The success of the wider allied war effort depended upon the successful balance of the many competing demands on the limited pool of resources available.

The personality considerations that influenced the directorate-general of military railways' position within the military hierarchy centred upon the relationship between Lloyd George's choice for DGMR and the incumbent officers in the QMG's department. On 18 September the secretary of state for war offered Geddes the role of DGMR in a letter that made clear that

---

[73] UWMRC, Granet papers, MSS. 191/3/3/62, Attachment A, Brade to Geddes, 18 Sept. 1916.

[74] Grieves, 'The transportation mission', p. 67.

[75] UWMRC, Granet papers, MSS. 191/3/3/62, Attachment A.

the new directorate would take over responsibility for the directorate of movements from the QMG.[76] Therefore, Geddes was scheduled to become Stuart-Wortley's superior. The latter's thoughts on the matter were gathered by Lord Derby in what was 'rather a painful interview':

> He feels the position very strongly and I have great sympathy with him, for he has done his work well so far as it goes ... I explained to him that everything had been harmoniously settled in France and that the corollary for the new appointment that is to be made there was the appointment of somebody with a corresponding post in this Office, a man who would take in several departments, and at the same time would be on such terms of equality with the corresponding holder of the post in France as would enable him, with the help of Haig, to insist on his views being carried out.

> Wortley took it very well from a personal POV. He told me that under no circumstances could he work under Geddes and that he should immediately resign.[77]

Lloyd George was prepared for the DOM's threat, which was backed up by expressions of opposition to Geddes's appointment from two members of the Army Council. The secretary of state's response was to issue Geddes with the honorary rank of major-general, a manoeuvre that both reinforced the directorate's status within the military hierarchy and solidified Geddes's authority in the army's command structure; below his nominal superior, Lieutenant-General Sir John Cowans, but above Brigadier-General Stuart-Wortley. The latter reiterated to Lloyd George in a personal meeting that he could not work under Geddes, and a compromise was fashioned. Following Geddes's acceptance, with 'some misgiving', of the post of DGMR on 21 September, the civilian took over the railway supply and IWT branches while Stuart-Wortley and the rest of his staff remained under the QMG's direct command.[78] The separation of his command, Stuart-Wortley admitted privately to Henry Wilson, was a necessity – his 'show had really got too big' by September 1916.[79]

Geddes's expertise was not just in demand in London following the transportation mission. One day after the civilian had accepted the post of DGMR, Haig informed Lloyd George that he wanted Geddes to head a directorate-general of transportation in France – created to manage the BEF's supply lines on the western front. Upon accepting the post of

[76] UWMRC, Granet papers, MSS. 191/3/3/62, Attachment A.
[77] PA, Lloyd George papers, LG/E/1/1/6, Derby to Lloyd George, 15 Sept. 1916.
[78] UWMRC, Granet papers, MSS. 191/3/3/51, memorandum to Granet, p. 7; Grieves, 'The transportation mission', p. 67.
[79] IWM, Wilson papers, HHW 2/84/34, Stuart-Wortley to Wilson, 7 Oct. 1916.

DGT, Geddes became responsible for the provision and maintenance of the logistics network that sustained and equipped the largest military force Britain had ever assembled in the field alongside the acquisition and supply of all the resources required by the transportation services behind all the nation's globally dispersed expeditionary forces.

The establishment of the two directorates, and the appointment of Geddes at their heads, resulted in a remarkable concentration of power over the army's future in the hands of a civilian. His possession of a senior military rank was not enough to silence the 'whispering staff officers' who perceived Geddes's appointments to be 'evidence of the threat which Lloyd George posed to the autonomy of [the] military high command'.[80] Cowans remarked to Haig after dinner on 14 October that the secretary of state for war had 'imported an element of distrust into the W[ar] O[ffice] so that one wants "eyes in the back of one's head" in London', and stated his belief that Lloyd George sought to place 'civilians into the military machine wherever he possibly can to replace soldiers'.[81] The abolition of the post of IGC and Sir Frederick Clayton's subsequent departure from France did nothing to alleviate similar fears among the soldiers on the western front, and his removal was later portrayed as a consequence of his opposition to Lloyd George's desire to employ Chinese and African labour behind the lines.[82] The truth was far more prosaic. In late September Haig had noted that Clayton was 'anxious to retain control of the ports' in the face of the impending restructure at GHQ.[83] However, within a fortnight he had accepted that there was 'not room for an IGC and QMG in France, and that the proposed amendments with the introduction of Geddes as DGT would work well, and that he would do all he could to assist with them prior to his return home. He was tired, his health was failing and he wanted/needed to go home'.[84] The strain of active service had caught up with the sixty-one-year-old Clayton, and he returned to Britain to receive the recognition he felt he had earned earlier in the war in the 1917 New Year's honours.

Haig, while acknowledging the unique nature of Geddes's position, championed the civilian's role from the outset. He recognized how important the application of expertise was to the solution of the complex problems the expanding war had generated, and met frequently with the railwayman to discuss transportation matters. Haig believed explicitly in the promotion

[80] Grieves, *Sir Eric Geddes*, p. 31.
[81] NLS, Haig papers, Acc. 3155/108, diary entry, 14 Oct. 1916.
[82] P. Fraser, *Lord Esher: a Political Biography* (London, 1973), pp. 332–3.
[83] NLS, Haig papers, Acc. 3155/108, diary entry, 29 Sept. 1916.
[84] NLS, Haig papers, Acc. 3155/108, diary entry, 10 Oct. 1916.

of the best man for the job in the BEF, regardless of their background or previous military experience:

> There is a good deal of criticism apparently being made at the appointment of a civilian like Geddes to an important post on the Headquarters of an Army in the Field. These critics seem to fail to realize the size of the Army, and the amount of work which the Army requires of a civilian nature. The working of the railways, the upkeep of the roads, even the baking of bread and 1,000 other industries go on in peace as well as in war. So with the whole nation at war, our object should be to *employ men on the same work in war as they are accustomed to do in peace.*[85]

In the context of an industrialized war that demanded the mobilization and coordination of the British empire's human and material resources, Haig and Lloyd George both understood that the inefficient use of those resources to placate the sensibilities of the 'military trade union' was incompatible with the goal of securing victory over a determined and organized enemy. The employment of a 'civilian who was unafraid of large-scale planning and had access to the necessary resources' was far more logical than the continued use of soldiers who were handed transportation work 'merely because they [were] generals and colonels'.[86] The commander-in-chief's enlightened attitude, coupled with the political backing of the secretary of state for war in London, provided Geddes with the support he required to establish functioning directorates on both sides of the Channel.

However, the powerful support of the British war effort's most prominent figures does not entirely explain the scale of Geddes's achievement. The directorate-general of transportation's organizational chart provided fourteen departmental heads – each with their own hierarchical management structure and units dispersed throughout the BEF's rear areas – with direct access to the DGT. To coordinate the various forms of transport under his control, to balance the conflicting priorities and competing demands of these groups, and to direct them towards the realization of Haig's strategic goals called for a man of exceptional organizing capacity. Geddes combined the fulfilment of these responsibilities in France with equally monumental duties in London, where he commanded the department charged with providing sufficient personnel and equipment to satisfy the seemingly insatiable demands of Britain's global war. The tasks that confronted the DGT and DGMR in October 1916 were Herculean. Geddes was a

---

[85] NLS, Haig papers, Acc. 3155/108, diary entry, 27 Oct. 1916. Emphasis in original.
[86] Grieves, *Sir Eric Geddes*, p. 32; NLS, Haig papers, Acc. 3155/108, diary entry, 27 Oct. 1916.

'Hercules'.[87] But he did not undertake his labours alone. In the aftermath of the battle of the Somme, Geddes was able to draw the skills of Britain's transport experts even more deeply into the prosecution of the First World War.

The population of the transport directorates

Geddes's key appointments to the directorates in London and France demonstrate his appreciation of the need for civilian and military elements to exist in close harmony. Working relationships on both sides of the Channel heavily influenced his decision to install Ralph Wedgwood as director of docks. The responsibility for ships despatched from Britain remained with the DOM until the vessels arrived at their destination port. Therefore, the docks directorate on the western front received goods straight from the care of Stuart-Wortley's department. The two directors had to work collaboratively to agree the composition of traffic to be moved across the Channel each month and to ensure the maintenance and improvement of communications either side of the sea.[88] In October 1916 Geddes reported that telegraphic advices received in France contained incomplete details as to the cargo on board each ship, did not state the departure time of the vessel and often did not arrive at the relevant dock before the ship had crossed the Channel – a problem encountered by Francis Dent eighteen months earlier at the Bassin Loubet.[89] To rectify such inefficiencies it was imperative that the working relationship between the principal officers involved in the operations of the docks was not hampered by the personal animosity that had characterized Geddes's interactions with Stuart-Wortley.

In Wedgwood, Geddes identified a civilian with whom Stuart-Wortley had 'always got on well' and a man with the necessary professional experience to take on the job.[90] He was the first graduate of the North-Eastern Railway's traffic apprenticeship scheme and had succeeded Geddes in the role of chief goods manager in 1912. He had been released for service with the railway transport establishment soon after the outbreak of war but,[91] like Brigadier-General Philip Nash and Lieutenant-Colonel J. George Beharrell – who were issued with honorary military ranks and appointed deputy DGT and assistant DGT (statistics) respectively – Brigadier-General

[87] Pritchard, *History of the Royal Engineers*, v. 617–18.

[88] S. Fay, *The War Office at War* (London, 1937), p. 23; 'Directorate of inland waterways and docks', *Royal Engineers Journal*, xxix (1919), 338–64, at p. 354.

[89] UWMRC, Granet papers, MSS. 191/3/3/102, memorandum by Geddes, p. 10.

[90] UWMRC, Granet papers, MSS. 191/3/3/51, memorandum to Granet, pp. 7–8.

[91] R. Bell, *Twenty-Five Years of the North Eastern Railway, 1898–1922* (London, 1951), p. 40. Wedgwood was recalled to Britain for service in the Ministry of Munitions in 1915.

Wedgwood was yet another railwayman with scant military experience parachuted into a senior military position in the autumn of 1916. The trend led Lord Northcliffe to observe cynically that 'we have brought to France a considerable portion of industrial England'.[92]

Northcliffe was not the only observer to be perturbed by the outflow of prominent railwaymen to France. The departure of Nash, Beharrell and Geddes from the Ministry of Munitions was felt keenly, if somewhat melodramatically, by Lloyd George's successor as minister. As Edwin Montagu wrote to Lloyd George on 11 October 1916:

> To meet your wishes, and with tears in my eyes, tears which have been flowing ever since, Geddes left the Ministry … When Geddes left this Ministry he took with him Nash and Beharrell, and since then I can hardly bear to look at War Office correspondence, for almost every day, if you will excuse a slight exaggeration, I receive a request for the service of some new man to be sent somewhere or other, sometimes China, sometimes France. By a curious coincidence they are nearly always NER men, and it looks as though we shall be left without a railway man anywhere about.[93]

Just two days later, and 'despite the fact' he found it 'very difficult to spare him', Montagu agreed to Wedgwood's release.[94]

The 'curious coincidence' to which Montagu referred was a consequence of the North-Eastern's progressive approach to management before the First World War. Geddes, Beharrell and Wedgwood were all graduates of the company's managerial apprenticeship programme, and had proven themselves adaptable to the diverse challenges of wartime administration. The directorate-general of transportation's establishment permitted these senior executives to refocus their energies from munitions production to a more recognizable challenge: the provision of a reliable and efficient transportation system. However, the North-Eastern was far from the only British railway company to contribute personnel to the new military transport hierarchies created in France and London.

The discussions that surrounded the appointment of Sir Guy Granet as Geddes's deputy in London illustrate the delicate balance between civilian and military demands that the British war effort could not irrevocably upset. As Geddes anticipated that the scale of the task in France was likely to occupy most of his time, he sought out a highly qualified man to oversee affairs at the War Office. In Granet he doubtless saw many of his own qualities. Both had experience outside the railway industry – Granet having

---

[92] Quoted in Grieves, 'The transportation mission', p. 68.
[93] PA, Lloyd George papers, LG/E/2/19/8, Montagu to Lloyd George, 11 Oct. 1916.
[94] PA, Lloyd George papers, LG/E/2/19/9, Montagu to Lloyd George, 13 Oct. 1916.

initially trained as a barrister – and both had experienced a rapid rise to positions of seniority. Just six years after entering the railway industry Granet took over as general manager of the Midland Railway, an 'undertaking rather living on its past reputation'.[95] The company had become known for the 'easy-going regard to the virtue of punctuality' embraced by its 66,000 employees.[96] However, assisted by a good team of senior managers – which from 1907 included the future ROD commanding officer, Cecil Paget, as chief operating officer – Granet rapidly cultivated a systemic change in the Midland's organizational culture that was officially acknowledged with the award of a knighthood in 1911. His readiness to employ new methods, such as the train control system discussed below, and his high standing within the profession doubtless encouraged Geddes to choose Granet as his deputy in October 1916. Yet Granet was also already well known within the War Office thanks to his membership of the ERSC and REC, and Sir John Cowans offered his 'hearty approval' to Granet's appointment.[97] Even Stuart-Wortley found the Midland's general manager to be a 'nice fellow',[98] a further demonstration that his animosity towards Geddes was fuelled by personal dislike more than professional jealousy.

The soldiers' enthusiasm for Granet's appointment was not shared by his employers. As the correspondence between Lloyd George and the Midland's chairman demonstrates, Geddes's request for Granet's services created substantial difficulties for a company that had already endured serious privations thanks to the war's incessant demands. In response to Lloyd George's appeal for Granet to be released, George Murray Smith wrote:

> I cannot refrain from telling you that the Directors were very reluctant to release Sir Guy Granet from his duties. Apart from the difficulties we are experiencing from the absence of so many of our chief and subordinate officers, who are either serving in the Munitions Department, or who are fighting, the Assistant General Manager is only just recovering from a serious breakdown caused by overwork during Sir Guy Granet's absence at the Import Restrictions Department under the Board of Trade.[99]

---

[95] UWMRC, Granet papers, MSS. 191/10/1/40, 'A maker of railway history', *Railway Gazette*, 22 Oct. 1943 (press cutting).

[96] C. Hamilton Ellis, *The Midland Railway* (London, 1953), p. 144.

[97] UWMRC, Granet papers, MSS. 191/3/4/2, Geddes to Cowans, 20 Oct. 1916.

[98] IWM, Wilson papers, HHW 2/84/68, Stuart-Wortley to Wilson, 25 Oct. 1916.

[99] UWMRC, Granet papers, MSS. 191/3/4/9, Murray Smith to Lloyd George, 19 Oct. 1916. The Midland had placed Granet at the Board of Trade's disposal in March 1916.

Lloyd George's appreciation of the company's 'patriotic action' and his 'regret' at the Midland's 'inconvenience' did nothing to ameliorate the pressures under which the railway operated during the war.[100]

The Midland, like many of its colleagues, found it difficult to replace highly skilled officials such as Granet easily during the war. The absence of his experience and ability while on governmental duties created further discomfort for the railway servants who remained in post as the demands on the British network grew in line with the expanding war effort. In the heaviest year of the war, the Midland carried eighteen million more passengers and 3,220,000 tons more goods than it had in 1913, despite having lost 29 per cent of its male staff to the armed forces.[101] Under such testing conditions, the Midland's decision to permit Granet's release for service in the directorate-general of military railways underlines the continued existence of the cooperative spirit fostered between the railways, government and armed forces prior to the outbreak of the war.

The accommodating responses to governmental requests from railway companies such as the Midland and the North-Eastern were not matched by all of Britain's transport enterprises. For the Port of London, domestic requirements prevailed over the demands of the western front. Geddes sought to employ a man with 'practical knowledge', particularly of the mechanical engineering aspects of dock work, to act as a deputy to Wedgwood in France and help improve the Channel ports' throughput rates.[102] He considered Cyril Kirkpatrick, the Port of London's chief engineer – and a future president of ICE – to be the perfect candidate as he was 'a very strong man and a pusher'. Kirkpatrick and Wedgwood knew one another from the former's tenure as city engineer and town surveyor in Newcastle upon Tyne before the war, and Geddes believed Kirkpatrick to be 'quite glad' of the opportunity to go to France. However, the Port of London refused to release him, as he was engaged on the construction of what became the King George V Dock on the Thames. Undeterred, Geddes wrote to Lloyd George that 'if the ports over here are to be worked satisfactorily it is essential that we should have not the third or fourth class men from the British ports but the best'. The port continued to resist, even after the secretary of state for war despatched a letter to its chairman that stressed the 'national importance' of Kirkpatrick's proposed role 'to help forward to a satisfactory solution the vital question of transportation in

[100] UWMRC, Granet papers, MSS. 191/3/4/7, Lloyd George to Murray Smith, 20 Oct. 1916.

[101] E. A. Pratt, *British Railways and the Great War: Organisation, Efforts, Difficulties and Achievements* (2 vols., London, 1921), ii. 1048–50.

[102] PA, Lloyd George papers, LG/E/1/5/16, Geddes to Lloyd George, 19 Nov. 1916.

France'.[103] Clearly then, the later assertions of the official historian – that Geddes received everything he desired following his appointment – were misguided. The ongoing requirements of the domestic transport industry limited Geddes's access to the best men that civilian enterprise could provide.

From the winter of 1916–17 onwards the higher organization of the BEF's transport requirements was administered by a synthesis of civilian and military expertise. The influx of civilians into military roles did not lead to the wholesale replacement of soldiers. Where the incumbent – whether a general, colonel or otherwise – had proven themselves capable of discharging their duties, Geddes understood the benefits of retaining their services. Following Mance's performance on the transportation mission he returned to the War Office, was rewarded with a promotion to brigadier-general, and handed responsibility for sourcing the material and personnel required for the army's enlarged road, railway and light railway departments. Colonel Collard, who had acquired responsibility for the provision of men and material for the army's IWT services in January 1915, also kept his job.[104]

The explanation given to Granet for the two soldiers' retention demonstrates Geddes's appreciation of the crucial role to be played by military experience within the directorate-general of military railways. 'Our chief difficulty', he wrote, 'will be to get things "through" the War Office'. Mance and Collard understood the army's bureaucracy and were 'very wise' to the 'minor tricks of the trade' that had to be deployed in aid of the directorate's goals. According to Geddes, once a paper reached the War Office it passed beyond 'the wit of man to get it out again'. It was 'only by knowing the ropes and knowing where the snags' were, 'and how either to get round them or knock them out of the way', that Geddes believed anyone could 'get anything done at all'.[105] In addition, in recognition of the multitude of concerns with which both he and Granet were likely to be bombarded as the directorate evolved, Geddes stressed that Mance and Collard were able to run their own departments and work confidently without the need for close supervision from above.[106] Mance remained in post for the rest of the war and ultimately became a highly respected author on international transportation matters, while Collard was taken to the Admiralty by Geddes in May 1917 and appointed as deputy controller of auxiliary shipbuilding. Geddes was not alone in admiring Collard's talents. Sir Sam Fay recalled in his memoirs that Collard was 'an extraordinary

[103] PA, Lloyd George papers, LG/E/1/5/18(B), Lloyd George to Devonport, 27 Nov. 1916.
[104] UWMRC, Granet papers, MSS. 191/3/3/51, memorandum to Granet, pp. 1–3.
[105] UWMRC, Granet papers, MSS. 191/3/3/49, Geddes to Granet, 19 Oct. 1916.
[106] UWMRC, Granet papers, MSS. 191/3/3/51, memorandum to Granet, p. 5.

man, full of energy, very able, and prepared to take on anything from the construction of a battleship to the manufacture of a watch'.[107]

Geddes was also keen to retain Stuart-Wortley, at least in the short-term. Regardless of the obvious disdain the DOM had shown towards him, three factors combined to persuade Geddes not to immediately seek the replacement of Stuart-Wortley with a more compliant personality. First, over the summer of 1916 comments that questioned the veracity of placing civilians in key positions of authority began to appear in the pages of the Northcliffe press. Lord Northcliffe and Lloyd George had 'mysteriously drifted apart' earlier in the year,[108] and the former argued in articles for *The Times* that civilians should 'leave it to the service chiefs to decide strategy and the soldiers to die in battle'. As Lloyd George was facilitating Geddes's mission to France in August, Northcliffe warned that 'we must make changes [to the command structure of the army] with caution'.[109] On a related note, in order to ensure the smoothness of operations while the directorate-general of military railways was bedded in, Geddes was keenly aware of the need to maintain the good will of the professional soldiers in the War Office – many of whom were longstanding colleagues of Stuart-Wortley's. The king was 'glad to hear' that Stuart-Wortley remained as DOM in early October 1916, 'and that he and Sir Eric Geddes [were] working in complete harmony' despite their personal animosity.[110] At a more practical level, the backing of Sir John Cowans – Stuart-Wortley's most fervent supporter – was critical to the project's overall success. As Fay discovered when he eventually took over as DOM in early 1917, the removal of Cowans's friend elicited an emotional response from the QMG:

> When I saw General Cowans … he was angry and called me a damn fool. He said I could not carry on the job, that it was a military post, that the tentacles of the Director of Movements were all over the War Office and could not be moved from the building, although they were overcrowded … He reminded me that he had held the position ten years before Stuart-Wortley, and knew something about it.[111]

Cowans's outburst was highly uncharitable towards one of the pre-war British railway industry's most respected figures. But it also demonstrated the second reason why Geddes was loath to dispense with Stuart-Wortley's

---

[107] Fay, *The War Office at War*, p. 167.

[108] J. M. McEwen, 'Northcliffe and Lloyd George at war, 1914–1918', *Hist. Jour.*, xxiv (1981), 651–72, at p. 657.

[109] 'The army behind the army', *The Times*, 7 Aug. 1916, p. 7.

[110] PA, Lloyd George papers, LG/E/2/16/3, Stamfordham to Lloyd George, 5 Oct. 1916.

[111] Fay, *The War Office at War*, p. 26.

services straight away. Put simply, the latter's experience and understanding of his role made him temporarily indispensable. The process of replacing him threatened the directorate's efficiency, and had to be handled carefully. As Fay himself acknowledged after shadowing Stuart-Wortley for a week before he took over, nobody could have 'run the show' as well as the outgoing DOM had to that point.[112]

Finally, Geddes was more interested in the creation of efficient, functional departments than the settling of any personal scores. Stuart-Wortley was not removed because of the animosity between him and Geddes, but as a result of a decision made by Lloyd George's replacement as secretary of state for war. Had the organizational fudge created to accommodate Stuart-Wortley's desire not to serve under Geddes worked then it would doubtless have remained in place for the remainder of the war. However, Lord Derby was convinced that the pseudo-subordination had not been a success. He announced to Haig that Geddes recognized how Stuart-Wortley 'had played the game ... and nobody could have behaved better', but the separation of the directorate of movements had 'prevented things going smoothly'.[113] Stuart-Wortley's sustained refusal to serve under Geddes saw him replaced as DOM by Sir Sam Fay in early January 1917. Following an unsuccessful stint on the western front, where he served briefly in command of a brigade and even more briefly as a divisional commander, Stuart-Wortley ended the war as deputy QMG in Mesopotamia.

Alongside friction in London, Geddes's appointment as DGT and the corresponding restructure of the BEF's organization generated passionate opposition in France. Sir Ronald Maxwell, Haig's QMG, failed to reach agreement on the relationship between his department and Geddes's new directorate, and threatened to resign if the new organization was 'forced' on him.[114] On 30 October 1916 Haig made a personal intervention in an attempt both to assuage Maxwell's fears that Geddes had been sent to France by Lloyd George to replace him and to establish a 'workable scheme ... suitable to the personalities who had to work it'. The commander-in-chief held a conference with his senior staff officers, which included Maxwell and Geddes at separate times, 'in order to try and ascertain what [the QMG's] objections to the scheme really were'. To Haig it was soon evident 'that there was [sic] no solid grounds for disagreement' between the two men, and the chief of the general staff sketched out the boundaries between the QMG's and DGT's responsibilities that evening. Haig's support for

---

[112] Fay, *The War Office at War*, pp. 26–8.
[113] NLS, Haig papers, Acc. 3155/109, Derby to Haig, 27 Dec. 1916.
[114] NLS, Haig papers, Acc. 3155/108, diary entry, 30 Oct. 1916. Unless otherwise stated, all quotations in this passage are taken from this source.

Geddes's position, alongside his affirmation that the civilian had not been imposed upon the BEF by Lloyd George, were sufficient to convince Maxwell to withdraw his resignation. In addition, he 'said that he would tell his Directors to stop their criticism' of Geddes's appointment as DGT. That such an action was required illustrates the depth of military hostility towards the British war effort's new direction in the winter of 1916–17.

However, the existence of military apprehension with regards to the structural and personnel changes within the BEF and the War Office should not be taken as evidence of a concerted attempt to assert civilian dominance over the army. Geddes, with the full support of both Haig and Lloyd George, sought to merge the talents of Britain's transport experts with the bespoke knowledge acquired by soldiers through two years' practical experience of industrial war. His correspondence on the subject of Colonel M. C. Rowland, whose name was forwarded to Geddes for consideration in December 1916, illustrates the qualities he demanded from candidates for employment in the directorate-general of transportation. Upon the document outlining Rowland's skills and aptitudes, Geddes underlined the following: control of mechanical transport, rail and sea transport; record work; and recruiting.[115] Where professional soldiers had demonstrated their possession of such qualities they were retained. Where they fell short of the competencies necessary to discharge their duties – as in the case of the DRT, Brigadier-General John Twiss – they were swiftly removed. In November 1916 Geddes complained to Haig that Twiss had failed to pursue orders that had been placed for the railway equipment necessary to prepare the BEF for its intended offensive operations in early 1917. Furthermore, the DRT had relied upon 'one of Geddes's men' to identify the correct estimates for the force's requirements in terms of locomotives and miles of track. Twiss, Haig recorded, ought to either have argued against Geddes's projections or resigned. He had done neither, nor demonstrated sufficient mastery of the details of his brief to retain the confidence of his superiors.[116]

With two entirely new departments to populate and the majority of the army's most skilled administrators already employed either at home or abroad, Britain's transport companies were the most logical source of talent for Geddes to exploit in 1916–17. Suitably skilled civilians were identified, appointed and applied to the challenges of wartime transportation at both senior executive and junior management levels following his appointments as DGT and DGMR. Approximately one half of the technical officers

---

[115] TNA, ADM 116/1805, Sir Eric Geddes – private correspondence, Colonel M. C. Rowland: QMG: Union Defence Forces. statement of colonial service, 24 Dec. 1916.

[116] NLS, Haig papers, Acc. 3155/109, diary entry, 9 Nov. 1916.

under the DGT's control 'were furnished by the British railway companies or on recommendation of the Railway Executive Committee, and the other half were men from overseas employed on Colonial or foreign railways who offered their services'.[117] Brigadier-General Geoffry Harrisson, who oversaw light railway operations from February 1918 onwards, exemplified the latter. Harrisson served with the Royal Engineers in South Africa between 1901 and 1902, but had abandoned a civil engineering post with the LNWR to work in Argentina before the war. From 1907 onwards he had worked for the Argentine North-East Railway at Concordia, and the outbreak of the conflict occurred when he was building a railway in Brazil. Harrisson's experience of railway construction and military discipline made him an obvious candidate for service within the upper echelons of the transport directorate.[118]

Yet expertise was required throughout the organization, not just at managerial levels of the command hierarchy. The large-scale transport challenges that confronted the BEF from 1917 onwards necessitated the enlistment of huge numbers of men with the practical skills to undertake and supervise varied construction and operation duties effectively. Where Geddes lacked personal familiarity with the requirements of a role, he followed the template provided by the Ministry of Munitions earlier in the war and employed men who possessed the requisite skills and contacts. When Henry Maybury, the chief engineer of the road board, was appointed director of roads in France he was provided with a free hand to recruit suitable officers for the technical work of road construction. Around 2,600 men were selected from lists that comprised both serving officers and civilians, while a further 400 men were offered temporary commissions as officers in the Labour Corps. Maybury used his peacetime position to convene conferences of the chief officials of the local road authorities immediately after his appointment, and raised a number of complete companies of 250 men drawn from the same local area – a lines-of-communications-equivalent of the front-line 'Pals battalions'.[119] Already by 1 December 1916 Haig could record in his diary that 1,200 over-age men had been made available for road-related duties, and 1,800 'expert road men' were in the process of being enlisted.[120]

---

[117] 'Organisation and work of the transportation directorate', *Railway Gazette: Special War Transportation Number*, 21 Sept. 1920, pp. 14–20, at p. 18.

[118] S. Damus, *Who was Who in Argentine Railways, 1860–1960* (Ottawa, ON, 2008), pp. 236–7 <http://www.diaagency.ca/railways/WWW_sample.pdf> [accessed 13 Nov. 2014].

[119] 'Organisation and work', pp. 18–19; *The Work of the Royal Engineers in the European War, 1914–19: the Organization and Expansion of the Corps, 1914–18* (Uckfield, 2006), p. 31.

[120] NLS, Haig papers, Acc. 3155/109, diary entry, 1 Dec. 1916.

The commander-in-chief's diary entry that day also emphasized the scale of the railway recruitment process that took place after the creation of the directorate-general of transportation. Geddes had secured 12,000 railwaymen from the British railways 'to improve the BEF's capacity on the mainline railways' ahead of the 1917 campaigning season.[121] The wartime career of Company Sergeant Major L. W. Conibear provides just one example of the skills sets that ordinary railway servants contributed to the BEF's transport services in the second half of the war. An employee of the Great Western Railway at Bristol, Conibear joined the ROD in January 1917 and was in France by 4 February. Over the next five months he undertook a range of duties on board trains, including those of brakesman, guardsman and signalman, and was employed on clerical tasks such as organizing traffic and maintaining the orderly room. By July 1917, just six months after he had enlisted in the army, Conibear took over responsibility for all the administrative work in the Fifth Army's light railways department. He dealt with

> all personnel questions affecting eight Light Railway Operating Companies (over 2,000 men), leave, sickness, promotions, casualties, examinations and general routine. Traffic policy, new construction, signalling arrangements, pay, accounts … numerous telephonic and telegraphic enquiries in the absence of the Superintendent of the Line. [Collating] statistics appertaining to the general working of light railways as required by the Director of Light Railways.[122]

Conibear occupied this role until, following the dislocation and confusion caused by the German spring offensive in March 1918, his versatility proved invaluable to the BEF in the conflict's final months. After a period of 'considerable roaming' when the Fifth Army disintegrated, Conibear was placed in charge of sixty men attached to the Canadian Railway Troops and tasked with broad gauge reconstruction. He was responsible for the building of lines until the Fifth Army was reconstituted at the end of June, when he took on the job of central traffic controller. For the remainder of the war Conibear oversaw the 'movement of all power, wagons and traffic' in the Fifth Army, under the direction of the superintendent of the line.[123]

Conclusion

'Warfare', Haig wrote to Geddes in September 1916, 'consists of men, munitions and movement. We have got the men and the munitions,

---

[121] NLS, Haig papers, Acc. 3155/109, diary entry, 1 Dec. 1916p
[122] BLSC, Liddle collection, papers of Major L. W. Conibear, LIDDLE/WW1/GS0346, particulars of service with the colours, 23 July 1917.
[123] BLSC, Conibear papers, LIDDLE/WW1/GS0346, particulars of service.

but we seem to have forgotten the movement'.[124] The BEF had increased tenfold from the small, professional force that had left Southampton in the summer of 1914, while the firepower amassed behind the front line dwarfed that collected together in any previous British conflict. The quantity of shells, supplies and myriad stores consumed by the BEF in 1916 were insufficient to overcome the German resistance on the Somme. However, they were sufficient to illuminate the profound weaknesses in the transport infrastructure upon which any allied advance depended. As one corps' chief engineer (the future DGT, Major-General Sir Sydney Crookshank) admitted after the war, 'on the Somme the British Army was practically immobile'.[125] Had Haig's much desired break through occurred, the BEF was in no position to take advantage for much of the battle.

The army's largest military engagement in history to that point emphasized the need for a holistic examination of the BEF's road, rail and waterborne resources. Sir Eric Geddes undertook that investigation, and the character of the British war effort for the remainder of the war was shaped by his response to the unfolding crisis in Picardy. He argued for the centralization of transport policy on the western front, and accepted responsibility for the creation, population and direction of entirely new organizations in France and London. Upon his appointment as DGT and DGMR, the civilian railway expert obtained 'a position of most unusual authority and power'. The concentration of such remarkable control in the hands of a non-military actor engendered jealousy from professional soldiers on both sides of the Channel, emotions that coloured post-war interpretations of what took place after Geddes's transportation mission had been completed. As the Royal Engineers' history of the war observed, 'no QMG or Brigadier-General of Railways in France would ever have been allowed the power and the resources' showered upon Geddes in the autumn of 1916.[126]

Auckland Geddes provided a retort on his brother's behalf. He asserted that 'until experts, with experience of the transport problems – both rail and road – of crowded industrial England, were on the spot in charge of supply movement, fully adequate provision for the fighting men had proved impossible'.[127] This conclusion misrepresents the civil–military dynamic within the transport directorates created by his brother in France and London. From the very outset the directorate-generals of military railways

[124] Haig to Geddes, quoted in W. J. K. Davies, *Light Railways of the First World War: a History of Tactical Rail Communications on the British Fronts, 1914–18* (Newton Abbot, 1967), p. 27.
[125] Crookshank, 'Transportation with the B.E.F.', p. 194.
[126] Pritchard, *History of the Royal Engineers*, v. 614.
[127] Geddes, *The Forging of a Family*, p. 238.

and transportation were conceived and staffed in a way that took advantage of both military and civilian transport expertise. Geddes's recognition of the importance of technical and administrative experience was as manifest in his retention of talented soldiers like Collard and Mance as it was in his appointment of civilian railway managers such as Granet and Fay to responsible positions in the War Office.

The successful integration of civilian and military elements owed much to Haig's unequivocal support for Geddes. The motivations behind the commander-in-chief's backing of the civilian were twofold. First, as Keith Grieves identified, Geddes's employment symbolized Haig's acknowledgement of the 'forgotten interrelationship of strategy and transport' on the western front.[128] Second, Geddes's appointment was viewed as an opportunity for the BEF's senior field commander to influence the higher direction of the war. As Haig recorded in his diary after a conference with Lloyd George on the subject of light railway materials:

> LG promised to help me to the utmost of his power. The total cost will be under three million pounds, not much in comparison with our other expenses. The difficulty of provision is due to the present lack of steel, and in obtaining the material by next March. It is interesting to note how I have been striving to get a L[ight] R[ailway] organization ever since January 1915 when the First Army was formed. But it requires a civilian railway expert … to come on the scene and make a report to convince our government and War Office that such an organization is necessary.[129]

Haig's comment carries more than a hint of post-war observations about the government's willingness to act upon the advice of businessmen rather than soldiers. It also demonstrates his recognition that a war for human and material resources took place *within* the British war effort as well as between the belligerents on either side of no man's land. And for the remainder of the conflict, a union of military and industrial experience provided the means and methods by which the British army employed those resources to prosecute warfare on a truly industrial scale.

[128] Grieves, 'The transportation mission', p. 67.
[129] NLS, Haig papers, Acc. 3155/108, diary entry, 12 Sept. 1916.

III. Armageddon

7. 'By similar methods as adopted by the English railway companies': materials and working practices on the western front, 1916–18

Prior to wielding the spending 'axe' that bore his name in the early 1920s, Sir Eric Geddes had gained a reputation for being an 'improvident spender' of public money.[1] His approach was contrasted to the policies hitherto pursued in the War Office by Brigadier-General Richard Montagu Stuart-Wortley just a week after the directorate-general of military railways had been established. In a letter to Henry Wilson, the DOM emphasized that the civilianization of the War Office had been accompanied by pledges of financial support that had never been extended to the soldiers. Departments that had previously been staffed by small groups of officers were 'largely increased' and the incumbent departmental heads promoted to higher grades to reflect their expanding responsibilities. 'The way they waste money', Stuart-Wortley observed, 'is awful'.[2] Drawing upon his experiences of GHQ when the directorate-general of transportation was created, Brigadier-General Sir James Edmonds took much the same line after the war. In the official history's volume on transportation, he recorded that Geddes employed a 'very large staff of civilian engineers and officials' at the directorate's headquarters. Edmonds's observation was followed immediately by an unattributed quotation, which claimed that '"It has been said that at the outset the D.G.T. employed double the staff really needed for the work to be done, but that he did so in order to obtain 30 per cent increased output", and in this he was successful'.[3] Even soldiers with whom Geddes had fostered a strong working relationship before and during the conflict, such as Henry Mance, acknowledged that the civilians had operated with a liberty that had not been extended to the professional soldier. In the discussion that followed a lecture delivered at the Royal United Services Institution in 1921, Mance described how Sir Henry

---

[1] P. K. Cline, 'Eric Geddes and the "experiment" with businessmen in government, 1915–22', in *Essays in Anti-Labour History*, ed. K. D. Brown (London, 1974), pp. 74–104, at pp. 80, 99.

[2] IWM, Wilson papers, HHW 2/84/68, Stuart-Wortley to Wilson, 25 Oct. 1916.

[3] Edmonds's introduction in A. M. Henniker, *History of the Great War: Transportation on the Western Front, 1914–1918* (London, 1937), p. xiv.

Maybury had 'ransacked England and [taken] away all the skilled men and rollers and everything else connected with the roads and quarries that he could lay his hands on' upon becoming the BEF's director of roads.[4]

That a vast expansion in the size and capacity of the BEF's transportation services occurred in the second half of the war is beyond doubt. Prior to the establishment of the directorate-general of transportation, the existing railway and IWT units in France comprised 17,500 men of all ranks. Geddes's initial estimate of the personnel required to man his new light railways organization alone numbered 25,000, and by 1 January 1917 the DGT had his sights set on the deployment of 66,000 men (including those already in France) on transport duties behind the western front. Fresh proposals, which increased the paper strength of the directorate-general by a further 42 per cent, had been submitted to the War Office for sanction within four months.[5]

The composition of the human and material resources 'ransacked' from Britain and the empire, and the methods by which they were applied to the challenges of industrial warfare, are the subjects of this chapter. The acquisition of the resources necessary to increase the BEF's transport capacity from 1917 onwards owed much to Geddes's dual position, his contacts within the British government and the railway industry, and an acknowledgement among French and British leaders of the inapplicability of the pre-war arrangements made between the coalition partners. The transportation crisis of 1916–17 provided the catalyst for manpower and materials provided by Britain to become a far more integral component of the infrastructure and services operated in France than had been the case earlier in the war. The effective use of those resources drew upon the methods and expertise latent within the operations of an industrial economy. The provision of men and materials in quantities far above what had been made available to the military previously may have bred a resentment among officers that pervaded the post-war analysis of Britain's war effort. However, it also laid the foundations for the material-intensive warfare that helped bring the conflict to a successful conclusion.

The provision of British resources on the western front

The directorate-general of transportation acquired responsibility for the coordination of all aspects of the BEF's transport infrastructure over a vast

---

[4] Quoted in M. G. Taylor, 'Land transportation in the late war', *Royal United Services Institution. Journal*, lxvi (1921), 699–722, at p. 715.

[5] Henniker, *History of the Great War*, p. 222. The directorate-general's ultimate strength of 94,000 men was not achieved before the armistice.

area. Geddes took over the roads, railways and waterways from the Channel ports up to a point – known as the DGT line – that was 'roughly defined as the rear of the area under fire of the enemy's medium artillery', and beyond which authority devolved upon the armies and corps occupying the space.[6] The impact of industrial warfare upon the extant communications in France was stark on both sides of the DGT line in 1916. The munitions that Britain's transport experts had helped to produce in time for the battle of the Somme created a 'destroyed zone' of some three to six miles, which had to be traversed by the troops responsible for supplying their colleagues in the attack.[7] As the offensive ground on into autumn, the destructive power of the artillery was augmented by the deteriorating weather to produce a quagmire on the roads either side of the front line. Already by 5 October 1916, Haig recorded that the road between Montauban-de-Picardie and Guillemont had been closed 'owing to its breaking up ... The rain of the last few days has been very hard on the roads'.[8] A month later the 'soft state of the roads' made it impossible for lorries to carry the 1,400-pound shells fired by the 15-inch howitzers, leading Haig to observe that the BEF was 'fighting under the same conditions as in October 1914, i.e., with rifle and machine-gun only, because bombs and mortar ammunition cannot be carried forward as the roads are so bad'.[9]

Matters were no better on the roads maintained by the French army during the battle. During the entire Somme campaign over two million men and 371,000 tons of goods passed along the Amiens–Proyart road through Villers-Bretonneux on their way to the front. The road was essential to the supply of the French Sixth and Tenth armies in the vicinity of the Somme, and on 30 September alone it carried 38,000 men and 3,700 tons of material eastbound – twice as much as had passed along the fabled *voie sacrée* that sustained the defenders of Verdun. As the Somme drew to an inconclusive end, the smaller roads that branched off the Amiens–Proyart road lay broken up by the constant pounding of the ceaseless traffic. As winter set in both the French and British lacked the manpower and resources necessary to maintain all but the major traffic arteries behind their armies.[10] By November 1916 the Reserve Army's chief engineer admitted to Geddes that 'with three successive wet days motor lorry traffic must be discontinued'.[11]

---

[6] *History of the Corps of Royal Engineers*, ed. H. L. Pritchard (11 vols., Chatham, 1952), v. 616.

[7] UWMRC, Granet papers, MSS. 191/3/3/4, Geddes to Lloyd George, 15 Sept. 1916, p. 5.

[8] NLS, Haig papers, Acc. 3155/108, diary entry, 5 Oct. 1916.

[9] NLS, Haig papers, Acc. 3155/109, diary entries, 7 and 8 Nov. 1916.

[10] W. Philpott, *Bloody Victory: the Sacrifice on the Somme* (London, 2009), pp. 389–90.

[11] UWMRC, Granet papers, MSS. 191/3/3/102, memorandum by Geddes, p. 21.

Table 7.1. Principal road plant available in France, 1916–17.

Item of equipment	Number in France		Percentage increase
	1916	1917	
Steam rollers	85	170	100.00
Steam wagons	11	395	3490.91
Petrol rollers	—	35	—
Sweeping machines	57	175	207.02
Petrol lorries	—	235	—
Tarring machines	—	54	—
Water carts	72	132	83.33
Dump carts	91	930	921.98
Mud tumbler carts	16	132	725.00

Source: 'Railways and roads on the western front', *Railway Gazette: Special War Transportation Number*, 21 Sept. 1920, p. 27.

Following his investigations in the summer, Geddes doubted the BEF's ability to successfully maintain the road network during active operations, regardless of the weather. He told Lloyd George that the labour detailed to road repairs was inefficiently handled and the equipment available to the engineers was 'not the most suitable' for the task.[12] Maybury, as director of roads, was charged with rectifying these deficiencies. As he 'ransacked' England of vast quantities of road plant following his appointment (see Table 7.1), he became responsible for the upkeep of 'all roads of any military importance on the lines of communication'.[13] Progress was swift. By the end of April 1917 Haig felt assured enough by Maybury's efforts to record in his diary that the '10,000 workmen, road engineers, quarry men' and modern equipment procured by the director of roads meant he 'need have no further anxiety as regards roads on the western front'.[14]

Maybury's accumulation of manpower and equipment was a response to the BEF's progressive increase in responsibilities for road maintenance as the war continued. As with the pre-war agreement regarding the provision of stevedores to unload British vessels at the ports, the Franco-British arrangements for road maintenance collapsed under the pressures of

[12] UWMRC, Granet papers, MSS. 191/3/3/4, Geddes to Lloyd George, 15 Sept. 1916, p. 5.
[13] 'Railways and roads on the western front', *Railway Gazette: Special War Transportation Number*, 21 Sept. 1920, pp. 21–9, at p. 26.
[14] NLS, Haig papers, Acc. 3155/112, diary entry, 28 Apr. 1917.

industrial warfare. Throughout 1917 Maybury's directorate was immersed in a vast programme of road-building and upkeep that saw it construct, reconstruct or resurface almost thirteen million square yards of 'cours' by the end of the year – across a network that grew to comprise 3,267 miles of French roads.[15] However, despite the use of 2,340,000 tons of road stone in 1917 alone, 'demands for additional roads continue[d] to be received' at GHQ over the winter of 1917–18 as the BEF sought to solidify its position ahead of an anticipated German offensive in the spring. In January 1918 Haig was forced to issue a memorandum to his armies, warning that 'the available road stone is barely sufficient to maintain existing roads, and the present output cannot be increased to any considerable extent'.[16] GHQ's solution to the challenges of the war's final winter will be discussed further in chapter nine, but it is striking at this point to highlight that – just as Geddes had noted in September 1916 – limited quantities of crucial resources continued to constrain the development of the BEF's transport infrastructure in early 1918.

Attempts to relieve the pressure on the road network by the use of other means of transport were at an immature stage when Geddes had undertaken his mission to France. The DGT believed that a combination of 'intelligent organization, labour saving devices, and light railway' were required to economize labour on the road network and increase the speed with which the supply services maintained links to the troops on the far side of the 'destroyed zone'. He wrote to Lloyd George that 'plant must be secured from this country and the organization [to operate light railways] has to be created. We have little enough time to do it if it is to be ready by the spring and I am of the opinion that the matter should be taken in hand promptly and efficiently with executive authority, without one day's delay'.[17] Geddes was not the first man to recognize the potential value of light railways to the BEF. A policy for the operation of the isolated and disconnected systems behind the front line was first advanced in December 1915 by the QMG, Sir Ronald Maxwell, and a month later Haig mused in his diary that light railways could be used to 'save the roads' from excess wear over the winter months.[18] As British units had taken over portions of the front line from French troops over the spring of 1916 the 'usefulness' of the light railway systems constructed by the French 'quickly became

[15] TNA, WO 107/296, report of British armies, p. 26.
[16] Henniker, *History of the Great War*, p. 344.
[17] UWMRC, Granet papers, MSS. 191/3/3/4, Geddes to Lloyd George, 15 Sept. 1916, pp. 5–6.
[18] TNA, WO 107/69, work of the QMG's branch, p. 20; NLS, Haig papers, Acc. 3155/104, diary entry, 4 Jan. 1916.

apparent'.[19] In August Haig decided to 'inaugurate a complete system of light railways, and combine the systems already in existence under definite policy and control'.[20] However, discussions with individual armies over the form that policy should take led nowhere before Geddes arrived on the western front as DGT.

A lack of strong central coordination of the discussions from GHQ – Haig placed the DRT, Brigadier-General John Twiss, in charge – and the fact that the backdrop of the Somme overshadowed the light railways question led to stagnation. For individual army commanders the consideration of light railways was yet another intrusion upon the day-to-day business of running their armies. A month after he had received Haig's instructions, Twiss was unable to report any progress on the development of a universal policy for the employment of light railways on the western front.[21] Time, as both Geddes and the deputy QMG recognized, was not on the BEF's side. It was 'necessary', Colonel Woodroffe asserted, 'to apply all our efforts to developing a 60-centimetre system at the greatest possible speed in order to ensure that as much of the front area as possible is served by this means before the winter sets in'.[22]

Thanks to his involvement in the procurement of materials and the establishment of the directorate that oversaw its operations, the light railway network that emerged on the western front has come to represent Geddes's work as DGT. After both hearing Haig's views on the subject and inspecting the French light railways, the civilian recommended 'the purchase of a considerable mileage of track, viz. 1,000 miles; some 800 steam locomotives, 200 electric tractors, and some 3,000 wagons'. He stressed that 'no further time should be lost' in the procurement of the materials, and the equipment had been ordered even before he accepted the post of DGT.[23] However, light railway locomotives and tractors had a long lead time – which had combined with the lack of priority afforded to the medium by the BEF to retard the development of a coherent, widespread light railway policy in France. The records of Robert Hudson (a light railway equipment supplier in Leeds) from October 1914 to May 1916 show that nine of the eighteen engines built by the firm during the first half of the war were despatched to Mesopotamia, while the War Office discovered that other British manufacturers were fulfilling orders for French artillery

[19] W. J. K. Davies, *Light Railways of the First World War: a History of Tactical Rail Communications on the British Fronts, 1914–18* (Newton Abbot, 1967), p. 25.

[20] TNA, WO 107/69, work of the QMG's branch, p. 20.

[21] UWMRC, Granet papers, MSS. 191/3/3/4, Geddes to Lloyd George, 15 Sept. 1916, p. 3.

[22] IWM, Woodroffe papers, 3/38/1/2, 60 cm railways, 9 Sept. 1916.

[23] UWMRC, Granet papers, MSS. 191/3/3/4, Geddes to Lloyd George, 15 Sept. 1916, p. 3.

Table 7.2. Light railway construction in France and Flanders, 1917–18.

Quarter-year period	Miles constructed, 1917	Miles constructed, 1918
January to March	135	214
April to June	364	202
July to September	328	297
October to December	195	73
Total	1,022	786

Source: *History of the Corps of Royal Engineers*, ed. H. L. Pritchard (11 vols., Chatham, 1952), v. 665.

railways. In September 1915 an indent for ten locomotives and 200 wagons had been sent home from GHQ, for equipment that was expected to form a useful reserve but not considered urgent – an order based on the indent was not placed until January 1916. Two months later the original order was supplemented by a request for fifty more locomotives, 1,200 wagons and fifty miles of track, and the War Office decreed that British army orders should take precedence over any other orders placed with British manufacturers. However, deliveries of the materials requested before Geddes arrived in France were not completed until June 1917.[24] Consequently, when Sapper W. J. Hill, in peacetime an employee of the LNWR, arrived at Marœuil with the 19th Light Railway Operating Company early in 1917 he and his comrades found 'no motive power of any description, and only a few bogie wagons of French design'.[25]

Units like Hill's, formed to oversee the operation and maintenance of the network, witnessed a remarkable growth in motive power, carrying capacity and track mileage over the course of the year. Geddes's decision to order 1,000 miles of track and abundant quantities of locomotives, tractors and rolling stock in September 1916 – coupled with his ability to ensure sufficient attention was afforded to their provision – permitted the rapid expansion of the BEF's light railways organization. Before Geddes's arrival the BEF possessed just eighty miles of operable light railways. By April 1917 this had risen to over 200 miles, and before the end of 1917 the British had constructed over 1,000 miles of light railway track behind the front (see Table 7.2). Before the end of 1917, Hill reflected, the depot at Fosseux where he was employed on wagon repairs resembled 'an English railway yard ...

[24] Henniker, *History of the Great War*, pp. 68–9; Davies, *Light Railways of the First World War*, pp. 29–30.
[25] BLSC, Liddle collection, papers of Sapper W. J. Hill, LIDDLE/WW1/GS0767, recollections of France and the LRs during the Great War, 1914–1919, p. 10.

on, of course, a small scale'.[26] The yard contained British-built locomotive and wagon sheds for the repair and maintenance of largely British-supplied equipment, and the drivers were detailed for duty through a time office run 'by similar methods as adopted by the English railway companies'.[27]

Without light railways the bombardments of unprecedented ferocity that characterized the BEF's offensive operations in 1917 could not have been sustained at anything like the same intensity for the same duration. In September 'no less than 7,000 tons of ammunition were being carried daily' by light railways in support of the operations around Ypres,[28] on a network that was almost exclusively conceived, constructed and operated by Britain's transport experts. The BEF's light railways relied upon skills developed on imperial and global engineering projects alongside working methods pioneered on some of the nation's largest railways. However, in terms of one of its principal duties – that of providing relief to the overburdened French road network – the light railway organization created in 1917 was a failure.[29] In fact, despite the vast increase in the tonnage conveyed by light railways as the year unfolded (see Figure 7.1), demands for stone to repair and construct roads continued to grow. In January 1917 General Hubert Gough's Fifth Army received 405 lorry-loads of road stone. By July the same army required 1,000 lorry-loads, even though its light railways carried an average of 60,000 tons of stone (the equivalent of 1,350 lorries) each week.[30] Light railways, rather than removing the need for motorized transport entirely, merely shifted the traffic to new locations that – particularly in the form of marshalling yards and access roads – created their own considerable demands for new road-building. As will be seen, over the winter of 1917–18 this development had significant implications for BEF's transportation policies beyond the railhead.

Behind the railhead, British resources also became increasingly important in the second half of the war. Geddes's arrival coincided with the final abandonment of the pre-war Franco-British arrangements regarding the BEF's supply needs on the French main line railways. As the BEF had expanded in 1915 and 1916 the French authorities had 'urged repeatedly that more rolling stock should be imported' from Britain to help carry the growing volume of British traffic,[31] and the twin pressures of Verdun and

[26] BLSC, Hill papers, LIDDLE/WW1/GS0767, recollections of France, p. 22.
[27] BLSC, Hill papers, LIDDLE/WW1/GS0767, recollections of France, pp. 32–3.
[28] Davies, *Light Railways of the First World War*, p. 72.
[29] TNA, WO 158/852, director general of transport: history of light railways, 1916–1918, p. 3.
[30] Davies, *Light Railways of the First World War*, p. 68.
[31] Henniker, *History of the Great War*, pp. 170–1, 245. The first request for support from

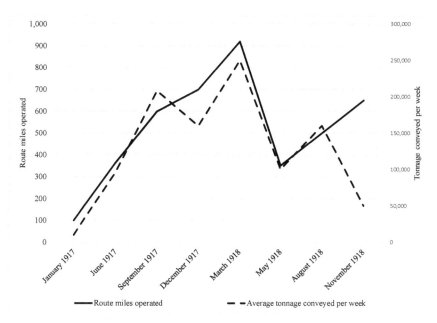

Figure 7.1. Route miles operated and average tonnage conveyed per week by the British Expeditionary Force's light railways, 1917–18.

Source: TNA, WO 158/852, director general of transport: history of light railways, 1916–1918, p. 19.

the Somme encouraged further modifications of British responsibilities on the western front. As early as 2 March 1916 Maxwell wrote to Sir John Cowans to advise his counterpart at the War Office that broad-gauge rolling stock 'in large quantities [was] urgently required in France'. Alongside the immediate placing of further large orders in Canada and the United States, the BEF's hosts 'demanded wagons from England' to ease the pressure on French resources.[32] By April orders for 13,000 wagons had been placed, but inter-allied conferences over the summer failed to reach a mutually acceptable conclusion as to the quantity of rolling stock required from British sources to fulfil the coalition's traffic requirements. In early October 1916 General Joffre submitted a formal request to GHQ for 'a large measure of assistance' in the maintenance and improvement of the Chemins de Fer du Nord.[33] The pre-war agreement, which had been

---

the French, made in March 1915, asked the British to supply between 2,000 and 3,000 wagons towards the projected ultimate requirements for the BEF's traffic of between 5,000 and 6,000 wagons.

[32] TNA, WO 107/15, IGC, general correspondence, Maxwell to Cowans, 2 March 1916.

[33] Unless otherwise stated, all quotations in this passage are taken from Henniker, *History*

tweaked and adapted in response to the changing conditions of the war, was 'considered at an end' by both armies following the receipt of Joffre's letter. Consequently, in addition to the various materials and labour-saving devices that Geddes had identified as crucial to the efficient operation of the BEF's transport services, the DGT also inherited responsibility for the provision of locomotives, rolling stock and personnel for the operation of trains on the French main lines. Furthermore, as 'it soon transpired that the railway situation [in France] was worse than the British had understood it to be', the French requested that the rolling stock delivery schedule agreed over the spring and summer of 1916 be expedited to sustain the network over the winter months.

The periodic nature of the French requests for British material support before October 1916 illustrate both the gradual erosion of the pre-war agreement in the wake of France's hammering in the first half of the war and the creeping increase in Britain's involvement in the transport infrastructure on the western front. The French had lost over 43,000 units of rolling stock during the initial German invasion – almost 12 per cent of the pre-war number of wagons in use on the French railways – while the withdrawal of railwaymen to replace losses in combat units had hindered the construction of replacements and maintenance of the remaining stock.[34] By November 1915 the British ambassador in Paris reported that a further 9,000 wagons were unavailable for service due to the absence of sufficient labour to affect the necessary repairs.[35] After the battles of Verdun and the Somme had taken a further toll on France's human and material resources, the railway network 'had fallen into very bad condition'.[36] As with light railway equipment, orders placed for new standard gauge rolling stock had a long lead time; the first wagons from the orders placed in March 1915 had taken thirteen months to arrive on the western front.[37] Faced with an impending crisis over the winter of 1916–17, Geddes recognized that the transport infrastructure could not withstand a prolonged delay while new orders were placed with – and fulfilled by – companies in Britain and North America whose productive capacity was already full. In early 1915 the ROD had commenced work in France with just seven engines loaned from the

---

of the Great War, pp. 246–7.

[34] *Les armées françaises dans la grande guerre: la direction de l'arrière* (Paris, 1937), pp. 29–30; J. H. F. Le Hénaff and H. Bornecque, *Les chemins de fer français et la guerre* (Paris, 1922), pp. 173–4.

[35] E. A. Pratt, *British Railways and the Great War: Organisation, Efforts, Difficulties and Achievements* (2 vols., London, 1921), ii. 643.

[36] TNA, WO 107/69, work of the QMG's branch, p. 2.

[37] Pritchard, *History of the Royal Engineers*, v. 598.

SECR and 'five or six machines which [Cecil Paget] had begged, borrowed or stolen' from the Midland Railway.[38] Eighteen months later the British railways were called upon to provide a far more substantial contribution to the BEF's material requirements.

To facilitate the delivery of railway equipment Geddes called upon his most prominent military and industrial contacts. On 19 November 1916 Sir Douglas Haig wrote to Sir Herbert Walker, the acting chairman of the REC, to request 'very large supplies of railway material, rolling stock, locomotives and personnel' from the British railways. Haig acknowledged the implications for the home railways of such an appeal, but felt that his demands could be

> more sympathetically met, and my needs more thoroughly appreciated, if you and the railway General Managers who form the Railway Executive Committee have had an opportunity of seeing for yourselves the difficulties which my Transportation Departments have to overcome. [...] I feel sure that when you have seen the conditions for yourselves, and have heard from Sir Eric Geddes, who will explain the situation to you when you are out here, of the difficulties which confront us you will realise that no effort, sacrifice, or inconvenience is too great to enable the Armies under my Command to be adequately equipped with transportation facilities.[39]

Sir Guy Granet supplemented Haig's statement at a REC meeting on 28 November, where he impressed upon his colleagues the magnitude of the situation faced by the BEF. However, as he was unable to furnish the committee with details of the precise nature of the aid required in France,[40] a delegation of REC members crossed the Channel to explore the matter with the DGT. The party left London on Saturday 9 December, discussed Geddes's estimates for the transport effort required to support the BEF's planned offensives in 1917, and lunched with the commander-in-chief at GHQ prior to their return home.[41] Over lunch, Haig took the opportunity

---

[38] TNA, WO 107/69, work of the QMG's branch, p. 16; L. S. Simpson, 'Railway operating in France', *Journal of the Institution of Locomotive Engineers*, xii (1922), 697–728, at p. 699.

[39] Haig to Walker, 19 Nov. 1916, quoted in Pratt, *British Railways and the Great War*, ii. 652.

[40] UWMRC, Granet papers, MSS. 191/3/4/39, Butterworth to Geddes, 30 Nov. 1916.

[41] A. J. Mullay, 'Letter from the Somme: the Railway Executive Committee and the military in World War I', *BackTrack*, xxii (2008), 220–3; Pratt, *British Railways and the Great War*, ii. 653–4. The delegation consisted of (general managers of their respective companies unless otherwise stated): John Aspinall, Lancashire and Yorkshire Railway; Sir Alexander Kaye Butterworth, North-Eastern Railway; Guy Calthrop, LNWR; Charles Dent, Great Northern Railway; Donald Matheson, Caledonian Railway; Henry Thornton, Great Eastern Railway; Sir Robert Turnbull, an LNWR director; Arthur Watson, superintendent

to reiterate 'the very great importance of efficient railways in modern war and our need for locomotives, wagons, cranes and personnel' to his guests. 'All agreed that they could help us and might do more for us than we had asked', Haig recorded later in his diary, 'the only thing required was an order from the government authorizing the railway companies to provide the material'.[42]

The combination of military and civilian expertise proved to be fruitful. Haig's personal intervention, coupled with the support of Geddes and Granet, helped clear an impasse that had existed between the Ministry of Munitions and the REC since the summer. In June 1916 the War Office had notified the latter that the ROD could soon require 'a considerable number of locomotives' to be despatched from Britain for service on the continent. Yet at the same time the railway workshops' focus on the production of warlike stores had reduced the output of locomotives in Britain, while the Ministry of Munitions was stressing the 'vital importance' of the companies being prepared to deal with increased volumes of traffic on the British railways. At a meeting between representatives of the ministry and the REC on 17 August the government promised to obtain the raw materials necessary for the LNWR to construct seventy locomotives.[43] However, by 11 November 'no tangible assistance whatever had been rendered to the companies' to permit the supply of locomotives to the BEF. Following the receipt of Haig's letter on 28 November the goods managers and superintendents of the line of Britain's major railway companies took matters into their own hands. Rather than await an allocation of raw materials, they identified locations in Britain where services could be reduced to free up stock for service overseas, and by 1 December Geddes could report to Haig that the companies had agreed to send 350 locomotives, 20,000 wagons and 320,000 sleepers to the western front.[44]

Following the REC's visit to France the export of British materials to the fighting fronts increased rapidly. Sixty-two locomotives from British sources were in traffic behind the western front by the end of 1916. A year later, 450 locomotives lifted direct from British railways provided the majority of the total of 753 locomotives available to haul supplies for the BEF.[45] By the end of the war the ROD possessed a fleet that was 'representative of almost

---

of the Lancashire and Yorkshire Railway; and Major Gilbert Szlumper, the REC's secretary.

[42] NLS, Haig papers, Acc. 3155/109, diary entry, 12 Dec. 1916.

[43] UWMRC, Granet papers, MSS. 191/3/3/69, Appendix D – Memorandum on transport facilities in the various theatres of war, 28 Oct. 1916, pp. 6–7.

[44] Pratt, *British Railways and the Great War*, ii. 643–46; NLS, Haig papers, Acc. 3155/109, diary entry, 1 Dec. 1916.

[45] Pritchard, *History of the Royal Engineers*, v. 621–2.

every railway in England', and of which over one-third had been received direct from service on the British railways.[46] The supply capacities of the expeditionary forces further afield were also augmented by the receipt of British vehicles. The BSF had received nineteen locomotives from England by early 1918, and by March of that year the British advance in Egypt and Palestine was supplied by a pool of vehicles that included twenty-six locomotives from the LSWR, twenty-five from the LNWR and three petrol engines from the Manning-Wardle Locomotive Company in Leeds.[47] According to *The Times*, the LSWR engines in Egypt operated 'with loads as heavy as they would have hauled between Southampton Docks and Nine Elms', and ran 'without cessation night and day, week in, week out'.[48]

The British railways' response to the army's demands had a significant impact upon the domestic railways. The withdrawal of locomotives, over 29,000 wagons and hundreds of miles of track engendered a series of measures designed to improve efficiency and reduce unnecessary travel in Britain as the material implications of industrial war became manifest in the second half of the war. Piecemeal actions that had been taken by individual railways before 1917 – such as the removal of dining cars, the slowing down of express services to conserve coal and the suspension of suburban services – were replaced by a comprehensive, nationwide effort to economize railway transport from 1 January 1917. Around 400 stations in Britain were closed, Sunday services were further reduced and fares were increased by 50 per cent to discourage non-essential journeys. In recognition of the pressures on precious raw materials the prime minister made a 'personal appeal' to the travelling public to 'cut down unnecessary travelling' in February, where he underlined that all the steel conserved by Britain's railway passengers could be directed into shipbuilding to meet the German submarine menace.[49] In the war's final two years, much of the REC's attentions were diverted into the question of reducing goods and passenger traffic not directly linked to the war effort.[50]

The import of railway equipment on a hitherto unprecedented scale after 1916 was mirrored by the improvement of transport facilities that took place at the BEF's Channel ports. As the entry point to the western front

[46] TNA, WO 107/69, work of the QMG's branch, p. 17; J. A. B. Hamilton, *Britain's Railways in World War I* (London, 1947), p. 171.

[47] 'Railways and the Salonica campaign', *Railway Gazette: Special War Transportation Number*, 21 Sept. 1920, pp. 110–18, at p. 117; 'The Palestine campaign', *Railway Gazette: Special War Transportation Number*, 21 Sept. 1920, pp. 119–28, at p. 126.

[48] Quoted in Pratt, *British Railways and the Great War*, ii. 666.

[49] 'Premier's appeal to the travelling public', *Railway Gazette*, 9 Feb. 1917, p. 174.

[50] On Britain's railways during the First World War, see Hamilton, *Britain's Railways in World War I*.

for the majority of the BEF's supplies, the effective operation of the ports was fundamental to the maintenance of fluidity in the force's distribution network. Geddes's initial enquiries ascertained that the ports had to deal with 'roughly 190,000 tons per week' to cover the BEF's requirements, and needed additional space on the shore to deal with the disembarkation of personnel, animals and the usual accoutrements that accompanied divisional formations.[51] On 14 September 1916 the IGC, Sir Frederick Clayton, had produced a statement that detailed how 198,000 tons per week could be handled by the ports. However, Clayton's projections provided no contingency for unpredictable occurrences such as the irregular arrival of ships, the closure of the ports for naval reasons or even poor weather. 'Experience has shown', Geddes wrote in November 1916, 'that we ought to be able to deal with each week's traffic in five days so as to provide a proper and safe margin, and in considering port capacity the average tonnage to be dealt with should not be more than 80 per cent of the maximum capacity'.[52] In Clayton's statement only Boulogne had been allocated a tonnage that met Geddes's criteria in its existing condition, while Dunkirk had been allocated a tonnage that exceeded the port's maximum capacity by 43 per cent (see Table 7.3).

It was not until the nature of the war effort required to dislodge the German army from French soil had revealed itself, and Britain's transport experts had taken a direct role in the organization necessary to sustain that effort, that the provision of equipment to boost the maximum capacity of the BEF's ports substantially increased. Shortly after the establishment of the directorate-general of transportation, Geddes complained both that the directorate inherited 'an insufficiency of shore gear, trays, skids, and other minor, but very essential equipment' and that the cranes in situ at the ports appeared 'to be somewhat below modern docks standards'.[53] To meet the requirements of the docks programme finalized by Geddes in March 1917,[54] the BEF increased the number of onshore cranes available to discharge goods from arriving vessels from 121 to 314. Only seven of the additional units were provided by the French, the rest were obtained from British sources.[55] By December 1918 the BEF also possessed thirty-six floating cranes for light lifts, three large floating cranes operated by the Royal Navy, two floating

[51] UWMRC, Granet papers, MSS. 191/3/3/69, Appendix D, p. 2. Geddes's estimates from this period are reproduced in Henniker, *History of the Great War*, p. 187.

[52] UWMRC, Granet papers, MSS. 191/3/3/102, memorandum by Geddes, pp. 4–5.

[53] UWMRC, Granet papers, MSS. 191/3/3/102, memorandum by Geddes, pp. 7–8.

[54] I. M. Brown, *British Logistics on the Western Front, 1914–1919* (London, 1998), p. 156.

[55] Henniker, *History of the Great War*, p. 236.

Table 7.3. Projected traffic allocations for the British
Expeditionary Force's Channel ports, November 1916.

	Percentage that the tonnage allocated bears to the capacity of the port	
	With existing equipment	With existing and additional equipment on order, November 1916
Dunkirk	143%	100%
Calais	100%	83%
Dieppe	100%	71%
Le Havre	100%	100%
Rouen	91%	87%
Boulogne	80%	80%
Le Treport	83%	83%
St Valery	100%	83%
Fecamp	125%	100%

Source: UWMRC, Granet papers, MSS. 191/3/3/102, memorandum by Sir Eric Geddes, 26
November 1916, p. 5.

electric power stations and six floating grain elevators.[56] The introduction of
new equipment contributed to the achievement of an average daily import
figure during 1917 and 1918 of 25,000 deadweight tons per day. The docks
directorate dealt with 4,178,000 tons of ammunition alone in the final two
years of the war, providing the ingredients with which the BEF's artillery
commanders concocted the firepower mixtures unleashed upon the German
lines in the pursuit of victory.[57]

However, the provision of modern equipment alone was not a panacea to
the challenges of port operation during the First World War. As Commander
Underwood noted in a report on the use of so-called labour-saving devices at
Boulogne, mechanical contrivances were not always applicable to conditions
on the Channel coast. Following an examination of the Bassin Loubet, he
stated that the port was not well suited for the employment of an automatic
discharge and stacking appliance for use on grain ships for three reasons.
First, the Bassin Loubet was a 'rather small dock' in which to operate such a
large piece of machinery. Second, the tidal and meteorological conditions at
Boulogne militated against the use of a floating grain elevator. Underwood
explained that 'one of the disadvantages at this port for floating elevators is

[56] Pritchard, *History of the Royal Engineers*, v. 621.
[57] Pritchard, *History of the Royal Engineers*, v. 664.

the rise and fall of the tide, which at its maximum is 27 feet. The height of any floating elevators would have to be so great, that in the high winds of winter there would be danger of craft capsizing. The wind here at times is so great that the cranes have to stop working'. Finally, he concluded, 'the length of time it would take to erect' the appliance and 'the inconvenience which would be caused during its erection' would not be offset by the potential benefits of the equipment's use.[58]

Underwood's report demonstrates that the British army, even before Geddes's arrival on the western front, considered the possibilities of mechanization behind the lines as well as at the front. The relative importance to the BEF's combat methodology of innovations such as the tank, poison gas and sound-ranging equipment have been debated by historians throughout the post-war period. Discussions over the extent to which the First World War armies utilized these tools effectively has overshadowed considerations of the provision of adequate quantities of mundane, civilian technologies such as locomotives, wagons, steam rollers and cranes to service those armies' demands. Geddes did not view such expansion as 'extravagance', but merely the logical corollary of the fact that the British were asked to take on a far larger share of the BEF's transport burden from October 1916 onwards. Without a sufficient transport infrastructure, founded upon the machinery that underpinned Europe's industrial economies, the weapons of the industrial war could not have been transported across the continent in the quantities required to create the firestorm that took place between 1914 and 1918. However, the provision of new equipment alone cannot explain the BEF's increased ability to supply the material war that emerged on the western front. As Geddes himself acknowledged from the moment he arrived in France, alongside new tools the force's fighting capabilities depended upon the efficiency with which those tools were utilized.[59] Britain's transport experts' influence stretched beyond the provision of roads and rails to the organizational systems that sustained the conflict.

The application of business methods on the western front

During the summer of 1917 the BEF attained its peak strength of just over two million troops. From that moment until the conflict's end the following autumn, the British contribution to the main European theatre of war gradually declined in numerical terms. That the BEF responded to this reduction in strength through the more effective deployment of

[58] TNA, WO 95/3970, IGC war diary, Underwood to Clayton, 12 Aug. 1916.
[59] UWMRC, Granet papers, MSS. 191/3/3/4, Geddes to Lloyd George, 15 Sept. 1916; IWM, Woodroffe papers, 3/38/1/2, notes and reports, 25 Aug. 1916.

the available manpower, and a higher dependence upon the machines of war, has been central to the so-called learning curve theory of British military improvement after the nadir of 1 July 1916.[60] Following the United States' entry into the war in particular, Britain's desire to win the war and the subsequent peace while incurring the lowest possible cost had to be reframed as a determination to secure victory before the costs became so great that Britain was unable to exert sufficient influence at the post-war bargaining table.[61] To do so, the British government had to ensure that the nation's dwindling manpower resources were employed – regardless of the character of their contribution – in the most efficient manner possible.[62] On the western front, Lloyd George emphasized in a meeting of civilian and military authorities in January 1918 that, 'in view of the difficulties that were arising ... in regard to the question of Man-Power, it was essential that there should be no idle men'.[63]

The effective coordination and supervision of men and equipment dispersed across a vast geographical area was critical to the economical use of the BEF's resources during the First World War. The speed with which roads were repaired, railway tracks laid, trains moved from depot to railhead and wagons unloaded for return to depot directly affected the pace and intensity with which the fighting troops were able to concentrate their force against the enemy lines. The battle of the Somme illuminated the shortcomings in the BEF's logistical foundations in the summer and autumn of 1916. As Ian M. Brown has demonstrated, the shortages of ammunition noted by commanders early in the battle were the result of longstanding tactical delivery problems rather than insufficient production levels – issues that were exacerbated by the voluminous increase in supplies from Britain once the offensive was under way.[64] Three potential solutions existed to remove the supply bottleneck that reduced fluidity on the western front. The first option, proposed by the IGC in August 1916, was for ships to be sent from Britain at a slower rate, thereby synchronizing their arrivals with the discharge rates at the docks and the railway network's ability to remove goods from the ports.[65] The second, reported to Lloyd George by

---

[60] D. Stevenson, *With Our Backs to the Wall: Victory and Defeat in 1918* (London, 2011), pp. 170–243; S. Bidwell and D. Graham, *Fire-Power: British Army Weapons and Theories of War, 1904–1945* (Barnsley, 2004); G. Sheffield, *Forgotten Victory. The First World War: Myths and Realities* (London, 2001).

[61] D. French, *The Strategy of the Lloyd George Coalition, 1916–1918* (Oxford, 1995), p. 291.

[62] Stevenson, *With Our Backs to the Wall*, pp. 259–63.

[63] TNA, CAB 24/39/101, Chinese labour in France. Report of meeting held, Friday 18 Jan. 1918, p. 2.

[64] Brown, *British Logistics*, pp. 123–4.

[65] TNA, WO 95/3970, IGC war diary, Clayton to Lloyd George, 2 Aug. 1916.

Geddes the following month, was 'that the factories must slow down!' Both were impracticable. The scaling back of munitions production was a 'moral and physical impossibility' in a nation increasingly geared towards a more total form of warfare,[66] while a reduction in the frequency of deliveries to France would simply shift the storage problem to Britain's ports – where Geddes had already recognized pressure on the available space before the Somme offensive commenced.[67]

To permit the BEF to undertake offensive operations that consumed munitions at a greater rate than had proven possible at the Somme, but without overloading the capacity of the transport network, that network had to be operated more efficiently. The labour force attached to the directorate-general of transportation, employed on duties across France and Flanders – and numbering 89,000 men by November 1918 – played a key role in the maintenance and improvement of the BEF's supply chain in the second half of the war. Yet the constant, direct visual observation of such a large-scale, dispersed workforce from the directorate's headquarters was impossible. Therefore, the implementation of managerial tools that stimulated improvement and the application of operating procedures that promoted fluidity throughout the transport network became important facets of the British approach to material-intensive warfare.

To Britain's transport experts the control of a large workforce employed on tasks beyond the direct oversight of senior managerial figures was a familiar challenge. The growth of large-scale businesses during the second half of the nineteenth century had created a series of unprecedented difficulties for the employers of labour to solve. By virtue of being executives of the first companies to experience such growth, railway managers were by necessity the forerunners in addressing the problems associated with handling 'large amounts of men, money, and materials within a single business unit'.[68] Unlike even the largest factories of the day, which could be observed in their totality within a relatively brief period of time, the major railways were operated by units that were spread over hundreds of miles and engaged on a wide variety of activities.[69]

Geddes appreciated from the outset that the efficiency of the human and material resources under the DGT's command could be assessed and improved through the application of business methods deployed by civilian

---

[66] UWMRC, Granet papers, MSS. 191/3/3/4, Geddes to Lloyd George, 15 Sept. 1916, p. 9.
[67] PA, Lloyd George papers, LG/D/5/2/4, memorandum on filling for week ending 10 June 1916, 17 June 1916, pp. 7–8.
[68] A. D. Chandler, 'The railroads: pioneers in modern corporate management', *Business History Review*, xxxix (1965), 16–40, at p. 16.
[69] Chandler, 'The railroads', p. 19.

industries to manage their peacetime endeavours. Nowhere was this more pronounced than in BEF's light railway operations from 1917 onwards. Alongside the locomotives, wagons and personnel necessary to meet the BEF's increased demands for transport capacity in the second half of the First World War, Britain's transport experts provided the management tools designed to squeeze the maximum fluidity out of the network. As Figure 7.2 illustrates, during 1917 the component parts of the BEF's light railway network – managed by individual operating companies – grew into a complex system. To ensure the maintenance of fluidity across the expanding network required the coordination of train movements, just as took place on the main line railways that connected the BEF to the Channel ports. However, unlike on the main lines – managed according to rules and regulations laid down by the French rail authorities – the light railway network presented the directorate-general of transportation with an opportunity to introduce working practices adapted from the latest innovations conceived by Britain's transport experts.

Auckland Geddes later suggested that his brother had taken the inspiration for the BEF's light railway operations from the Powayan Steam Tramway in India that Eric had helped run in the early 1900s.[70] However, the origins of the BEF's traffic management procedures could be found much closer to home. Before the outbreak of the First World War, operating methods were the subject of intense experimentation among Britain's competitive railway companies. The maintenance of fluidity and efficiency upon individual companies' networks demanded that traffic officers possessed a thorough understanding of the location and movements of the trains that circulated around the system, and major advances in the companies' surveillance capacities had taken place before 1914.[71] By the time Geddes became DGT both the Midland Railway and the Lancashire and Yorkshire had installed centralized control systems to help coordinate their main line operations,[72] and it was from the former that the BEF's light railway directorate drew its inspiration.

---

[70] A. C. Geddes, *The Forging of a Family: a Family Story Studied in its Genetical, Cultural and Spiritual Aspects and a Testament of Personal Belief Founded Thereon* (London, 1952), p. 238.

[71] R. Edwards, *Instruments of Control, Measures of Output: Contending Approaches to the Practice of 'Scientific' Management on Britain's Railways in the Early Twentieth Century* (Southampton, 2000); C. A. Williams, *Police Control Systems in Britain, 1775–1975: from Parish Constable to National Computer* (Manchester, 2014), p. 153.

[72] TNA, RAIL 491/815, train control office at Derby, 1914; J. A. F. Aspinall, *Train Control Arrangements: a Survey of the Comprehensive Control System Operating on the Lancashire and Yorkshire Railway* (Manchester, 1915).

The primary motivations for the Midland Railway's institution of a centralized train control system in 1909 – efficiency, flexibility, the economic use of rolling stock and the acquisition of timely information on the whereabouts of the company's resources – mirrored the BEF's priorities in 1917. The similarities between military and civilian railway operations did not stop there, either. The reliability of the Edwardian railway industry had created a logistical environment in Britain that encouraged firms to reduce their stock levels on-hand, confident that goods would be delivered swiftly by rail when required. This approach created concerns that any prolonged dislocation to railway services, whether the result of congestion or industrial action, would quickly starve manufacturers of raw materials and consumers of staples such as bread and milk. As Lloyd George, the then president of the Board of Trade, acknowledged in 1907, 'there was hardly a country in the world ... which depends so much upon the absolute promptitude with which goods are delivered'.[73] The military practice during the First World War was to ensure that stores were placed far enough away from the front

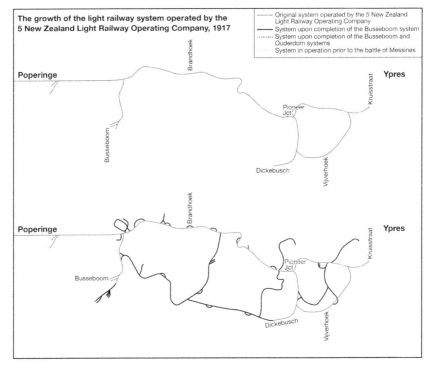

The growth of the light railway system operated by the 5 New Zealand Light Railway Operating Company, 1917

———— Original system operated by the 5 New Zealand Light Railway Operating Company
———— System upon completion of the Busseboom system
············ System upon completion of the Busseboom and Ouderdom systems
System in operation prior to the battle of Messines

[73] TNA, RAIL 1053/258, railway dispute: conference between David Lloyd George, president of the Board of Trade, and representatives of the railway companies, 25 Oct. 1907, p. 6.

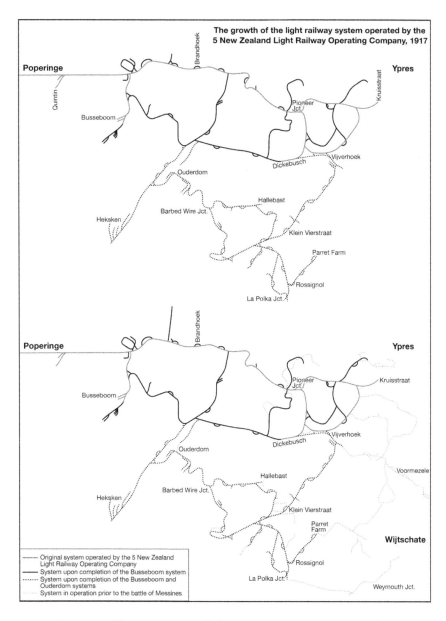

Figure 7.2. The growth of the light railway system operated by the
5 New Zealand Light Railway Operating Company, 1917.

Source: TNA, WO 95/4061/7, lines of communication troops. 5 New Zealand Light Railway
Operating Company. Map drawn by Cath D'Alton.

line to reduce their susceptibility either to hostile artillery or capture in the event of a surprise enemy advance. Therefore, the BEF's customers at the front line were also wholly dependent upon an effective transport network that delivered the goods reliably and frequently. A cut in supply, whether by enemy action, mismanagement or the withdrawal of labour, would produce the same results.

Congestion, which had brought about the near paralysis of the BEF's transportation network during the battle of the Somme, had been a significant problem for the Midland in the early part of the century. The use of railway sidings as makeshift depots had created difficulties for those tasked with unloading freight wagons on the Midland's lines, which increased the time required to unload individual trains and led to widespread delays to traffic across the network. The unpredictable nature of the traffic meant that engine crews were frequently forced to work shifts of fifteen hours or more, as replacement crews were allotted according to the timetables rather than the actual locations of the trains. A Board of Trade enquiry into the working hours of 18,354 engine drivers in Britain uncovered that the majority worked for between sixty and sixty-two hours each week in 1907, while 3,689 of those examined worked for more than sixty-six hours per week.[74] On the Midland alone a total of 24,760 cases of extended duty were recorded in the first six months of 1907, a situation that contributed to numerous cases of illness-induced staff absences and inspired the company to amend its operating procedures.

The impressive results obtained by the Midland – coupled with the fact that its chief architect, Cecil Paget, was already in France at the head of the ROD – make Geddes's decision to adopt the train control system developed after 1907 for the BEF's light railway network unsurprising. The system involved the creation of district offices throughout the Midland's network, each linked by telephone to a central control office in Derby. The central office received regular updates on the whereabouts and composition of the trains on the system, which allowed traffic controllers to identify stations where trains were detained for unnecessarily long periods in order to focus improvements on the area.[75] Following successful trials on sections of the line, Paget, the railway's general superintendent, authorized the extension of the train control system to cover the entire 1,400-mile network despite the reservations of many within the railway industry. 'Quite a number of able

---

[74] *Earnings and hours enquiry. Report of an enquiry by the Board of Trade into the earnings and hours of labour of workpeople of the United Kingdom. VII. Railway service in 1907* (Parl. Papers 1912 [Cd. 6053], cviii), pp. 188–9.

[75] On the early development of train control systems on the British railways, see P. Burtt, *Control on the Railways: a Study in Methods* (London, 1926), pp. 95–101.

railway men', the company recalled, suggested that the existing methods of control in the industry could not be improved upon and expected the system – based upon the creation of a real-time record of all traffic on the network within a central control office – to fail.[76] However, between 1907 and 1913 Paget's system proved remarkably successful. As Figure 7.3 illustrates, the weekly average hours of traffic delays on the Midland fell by more than 64 per cent despite a 10 per cent increase in the tonnage of goods conveyed over the same period. From well in excess of 20,000 cases of men working for excessively long hours in early 1907, four years later there were none.[77]

As the BEF inherited relatively few operable light railways from the French the installation of a new operating procedure was relatively straightforward. The French practice of working a section of line – typically between twenty and thirty miles in length – using the box-to-box system, was not particularly well established among British troops. Under the box-to-box system orders for light railway were issued direct from the department that required the transport, rather than from a central office with access to the most up-to-date information on the army's priorities.[78] The retention of such uncoordinated methods was impracticable once the BEF's light railways grew into an interconnected network. The threat of 'friendly generals' impairing the system's efficiency through the forceful imposition of their personal priorities, without due appreciation of the wider army's requirements, encouraged the creation of a centralized administration through which all transport requests could be filtered. Therefore, the new lines in the British areas were constructed with the central control system in place from the start, while extant lines were gradually converted from the box-to-box system. In August 1917 the 31st Light Railway Operating Company recorded that the central control system, to which they had been recently converted, was 'working quite satisfactorily'.[79] However, with the susceptibility of the telephone connections to artillery fire in mind, the box-to-box equipment was retained in place to act as a back-up during periods when the telephone network was inoperable.[80]

As on the Midland Railway, the control offices became a fundamental component of the BEF's light railway operations during the second half

---

[76] TNA, ZLIB 29/620, the train control system of the Midland Railway (reprinted from *Railway Gazette*), 1921.

[77] C. Hamilton Ellis, *The Midland Railway* (London, 1953), pp. 150–2.

[78] TNA, WO 158/852, history of light railways, p. 19.

[79] TNA, WO 95/4056/2, lines of communication troops. 31 Light Railway Operating Company Royal Engineers, diary entry, 16 Aug. 1917.

[80] BLSC, Hill papers, LIDDLE/WW1/GS0767, recollections, p. 51.

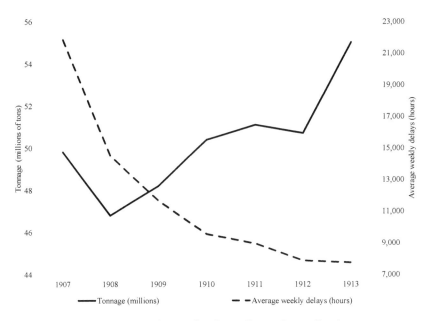

Figure 7.3. Delays to freight traffic on the Midland
Railway, average weekly hours, 1907–13.

Source: R. Edwards, *Instruments of Control, Measures of Output: Contending Approaches to the Practice of 'Scientific' Management on Britain's Railways in the Early Twentieth Century* (Southampton, 2000), p. 19.

of the war. Each of the BEF's five armies were served by a self-contained central control office based upon the Midland's template, at which requests for light railway transport from each corps were collated and prioritized. Large schematic diagrams were set up in each office, which displayed the army's portion of the light railway network and indicated the location of all motive power and rolling stock under the army's command. Information on the whereabouts of the army's resources was frequently updated by reports telephoned from numerous district offices, situated at the marshalling yards from which individual trains were made up for their journeys to the front. Within each district several stations or dumping points also contained reporting facilities that permitted local officers to keep the district offices informed of events out on the lines.

The 'ceaseless' surveillance from the army's central office produced an almost instant overview of the network's assets, which ensured the light railways organization could respond quickly to the demands of industrial warfare and fostered a more efficient use of the BEF's limited resources. Daily conferences among the technical staff allowed the officers responsible for allocating the available locomotives and rolling stock to do so according to the army's most recent priorities, minimizing the prospect of a district being left with insufficient stock on hand to service its daily traffic. In addition, central control ordered the movement of wagons around the system as necessary. On 31 July 1917 such a transfer of resources took place on the portion of line worked by the 12th Light Railway Operating Company. Based in Romarin, near Armentières on the French-Belgian border, the company received orders to 'transfer as many bogie wagons as could be put together quickly for ammunition work' to assist colleagues engaged on supplying the third battle of Ypres further north.[81]

Alongside creating flexibility over the allocation of resources, the central control system provided the BEF's light railways organization with the data to identify where rolling stock was being held under load for abnormally long periods. As on the peacetime railways, wagons left in sidings represented a reduction in the BEF's overall transport capacity and indicated the presence of uneconomical working practices. Every day the central control office produced a wagon register, which recorded the location of every piece of rolling stock in the BEF's possession and was drawn from the contents of train and engine shunting journals compiled in each district.[82] The journals contained information on every train that passed over the light railway network, 'often compiled in huts or dugouts under artillery fire' and forwarded to the district offices by telephone.[83] The data produced through this process became a core component of the comprehensive series of statistics used by the light railways directorate to improve working practices among the widely dispersed units under its command. The close observation of unloading methods between March and September 1917 brought about a reduction in the average turnaround time for a light railway wagon from 1.7 days to around 0.7 days, which – combined with a 200 per cent increase in the number of wagons in traffic during the same period – vastly increased the light railways' carrying capacity on the western front. From an average weekly traffic of 25,315 tons in March, by September the BEF's light railways carried an average weekly traffic of 210,808 tons to the front.[84]

A comprehensive system of data capture and statistical analysis lay at

[81] TNA, WO 95/4056, lines of communication troops. 12 Light Railway Operating Company Royal Engineers, diary entry, 31 July 1917.

[82] TNA, WO 158/852, history of light railways, p. 20.

[83] J. G. Beharrell, 'The value of full and accurate statistics: as shown under emergency conditions in the transportation service in France', *Railway Gazette: Special War Transportation Number*, 21 Sept. 1920, pp. 37–9, at p. 39.

[84] TNA, WO 158/852, history of light railways, p. 20; Davies, *Light Railways of the First World War*, pp. 73–4.

the heart of the directorate-general of transportation's supervisory practices, one similar to that described as an 'information infrastructure' by Lisa Bud-Frierman in her examination of business administration methods in the late nineteenth century.[85] In an interview with the American journalist Isaac Marcosson following his appointment as First Lord of the Admiralty, Geddes emphasized his belief in the utility of statistics as a managerial tool. When Marcosson asked him 'what single rule had been of most service to him' during the war, Geddes responded: 'The use of statistics. I statistise [*sic*] everything. Knowledge is power and statistics are the throttle valve of every business'.[86]

Geddes's appreciation of statistics provides a further example of the North-Eastern Railway's managerial culture's influence on his approach to the war. Sir George Gibb's progressive response to the challenging operating conditions in the turn-of-the-century railway industry inspired the business techniques that Geddes applied to the management of the directorate-general of transportation. Following a month-long tour of the United States in 1900, Gibb was convinced that the extensive collection and examination of statistical data could greatly improve the efficiency of work undertaken on the British railways. Throughout the pre-war period, within the trade press, to parliamentary committees and even in the discussion of papers delivered before the Royal Statistical Society,[87] Gibb passionately advocated the use of statistics for allowing

> a railway manager to test the work done in carrying passengers and merchandise on any part of the railway, to measure the work performed in relation to many important items of cost incurred in performing it, to compare period with period and district with district, to supervise local staff with a full knowledge of results, to control train mileage, and to enforce economy in working.[88]

Statistical information, disseminated throughout the company, was used to 'found judgments, to make policy decisions and to establish standards which … enable[d] officials to watch and control the effects of the steps being taken to improve working methods' on the North-Eastern.[89] In collaboration with

---

[85] L. Bud-Frierman, 'Information acumen', in *Information Acumen: the Understanding and Use of Knowledge in Modern Business*, ed. L. Bud-Frierman (London, 1994), pp. 7–25, at pp. 7–8.

[86] I. F. Marcosson, *The Business of War* (New York, 1918), p. 283.

[87] W. M. Acworth, 'English railway statistics', *Journal of the Royal Statistical Society*, lxv (1902), 613–64, at pp. 652–4.

[88] W. W. Tomlinson, *The North Eastern Railway: its Rise and Development* (Newcastle upon Tyne, 1915), p. 732.

[89] R. J. Irving, *The North Eastern Railway Company, 1870–1914: an Economic History* (Leicester, 1976), pp. 218–19.

the statistician George Paish, Gibb oversaw the establishment of a traffic statistics office in York in 1902 to 'pioneer and promulgate the use of new statistical concepts for operational measurement, control and efficiency'.[90]

The data-gathering system used by the BEF's transportation services was the brainchild of Geddes's assistant from the North-Eastern, J. George Beharrell. Shortly after the war he explained his procedures in an address to the annual meeting of the British Association for the Advancement of Science. At the outset of the lecture Beharrell directly responded to claims that 'the success of transportation in France was largely due to the fact that money was of no account and it was spent like water'.[91] He pointed out that shortages of manpower and materials on the western front compelled the DGT to ruthlessly pursue efficiencies throughout the organization. Drawing upon his own professional career at the North-Eastern Railway – where Geddes had given his abilities to collect, prepare and disseminate information from a variety of sources 'full play' within the goods department – Beharrell created bespoke, elaborate reporting systems for each of the departments under the DGT's command.[92] The statistics that were generated, Beharrell explained, 'told each responsible officer what he was doing, whether he was going back or forward, and how he compared with his opposite number in other places'.[93] In the IWT directorate 'the statistical and movements sections ... [were] provided [with] an efficient system whereby the work being performed by both personnel and craft could be carefully gauged, and immediate steps taken when necessary to speed up the work in sections where the best results were not being obtained'.[94] 'Without statistics', Beharrell claimed, it was 'impossible to know with any certainty that there ha[d] been any improvement' in the operations behind the front line upon which the fighting troops depended throughout the war. It was certainly impossible, he concluded, to accurately measure any such improvement.[95]

The constant flow of data from the DGT's workshops to its boardroom played a vital role in permitting administrative officers to ascertain the output levels that could be expected from units assigned to each task, and assisted them in the identification of inefficiencies across the western front.

[90] R. J. Irving, 'Gibb, Sir George Stegmann (1850–1925)', in *Dictionary of Business Biography: a Biographical Dictionary of Business Leaders Active in Britain in the Period 1860–1980*, ed. D. J. Jeremy (5 vols., London, 1984), ii. 543–5, at p. 545.
[91] Beharrell, 'The value of full and accurate statistics', p. 37.
[92] R. Bell, *Twenty-Five Years of the North Eastern Railway, 1898–1922* (London, 1951), p. 39.
[93] Beharrell, 'The value of full and accurate statistics', p. 37.
[94] TNA, WO 158/851, history of IWT, p. 22.
[95] Beharrell, 'The value of full and accurate statistics', p. 37.

The potential value of metrics had been recognized within the railway industry before the war, as the major companies grappled with the challenge of supervising and maintaining the productivity of a dispersed workforce. As George Boag advised in his *Manual of Railway Statistics*, 'by training the staff in the use of statistics an intelligent interest is aroused, and the figures will have already served their purpose if they have drawn the attention of the operating officers to the results of their own work'.[96] The knowledge that senior officers were liable to closely investigate sustained periods of inadequate performance acted as a stimulus for supervising officers to take a close interest in their men's working methods, to ensure that standards were rigorously maintained and improved, and to reflect upon their own contributions to the directorate's progress. Major-General John W. Stewart – a Scottish-born railwayman who had made a career building railways in Canada prior to the war – became infamous among the light railway construction units behind the western front for his 'unannounced and unexpected' site visits in 1917, undertaken to encourage the troops and their supervisors to build new lines as swiftly as possible.[97]

The existence of a pool of numerate administrators was critical to the maintenance of the directorate's programme. The 'rapid expansion of the clerical workforce' in Britain during the sixty years before the First World War provided the skilled labour required to convert Beharrell's concept into a reality. By 1914 over 5 per cent of occupied males in the British workforce were clerks, a larger middle-class occupational group than any other aside from employers and proprietors.[98] The complexities of administration in the twentieth century had been addressed by the state and a larger number of companies in Britain than in any other European nation before the war.[99] The so-called 'nation of shopkeepers' was short neither of 'numerical nous' nor the tools with which to harness, manipulate and analyse complex operational data as a foundation for decision-making. The pre-war British

[96] G. L. Boag, *Manual of Railway Statistics* (London, 1912), p. 42.

[97] G. W. Taylor, *The Railway Contractors: the Story of John W. Stewart, His Enterprises and Associates* (Victoria, BC, 1988), pp. 113–14.

[98] P. Scott and J. T. Walker, 'Demonstrating distinction at "the lowest edge of the black-coated class": the family expenditures of Edwardian railway clerks', *Centre for International Business History*, 2014 <https://assets.henley.ac.uk/legacyUploads/pdf/research/papers-publications/IBH-2014-04%20Scott%20and%20Walker.pdf?mtime=20170410170907> [accessed 8 Dec. 2014], p. 2.

[99] E. Higgs, *The Information State in England: the Central Collection of Information on Citizens since 1500* (Basingstoke, 2004), pp. 99–132; Y. Cassis, 'Big business in Britain and France, 1890–1990', in *Management and Business in Britain and France: the Age of the Corporate Economy*, ed. Y. Cassis, F. Crouzet and T. R. Gourvish (Oxford, 1995), pp. 214–26, at p. 216.

economy had embraced the 'quantifying spirit',[100] and the directorate-general of transportation applied the same ethos to the BEF's operations.

From 1917 onwards the BEF's workforce was relentlessly subjected to measurement in the pursuit of efficiency and economy. At the docks 'the rate of handling per man per hour was closely watched, each port being compared with its previous performance on similar cargoes'. Through a combination of 'better supervision and equipment', Beharrell recorded, in a year the tonnage handled per man per hour increased by 24 per cent across all the ports operated by the BEF. Behind the headline improvement lay a process of observation, measurement, comparison, adjustment and repetition – presented through an unambiguous, accessible comparison tool that helped identify 'the most suitable form of work' for each of the myriad units employed under the director of docks.[101] The tool Beharrell chose was the graph, which had 'exploded' into widespread use in the latter part of the nineteenth century.[102]

Plotting units' results alongside one another clearly illustrated the varied performance levels of individual groups within the category being measured, an activity that allowed busy commanders to distinguish at a glance where output was satisfactory and where causes for concern existed. Rather than pore over the raw data provided by each unit, senior officers received a visual depiction of the department's progress that could also be disseminated throughout the organization to help companies chart their own development. As the author of a report on the Chinese Labour Corps' work explained:

> By that ingenious method of making statistics intelligible to those who have no mathematics in their souls, the whole situation can be seen at a glance. [I] was shown one graph which dealt with the comparative results produced by the different types of labour in France, and another which compared, month with month, the total output of each type of labour in all the great dumps and workshops. In the first line the little blue strip which denoted the Chinese was more than holding its own, in the second there was a steadily increasing blue strip everywhere. A terrific amount of toil had gone into the making of those little strips, and the tale they told was cheering indeed.[103]

---

[100] J. Thompson, 'Printed statistics and the public sphere: numeracy, electoral politics, and the visual culture of numbers, 1880–1914', in *Statistics and the Public Sphere: Numbers and the People in Modern Britain, c. 1800–2000*, ed. T. Crook and G. O'Hara (Abingdon, 2011), pp. 121–43, at pp. 135–6.

[101] Beharrell, 'The value of full and accurate statistics', pp. 37–8.

[102] J. Thompson, 'Printed statistics and the public sphere', pp. 123–4.

[103] TNA, WO 106/33, the Chinese Labour Corps – recruitment and organisation – history of the corps, Chinese labour in France, p. 15.

This quote demonstrates how statistics were used to encourage workers as well as managers during the war. The tasks upon which those employed by the DGT laboured from 1917 onwards lacked a readily identifiable link to the BEF's overall goal of victory, and the newly created units did not possess a folklore of regimental histories upon which to draw for the inspiration of new recruits. Furthermore, many of the units under the DGT's charge were forced to accept the descriptor 'unskilled', a 'negative qualification' unlikely to engender a sense of pride and commitment within those to whom it was attached.[104] When assigned to the unskilled labour of road-making duties, Richard Smith has noted, 'men of all ranks in the British West Indian Regiment were keen to shed the stigma attached with labour' and stressed their commitment to martial customs.[105]

Through the concentration on targets, encouraged by the proliferation of easily understood graphs, the abstract notion that building a road or unloading lorries was fundamental to the prosecution of the war effort was superseded by the immediately recognizable, attainable benchmarks that were naturally created by daily, weekly and monthly records of achievement. The raw data have since been destroyed, but the surviving war diaries of units such as the 12th Light Railway Operating Company illustrate the unit's pride when it surpassed previous performances: 'The ammunition tonnage handled today – highest on record – 2,250 tons. Every man doing splendidly and the system working perfectly',[106] the company's diarist recorded on 11 June 1917. During the same week the diary noted several instances of German shelling, which caused damage to the company's yard and camp. Clearly, the act of recording respectable statistics despite the enemy's best efforts played a significant role in motivating the company's troops.

Individual units created targets, both for the unit themselves to beat in subsequent periods and to demonstrate their abilities in comparison to neighbouring companies and departments. As John Starling and Ivor Lee discussed in their account of British military labour during the war:

> It was found that the South African Native Labour Corps were most efficient when used as unskilled labour loading and unloading stores. In July 1917 'a party of the SANLC working on ammunition at Martainville created a record. Fifty natives, with 4 NCOs and 1 White NCO man-handled 700 tons of ammunition, loading it from dump to lorry and from lorry to [railway] truck in 345 man hours, an average of 2.03 tons per man hour.'[107]

[104] TNA, WO 107/37, work of the labour force during the war: report, 1919, p. 6.

[105] R. Smith, *Jamaican Volunteers in the First World War: Race, Masculinity and the Development of National Consciousness* (Manchester, 2004), pp. 87–8.

[106] TNA, WO 95/4056/12, Light Railway Operating Company war diary, diary entry, 11 June 1917.

[107] J. Starling and I. Lee, *No Labour, No Battle: Military Labour during the First World War*

The diary extract quoted by Starling and Lee emphasizes the dual purpose of statistics as a management tool. First, the data provided senior officers located far from Martainville with quantitative evidence of the unit's output. Second, the generation of numerical indicators of the unit's work provided a stimulus for groups of men from diverse backgrounds to contribute to the war effort of a nation to whom many of the labourers possessed no patriotic attachment. For the different tribes represented within the South African Native Labour Corps, the war was less about defeating the German empire and more about proving their strength, fitness and stamina in a friendly competition with their fellow tribesmen.

However, competition created temptations to subvert the measurement process and distort the information collected by the directorate. As Brigadier-General John Charteris observed in early 1918:

> There is much rivalry between the light railways and the standard gauge in the forward areas. Each seeks to justify its supremacy by graphs showing the number of men and tons carried every day. If rumour is to be believed, the light railways will stop any of their trains wherever a body of troops appears and almost beseech them to take a lift anywhere up and down the line, so that they can record them on their graph.[108]

Charteris's somewhat light-hearted remark reveals the key limitation regarding the implementation of a data capture and performance monitoring system within the BEF during the First World War. The generation of accurate records relied upon accurate measurement, which was not always forthcoming. As the rumoured system-gaming activities of the light railway operating companies demonstrates, not all of those who participated in the BEF's information-gathering network were entirely honest or willing contributors.

The competence and good faith of those who participated in Beharrell's data-gathering system were fundamental to its success. However, examples of both incompetence and deception can be ascertained from the surviving records. In the first instance, even an educated man like the philosopher and academic Alexander Lindsay – who spent much of 1917 engaged in the accumulation of data and production of graphs – struggled to attain the precision required to submit accurate returns, and recorded his battles to master the intricacies of the process in a series of letters to his wife.[109] Elsewhere, among professional officers concerned with self-preservation and their future career prospects, the urge to 'present the best aspect' of a

---

(Stroud, 2009), p. 231.

[108] J. Charteris, *At G.H.Q.* (London, 1931), p. 283.

[109] Keele University Special Collections and Archives (KUSCA), papers of A. D. Lindsay, LIN 149, Lindsay to Erica Lindsay, 22 Jan. 1917; 28 Feb. 1917; 5 and 14 March 1917.

company's work in the official records rather than an honest and accurate account of its performance was an ever present temptation.[110] One unnamed 'old colonel', described by Lindsay as 'the bad sort of army man', was more interested in ensuring that his back was covered in case things went wrong than in using statistics to guide improvements in his unit's performance.[111]

Furthermore, Beharrell's system had to contend with the hostility of individuals unwilling to engage with the bureaucracy that it generated. While the conflict acted as what Lindsay referred to as a laboratory for an 'elaborate experiment in the organization of labour',[112] to many of those charged with accumulating the raw data for the 'carefully planned ... forms [and] graphs', Beharrell's time-consuming approach was considered an unnecessary additional burden in a war of unprecedented intensity.[113] Lieutenant-Colonel Bryan Fairfax, commander of the Chinese Labour Corps and ultimately responsible for a workforce of almost 100,000 men, was one such officer. A dugout who had been recalled from the reserve list in August 1914, Fairfax had served in China during the Boxer Uprising and as inspector of Chinese labourers in the Transvaal. Although a far cry from the 'Oxford men' like Lindsay who found themselves employed on administrative duties,[114] Fairfax was by no means the stereotypically insular and obstinate soldier. In South Africa he was described as 'an efficient and conscientious and trustworthy' officer who possessed 'in a high degree the quality of tact, so necessary in handling men' by the superintendent of the foreign labour department in Johannesburg, while his application for the post of king's messenger in 1914 was supported by a host of character references that underlined his aptitudes.[115]

For Fairfax, 'the war ... was not an exercise in scientific labour management, but a life and death struggle in which men must be exploited

---

[110] TNA, CAB 24/58/73, report upon an enquiry into the management of labour and the control of works in the British army zone in France, 12 June 1918, pp. 3–4.

[111] KUSCA: Lindsay papers, LIN 149, Lindsay to Erica Lindsay, 15 Aug. 1917.

[112] A. D. Lindsay, 'The organisation of labour in the army in France during the war and its lessons', *Econ. Jour.*, xxxiv (1924), 69–82, at p. 69.

[113] N. J. Griffin, 'Scientific management in the direction of Britain's military labour establishment during World War I', *Military Affairs*, xlii (1978), 197–201, at p. 198.

[114] 'G.S.O.', *G.H.Q. (Montreuil-Sur-Mer)* (London, 1920), p. 168.

[115] BLSC, Yorkshire Archaeological and Historical Society collection, Personal file of Captain Bryan Fairfax, YAS/MD335/10/1, Jamieson to private secretary to H.E. The Lt. Governor, Pretoria, 25 Sept. 1905; Jamieson to the attorney general, 1 June 1908. The testimonies written in support of Fairfax's application to the post of king's messenger in 1914 emphasized his 'good character', 'judgment and tact', 'good manner', 'leadership qualities', 'methodical' approach, and 'common sense'. See, e.g., testimonies written by Heath, 12 May 1914; Lyttleton, 12 May 1914; McMahon, 12 May 1914; Paget, 13 May 1914; Pratt, 14 May 1914; Scarborough, 15 May 1914.

regardless in order to secure victory'. He was, according to Nicholas Griffin, 'a production-oriented traditionalist who preferred only to see the "bottom line" without dwelling on the means of reaching it', and he condemned the 'repeated requests for graphs demonstrating job performance' that he received as 'irksome' and 'futile'.[116] So long as the work required 'got done', Fairfax argued, the accumulation and assessment of data that recorded the precise nature of his units' performance were a waste of time. He was unwilling to countenance the potential benefits of attempts to coordinate and distribute the BEF's resources in the most efficient and systematic manner, and instead was preoccupied by the 'paper-mongering' that Beharrell's methods demanded. Even Lindsay, who was relatively open-minded with regards to the benefits of constant measurement, admitted that the administrative requirements of the system were incredibly bloated and generated data that was not necessarily consulted by busy senior officers. He wrote in April 1917 that he was 'waging a great war on unnecessary or rather meaningless forms but I am not sure that I shall win. Some people like returns as such even though they never mean to act on them'.[117]

However, others liked statistical returns so much that they became reluctant to engage with alternative sources of information on their working practices. In the directorate of docks, the plethora of quantitative data generated by the relentless measurement of the troops under Ralph Wedgwood's command created an image of remarkable progress. The twin pressures of the German submarine campaign and, from mid 1917 onwards, of the need to apportion shipping capacity for the transport of American troops exacerbated the need for the docks under Wedgwood's control to be operated with the maximum efficiency. The returns compiled within the directorate indicated that it had responded to the challenge remarkably well. Between January and September 1917 the volumes discharged from vessels at the ports under the directorate's authority rose from 12.5 tons per hour in port to 25.8 tons per hour. In the first half of 1918 the figures 'showed a continuous increase', which reached a peak of 34.4 tons per hour in port in July. Over an eighteen-month period Wedgwood oversaw a 175.2 per cent increase in the docks directorate's discharge rate, which contributed to a significant decrease in the number of ship-days lost by vessels awaiting a berth. From a peak of 100 days per week in February 1917, by March 1918 just 5.8 ship-days per week were lost at the Channel ports.[118]

---

[116] Griffin, 'Scientific management', p. 199.
[117] KUSCA: Lindsay papers, LIN 149, Lindsay to Erica Lindsay, 17 Apr. 1917.
[118] TNA, WO 107/296, report of British armies, pp. 16–17.

Behind the numbers a different story emerged, thanks to the investigative efforts of officers under the controller of labour, Colonel Edmund Wace.[119] As British manpower resources became increasingly stretched in the second half of the war – and the number of tasks devolved upon the British that had hitherto been the responsibility of the French grew – the demand for so-called 'unskilled' labour to dig, load, carry and build for the BEF outstripped supply. By November 1918 the labour controller's office was responsible for administering 385,000 men on the western front, and a core component of Wace's job was to flag up areas of the British war effort that were being run inefficiently. Under the terms of reference issued to the labour organization upon its establishment, Wace's officers provided a consultancy role rather than an executive function. They were unable to directly intervene to address examples of what they considered to be inefficient working practices. Instead, they recorded their observations and submitted them to the department or directorate responsible for employing the labour under investigation.[120] Over the summer of 1918 Wedgwood's directorate was the subject of a series of scathing memoranda, written by men with 'practical experience of dock labour in civil life' who had been drafted into the Labour Corps thanks to their specialist knowledge.[121] In July, when the directorate recorded its highest discharge rates of the war, the assistant controller of labour submitted the following observation of a morning's work at the port of Boulogne:

> On July 18th Docks assigned six Chinese [labourers] to each of 14 railway trucks in the rear of Transit Hangars L. & K. to accept from Horse Transport ammunition discharged from the steamship *Hansa*. At 11.00 a.m. [the] Chinese were ready in Trucks. At 11.20 a.m. the first load arrived at trucks; by 11.45 a.m. six out of the fourteen had received a load; by 12.20 p.m. a load had been delivered to all but one of the trucks. In other words, one gang of six men had been waiting one hour without doing any work.[122]

Two days earlier an even more stark example of inefficiency had been recorded at the same port: roughly ninety Chinese labourers were discovered 'lying asleep awaiting their turn to work' by Labour Corps observers.[123] The memoranda were forwarded to the QMG with the recommendation that labour control should be removed from the

---

[119] On the work of the various labour organizations established on the western front, see TNA, WO 107/37, labour force report; Starling and Lee, *No Labour, No Battle*, pp. 77–162.

[120] TNA, WO 107/37, labour force report, pp. 96–7.

[121] TNA, WO 107/37, labour force report, p. 99.

[122] TNA, WO 107/37, labour force report, pp. 99–100.

[123] TNA, WO 107/37, labour force report, p. 100.

directorate of docks. However, Wedgwood categorically refused to accept a proposal that the allocation of labour at the ports should be transferred away from his jurisdiction.

The Labour Corps' observations illuminated the key challenge of labour organization, which the BEF failed to adequately rectify over the course of the war. The actual labour needs of individual services, such as the directorate of docks, varied from day to day (and, indeed, from hour to hour). However, the various departments within which unskilled labour was employed 'wanted as much labour as [they] could get' throughout the conflict – ensuring that their individual requirements were met at all times was paramount, according to Lindsay. Therefore, even during periods where their own demands for labour were not so pronounced, the departments proved 'very reluctant' to release unskilled labour for employment elsewhere on the western front.[124] Whereas in a civilian business during a period of slack the wage costs of unproductive workers compelled employers to lay off unnecessary employees to protect profit levels, the absence of a profit margin in the BEF gave individual officers little inclination to willingly surrender labourers for service in other departments. The fear that they would not receive 'their' labour back when required appeared to supersede all other considerations.

As demonstrated earlier in the war with regards to individual units' reluctance to embrace canal transport to relieve pressure on the overburdened railway network, departmental desires to retain unskilled labour eclipsed considerations of the BEF's priorities as a whole. Consequently, the acceptance of any downgrade – however temporary – in the allocation of manpower was something to be fiercely resisted and combatted by the submission of 'extravagant' demands for labour. As Lindsay recalled after the war:

[hearing] a high official ... say, 'If no ships came into my ports for thirty days, I would whitewash all my buildings and relay all my track sooner than let another damned department have a single man of mine'. He was no doubt an extreme example, but there was a trace of that spirit in most administrative services.[125]

Individual departments essentially competed with one another for the finite resources available, rather than accepting a number of labourers allocated according to the BEF's needs. There was, noted Brigadier-General Lawson in a report into the number of men employed behind the lines, 'not one army but many armies' – for whom mutual help was anathema.[126]

[124] Lindsay, 'The organisation of labour in the army', p. 72.
[125] Lindsay, 'The organisation of labour in the army', p. 72.
[126] TNA, WO 106/362, physical categories and number of men employed out of the fighting area in France; Report by Lt. Gen. H. M. Lawson CB, 16 Jan. 1917, p. 19.

Far from being alleviated by the influx of civilians into positions of authority, the entrenched attitudes of individual departments were heavily reinforced. Ralph Wedgwood had been specifically appointed as director of docks because of his working knowledge of port operations, honed by years of experience at the North-Eastern Railway. Unsurprisingly, he felt little compulsion to consult labour officers over the employment of men at the docks, and was unwilling to surrender control over the manpower that he considered integral to the continued functioning of his directorate. In October 1918 Wedgwood did agree to the constitution of a committee that comprised himself, the controller of labour (Wace), a representative of the QMG's office and the principal naval transport officer, who was responsible for the naval aspects of operations at the docks. However, the committee's terms of reference explicitly stated that no 'fundamental changes' to labour allocation at the ports were to be considered, which meant that Wedgwood would retain overall control of labour distribution within the docks regardless of the committee's deliberations. Ultimately, the committee provided a platform for representatives of the various departments engaged at the docks to air their views and resulted in: the institution of weekly meetings at which issues could be discussed; an acceptance from the navy that labour could be released for service away from the ports in quiet periods; and an agreement that statistics compiled by each department would be made available to the others for consultation.[127] The agreement, Wace acknowledged – particularly in terms of the establishment of weekly meetings – did 'undoubted good' in what remained of the war.[128]

Wedgwood's protectionist attitude towards his authority over the directorate of docks' work demonstrates that the introduction of Britain's transport experts into positions of seniority within the BEF brought its own complications. The fact that there appears to have been a degree of confidentiality attached to the circulation of statistics between departments illustrates the continued existence of compartmentalized thinking within the BEF during the First World War. Lindsay ascribed this in part to the 'bad competitive habits' of the 'capitalists' drawn into the BEF following Geddes's appointment as DGT.[129] Wedgwood's conception of his own duties and the extent of his own jurisdiction made him resistant to accepting criticism of the working practices for which he was ultimately responsible, particularly when those working practices had helped contribute to an unambiguously vast improvement in the discharge rates achieved at the ports under the

[127] TNA, WO 107/37, labour force report, Appendix Z, 2 Oct. 1918.
[128] TNA, WO 107/37, labour force report, p. 104.
[129] Lindsay, 'The organisation of labour in the army', pp. 78–9.

BEF's control. The result, Wace noted pointedly after the war, was that 'the attitude assumed' within Wedgwood's directorate 'appeared to be that so long as the quick turnaround of ships was secured, no other consideration had any weight'.[130]

Wedgwood's actions in the latter part of the war undermined Lloyd George's 'rhetoric that the great feats of wartime organization were achieved by civilian experts', unaided if not actively hindered by insular and self-preserving soldiers.'[131] Civilianization was not a panacea to the organizational difficulties faced by the BEF that Lloyd George later proclaimed it to have been. The treatment of individual departments as personal fiefdoms was not an accusation that could be levelled solely at the military professionals. Both Fairfax and Wedgwood, the career soldier and the civilian specialist, pointed to their results to justify their approaches to the tasks for which they were responsible. Yet while the former saw the encroachment of civilian business methods as an unnecessary distraction from his duties, the latter drew upon the unequivocal message of progress that emerged from the quantitative analysis of his work to bolster his position and defend the utility of his methods.

Conclusion

Without access to the raw data it is impossible to assess the extent to which the directorate-general of transportation's data-gathering and analysis methods provided the catalyst for the operational improvements charted over the course of 1917 and 1918. The surviving records certainly demonstrate that the *esprit de corps* generated by the achievement of strong results provided a boon to productivity among the units under the directorate's control, while the unequivocal clarity of numerical representations of work done made the identification of inefficiencies easier. However, it is unclear how much of the increases made in, for example, the discharge rates at the ports were due to better working and supervisory practices enacted by Ralph Wedgwood and how much was the result of practice-based improvements and the introduction of new equipment to augment the existing capacity of the docks. As Theodore Stewart recognized after the war, men engaged on the tasks that underpinned the BEF's transport infrastructure simply got better at their jobs the more they did them. Stewart noted that men for whom the duties of road making and repairing were 'entirely new' in July 1916 had,

---

[130] TNA, WO 107/37, labour force report, p. 99.

[131] K. Grieves, 'The transportation mission to GHQ, 1916', in *'Look to Your Front!' Studies in the First World War by the British Commission for Military History*, ed. B. Bond et al. (Staplehurst, 1999), pp. 63–78, at p. 74.

through repetition and training, 'gained quite a reputation' for the work within six months.[132] As the general principle within the BEF was to retain men on the same class of work to help improve their skills and efficiency, it is unsurprising to see that they became more efficient as the war progressed.

Eric Geddes's powerful dual roles of DGT in France and DGMR in London, coupled with the unequivocal support he received from his superiors, insulated him from the constraints under which the BEF's professional soldiers had operated before the battle of the Somme. As Henniker reflected in the official history:

> The realisation of such great programmes for the provision of men and materials would have been impossible under the conditions existing during the first two years of the war. Under the earlier conditions demands from France to the War Office for transportation personnel and material were met with the answer 'the man-power situation does not permit', 'there is no labour available', 'the Ministry of Munitions will not allocate the steel', 'the Board of Trade say that the rolling stock cannot be spared', 'the Admiralty say they cannot find the shipping'.[133]

The evolving understanding among Britain's military and political leadership of the character of industrial, material-intensive warfare – and of the implications for the transport infrastructure upon which the conduct of colossal operations depended – weakened the constraints that Henniker recalled bitterly after the war. Geddes operated outside the previously rigid hierarchies that had subordinated the BEF's transport requirements on the western front to those of the artillery, the infantry and the needs of domestic industry. However, the freedom he and his colleagues enjoyed within the BEF was not limitless. Over the final two years of the war, Britain's transport experts were exposed to the constraints and compromises necessary for the successful prosecution of a global, coalition war effort.

---

[132] T. Stewart, 'With the Labour Corps in France', *Royal United Services Institution. Journal,* lxxiv (1929), 567–71, at p. 567.
[133] Henniker, *History of the Great War,* p. 226.

8. The balancing act: Britain's transport experts, the global war effort and coalition warfare, 1916–18

Lawson Billinton had seen little of the world beyond southern England before 18 March 1917 when, as a temporarily commissioned lieutenant-colonel in the Royal Engineers, he departed the country 'for a destination unknown'.[1] By the time he set foot on British soil once again the following June, the LBSCR's locomotive engineer had discussed the condition of Romania's railways with King Ferdinand I; inspected the oil fields and refineries of Baku; drawn his revolver upon the stationmaster at Kharkov; diverted the journey of an American ship in the Sea of Japan; and circumnavigated the globe. Billinton's wartime travels underline the First World War's global dimensions, and illustrate how civilian expertise could be applied to the wider allied war effort. From late 1916 until the end of the war Britain's transport experts were exposed to the peculiar demands of an international alliance.

This chapter investigates the manner in which Britain's transport experts navigated the challenges of multifaceted coalition warfare following Sir Eric Geddes's appointment as DGT and DGMR. Both on and beyond the western front Britain's transport contribution to the allied war effort expanded as the relative strengths of its partners waxed and waned under the sustained pressure of the ongoing conflict. From working in relatively small-scale roles with limited, localized objectives during the first half of the war, the second half of the conflict saw Britain's transport experts become immersed in the solution of international supply challenges that demonstrated the global interconnectivity of the various fronts and belligerents that contested the First World War.

The deployment of Britain's transport experts, both to new theatres and to work alongside different allies, illustrates two things: that the transportation problems caused by modern, material-intensive warfare could not be tackled by nations working in isolation from one another; and that the complexities of alliance warfare stretched beyond the machinations of the political and military high commands. The continued difficulties of supplying a worldwide war effort, with a voracious appetite for finite

---

[1] K. Marx, *Lawson Billinton: a Career Cut Short* (Usk, 2007), p. 78.

resources and raw materials, necessitated the consideration of supply and transportation questions at a supranational level. The war's demands required the application of British technical expertise to the solution of inter-allied transport problems, within the delicate balance of a fragile partnership. In the final two years of the war, Geddes, Billinton, and their contemporaries from across the British empire faced fresh obstacles to the execution of their duties. Their experiences of coalition warfare illuminate the challenges – cultural, political and strategic – that were an ever-present feature of the conduct of modern warfare alongside sovereign states with their own priorities and national interests.

Defining the global requirements of the British war effort

In his initial memorandum to Sir Guy Granet about the nascent directorate-general of military railways, Geddes acknowledged the distinctiveness of the directorate-general of transportation in France compared to the extant hierarchies in Britain's other expeditionary forces. When the latter wrote to the former in October 1916 he was, as the 'inspecting officer' for transportation on the western front, in the process of preparing 'a very definite statement of requirements in tons per week, working up to the maximum, together with the provision of locomotives, rolling stock, permanent way, personnel and so on, necessary to meet [the BEF's] requirements. We have, however, no similar organization in the other theatres'.[2] Before Geddes could be satisfied that the transport facilities behind Britain's multiple forces were sufficient to support the demands of the troops in their respective theatres, he appreciated the need for the War Office to obtain 'a very clear and definite statement from each theatre of war as to what is wanted, together with the date upon which it is required'.[3]

Geddes was reluctant to base his assessment of the needs of Britain's various expeditionary forces upon the judgments hitherto put forward by soldiers. As he explained to Granet, 'the soldier, as opposed to the civilian, asks for less than he really ought to have and – in my private opinion – this is due to the way in which he has been made to cheespare, and the fear he has of the Treasury, on account of the lean years before the war'. Geddes believed that Britain's soldiers had 'consistently put forward demands far below the real needs of the situation' throughout the war – an activity that had contributed greatly to the deficiencies he had witnessed behind the western front. Consequently, he felt unable to accept that 'the very modest demands' from Salonika, Egypt, Mesopotamia and East Africa were an

---

[2]   UWMRC, Granet papers, MSS. 191/3/3/51, memorandum to Granet, pp. 3–4.
[3]   UWMRC, Granet papers, MSS. 191/3/3/51, memorandum to Granet, p. 5.

accurate reflection of those theatres' resource requirements. The solution Geddes proposed was to replicate the transportation mission of August–September 1916; he suggested that 'we ... carefully select the best man we can get and send him out, with a small secretarial staff, to consult with the Administrative directors and commander-in-chief, in Salonika and Egypt, and another man to do the same thing in Mesopotamia and East Africa'.[4] Sir Francis Dent's investigations in the former, discussed in chapter four, represented the first of a series of investigations undertaken by civilian experts to identify the transport implications of Britain's global war effort.

Dent's assignment demonstrates Geddes's recognition both of the interdependence of Britain's myriad commitments to the fighting and the need to apportion resources on the basis of an overarching, long-term strategy. Railway wagons despatched to France could not be made available to serve General Murray's troops in Egypt, while every yard of light railway track sent to the western front was a yard of track that could not be laid behind General Milne's forces at Salonika. Ahead of Dent's departure for the Mediterranean, the Army Council outlined the constraints that prolonged, material-intensive warfare had placed upon the British empire's ability to provide for its armies in the field. In letters sent to both Murray and Milne the council emphasized

> that it may be impossible to meet urgent demands from the various theatres of war, and to decide upon the relative urgency of these demands, unless the Army Council is provided with the fullest information on the subject as far as possible in advance of the date when the need becomes acute. The present situation is that the supply of raw materials and the industrial capacity available is over-taxed ... The Army Council therefore has decided that it is more desirable that a complete survey of the requirements of the various theatres of war for transportation material should be made by experts who in consultation with the General Officers Commanding-in-Chief in the various theatres of war would consider the situation as a whole so that as far as possible provision may be made to meet the future demands of the various campaigns.[5]

For Murray in particular news of the 'present situation' came as no surprise. Arthur Webb, the Egyptian State Railways' agent in England, had already reported his 'great difficulty in obtaining ordinary stores for the maintenance of the Railway' on 4 October – a challenge greatly exacerbated

---

[4] UWMRC, Granet papers, MSS. 191/3/3/51, memorandum to Granet, p. 4.
[5] TNA, MT 23/677/9, mission of Sir Francis Dent to Egypt and Salonika to investigate land transportation questions. Request to naval transport staffs to afford him all possible facilities, Cubitt to Milne, 24 Oct. 1916; Cubitt to Murray, 24 Oct. 1916.

by the voluminous orders for railway materials placed by the BEF in the wake of Geddes's mission to the western front.[6]

Whereas Sir Douglas Haig – as the field commander of the BEF – advocated strongly and consistently throughout his tenure for the concentration of British resources within his theatre of operations, his first DGT on the western front had to take account of his wider brief. As DGMR, Geddes was tasked both with the fulfilment of the British army's global transportation needs and with coordinating the War Office's policy to increase the exploitation of local resources within the extra-European campaigns to preserve British shipping capacity. The increasing success of German and Austrian U-boats in the Mediterranean during 1916 reached its peak in the last quarter of the year, as 248,018 tons of cargo were lost between October and December – the period in which Dent's investigation took place.[7] Following Dent's advice that Britain – both for shipping and industrial capacity reasons – could not fulfil all of the EEF's demands for the materials considered necessary to construct new strategic lines, Murray's staff placed orders in India and the shipment of 822 miles of track from the sub-continent commenced on 1 January 1917.[8] By July of that year, when Brigadier-General John Stewart undertook a second investigation of circumstances in Egypt, the main line behind the EEF had progressed some 138 miles towards Palestine from its base on the Suez Canal.[9]

Stewart's visit to Egypt reflected the EEF's amended role following Lloyd George's appointment as prime minister, rather than any deficiencies with Dent's work the previous year. In early 1917 Murray was ordered to abandon his policy of aggressive defence in the desert to the east of the canal, and to use the summer months to prepare for 'large scale operations' later in the year. The single-line from the canal at Qantara, originally constructed to supply the troops engaged in forward defensive positions, became the principal supply artery for the EEF's proposed advance into Gaza and beyond.[10] Murray claimed in early May that the existing infrastructure was 'barely sufficient' to sustain the existing force, and predicted that the 'railway would be strained to its limit' following the arrival of the 74th and

[6]   TNA, WO 95/4379, branches and services. Deputy quarter-master general, diary entry, 4 Oct. 1916.

[7]   K. C. Ulrichsen, *The Logistics and Politics of the British Campaigns in the Middle East, 1914–22* (Basingstoke, 2011), pp. 64–5.

[8]   TNA, WO 95/4389, DRT war diary, diary entries, 9 and 14 Nov. 1916; WO 95/4379, deputy QMG war diary, diary entry, 29 Dec. 1916.

[9]   TNA, WO 106/720, Stewart (railway) commission: precis of report, July 1917, p. 2.

[10]   M. Carver, *The National Army Museum Book of the Turkish Front, 1914–1918: the Campaigns at Gallipoli, in Mesopotamia and in Palestine* (London, 2004), pp. 94, 96.

75th divisions.[11] He had pressed for authorization to double-track the line in April but stressed to London that the Egyptian State Railways could not supply further material. However, as over two million tons of British, allied and neutral shipping had been sunk between February and April 1917 – and a further 320,000 tons had been damaged by enemy action following Germany's adoption of unrestricted submarine warfare – the effective allocation of the available shipping tonnage exercised British strategic planners even more thoroughly than had been the case earlier in the conflict.[12] Consequently, London had to be satisfied that the requested materials were fundamental to the success of military operations.[13]

According to L. S. Simpson, who accompanied Stewart on the mission, such was the importance attached to the investigation that Geddes originally intended to lead it himself.[14] However, in May 1917, Lloyd George appointed the latter as controller of the Royal Navy in response to 'the mismanagement of resources, particularly of supply', which the prime minister perceived to be rife within the Admiralty.[15] Therefore, leadership of the mission passed on to the man said, according to Haig, 'to be about the ablest builder of railways in the world'.[16] Stewart was tasked to examine whether the EEF's intended operations were feasible given the existing capacity of the railway line, and to identify the improvements and equipment required should the force advance to Jaffa, Haifa, Beirut, Tripoli, Homs and Aleppo (the latter being almost 600 miles from Qantara).[17]

Stewart's report combined encouragement for the immediate future of operations in Palestine with forecasts of the theatre's resource requirements should the EEF press the Ottomans into a significant retreat. He acknowledged that sea transport offered a 'useful supplement to the railway', but warned against further dependence being placed upon such

---

[11] *History of the Corps of Royal Engineers*, ed. H. L. Pritchard (11 vols., Chatham, 1952), vi. 276–7.

[12] D. Stevenson, *1917: War, Peace, and Revolution* (Oxford, 2017), pp. 67–87; K. C. Ulrichsen, *The First World War in the Middle East* (London, 2014), pp. 44–5.

[13] Ulrichsen, *The Logistics and Politics of the British Campaigns*, p. 56.

[14] L. S. Simpson, 'Railway operating in France', *Journal of the Institution of Locomotive Engineers*, xii (1922), 697–728, at p. 711. Simpson and Stewart were accompanied on the mission by Colonel William McLellan, a Scottish electrical engineer and partner in the consultancy firm Merz and McLellan. Prior to the war the firm had worked alongside the North-Eastern Railway on the electrification of local lines in the Tyneside area.

[15] K. Grieves, *Sir Eric Geddes: Business and Government in War and Peace* (Manchester, 1989), p. 41.

[16] NLS, Haig papers, Acc. 3155/110, diary entry, 17 Feb. 1917.

[17] Simpson, 'Railway operating in France', p. 712.

an 'unsecure source of supply'.[18] Consequently, he confirmed that – unlike in France, where the existence of roads, light railways and IWT provided the BEF with a range of transport options – the EEF had to rely almost exclusively upon the construction and operation of the railway line across the desert. Stewart concluded that the force's seventy-seven locomotives and 1,300 wagons were 'ample for present and sufficient for immediate prospective requirements', and that the existing personnel in Egypt were capable of running sixteen trains per day over the line from Qantara to Jaffa once ongoing construction work was completed. Therefore, he considered it 'unnecessary to double the track or to send out additional rolling stock' provided the EEF was 'not largely increased' and halted at Jaffa. However, if the force advanced beyond the Jerusalem–Jaffa line, Stewart advised, the doubling of the line at least as far as Rafa would become necessary and he recommended that one hundred miles of track be made available to the Egyptian authorities to 'meet unforeseen eventualities'.[19]

Alongside providing London with an unequivocal, independent statement of the theatre's potential requirements, Stewart's report illustrates that Britain's transport experts did not always agree with each other's methods and projections. Murray's DRT in Egypt was the highly experienced Brigadier-General Sir George Macauley, who had become the Egyptian State Railways' general manager after a career in the Royal Engineers that included active service in Kitchener's Sudanese campaign. Stewart believed, regardless of Macauley's knowledge of both civilian and military railway operations in the region, that the latter's estimate of the rolling stock required to service the EEF's advance into Palestine was excessive. Macauley had based his forecasts upon a prospective locomotive mileage per day of forty-three miles, which broadly equated to the extant figures recorded in France during April 1917.[20] Stewart asserted that a locomotive mileage of 'not less than' sixty-six miles per day 'should be assumed and worked on in all estimates of rolling stock requirements for any extensive railway developments in Palestine'.[21] It is unclear whether Stewart drew his figures from the mileage that the directorate-general of transportation considered to be an achievable target on the French railways. If so, his projection proved to be somewhat unambitious, and Macauley's figure remarkably conservative

---

[18] TNA, WO 106/720, Stewart commission report, p. 5.

[19] TNA, WO 106/720, Stewart commission report, p. 4.

[20] TNA, WO 106/720, Stewart commission report, p. 4; J. G. Beharrell, 'The value of full and accurate statistics: as shown under emergency conditions in the transportation service in France', *Railway Gazette: Special War Transportation Number*, 21 Sept. 1920, pp. 37–9, at p. 38.

[21] TNA, WO 106/720, Stewart commission report, p. 4.

for the Egyptian theatre. Whereas the highest figure for locomotive mileage recorded by the ROD in France was just over fifty-five miles per day, the average goods engine on the Egyptian front 'ran at least 1,000 miles and sometimes 1,400 miles a week' during March 1918.[22]

General Edmund Allenby's appointment as commander-in-chief of the EEF on 27 June 1917 changed the EEF's outlook once again, and forced the War Office to confront the hitherto hypothetical additional resources listed in Stewart and Macauley's estimates. Before the new commander arrived in Cairo, Lloyd George 'promised to deliver all the men and resources Allenby might consider necessary' to realize the prime minister's ambition of 'Jerusalem by Christmas'. If the soldier 'cheespared' and the enterprise failed, Lloyd George warned, responsibility for the failure would fall upon Allenby alone.[23] The latter took the prime minister's advice seriously, and demanded another infantry division, five squadrons of aircraft, more artillery and additional engineer, signals and medical units to augment his forces for the assaults on Gaza and Beersheba.[24] The demands of Third Ypres meant that Allenby's requests were not met in full, but the War Cabinet insisted that he should 'strike the Turks as hard as possible' by the autumn.[25] On 21 July he received the authorization to double the line from Qantara to Rafa that Murray had requested in May, and the arrival of two more railway construction companies – which doubled the number attached to the EEF – permitted construction to proceed at a rate of one mile per day as preparations for the advance intensified.[26]

The railway behind the EEF was fundamental to the success of Allenby's 1917 and 1918 campaigns. The artillery support for the assault on Gaza from 31 October comprised a concentration of guns the equivalent of that assembled by the BEF on 1 July 1916, and it fired the heaviest bombardment to take place outside Europe in the entire war.[27] As Rob Johnson has observed, Allenby resisted pressure from London to attack before his forces were at full strength and prepared.[28] His methodical approach ensured the EEF achieved both numerical and material superiority over the defending Ottomans, which precipitated the latter's abandonment of Gaza on 6

---

[22] Beharrell, 'The value of full and accurate statistics', p. 38; 'The Palestine campaign', *Railway Gazette: Special War Transportation Number*, 21 Sept. 1920, pp. 119–28, at p. 127.

[23] L. James, *Imperial Warrior: the Life and Times of Field-Marshal Viscount Allenby 1861–1936* (London, 1993), p. 111.

[24] Carver, *The National Army Museum Book of the Turkish Front*, pp. 207–8; James, *Imperial Warrior*, p. 118.

[25] R. Johnson, *The Great War and the Middle East: a Strategic Study* (Oxford, 2016), p. 193.

[26] Pritchard, *History of the Royal Engineers*, vi. 277.

[27] Ulrichsen, *The Logistics and Politics of the British Campaigns*, p. 73.

[28] Johnson, *The Great War and the Middle East*, p. 199.

November. By the middle of the month the force had advanced sixty miles, despite considerable logistical difficulties.[29] 'Appalling weather conditions' rendered the tracks and roads in the Judean hills almost impassable. However, thanks to the 'untiring work' of the EEF's labourers, the necessary repairs to the existing rail and road networks around Jerusalem were completed in time for Allenby to press on and deliver Lloyd George's 'Christmas present to the nation' on 9 December.[30] When the advance recommenced the following September the double-track to Rafa carried more than 2,000 tons of supplies per day, while at its peak the 5,500-strong ROD in Egypt and Palestine operated 169 locomotives, 2,573 wagons, fifty passenger coaches and ninety-eight hospital coaches. While mechanical transport and coastal shipping took on an increased role after the EEF advanced beyond Damascus in October 1918, the 627 miles of standard-gauge track laid under Macauley's direction 'were probably the most important single factor in achieving the superiority in numbers and material resources which made the [EEF's] final campaign so decisive'.[31]

Closer to home, an even longer railway line consumed the attentions of Britain's transport experts from early 1917. The establishment of a 1,460-mile-long overland route from the French Channel coast to the heel of Italy, conceived to improve communications between Britain and the eastern Mediterranean, exemplified the increased exposure of British expertise to transport operations in continental Europe after the battles of 1916. Before 1914 the normal route for passengers travelling between Britain and India had involved an overland journey from the Channel to Marseille or Brindisi, and railway services had been synchronized with the departure and arrival times of ships at the two ports. As Britain's war effort expanded around the globe – bringing with it demands for the movement of military personnel, government officials, nurses and civilian specialists to theatres outside western Europe – the French and Italian authorities began to complain about being 'crowded out of their own trains' by British travellers.[32] The British and French governments had agreed to the establishment of a new line of communications for the transport of troops from the western front to Salonika, via the Italian and Greek railways, at an inter-allied conference on 20 October 1916.[33] However, activity only really began on 7

[29] Stevenson, *1917*, p. 345.

[30] Ulrichsen, *The Logistics and Politics of the British Campaigns*, pp. 44, 73.

[31] D. Stevenson, *With Our Backs to the Wall: Victory and Defeat in 1918* (London, 2011), p. 235; Pritchard, *History of the Royal Engineers*, vi. 410.

[32] A. M. Henniker, *History of the Great War: Transportation on the Western Front, 1914–1918* (London, 1937), p. 287.

[33] TNA, CAB 28/1, papers I.C. 0–12, Conclusions of the Anglo-French conference held at Boulogne, 20 Oct. 1916, p. 3.

January 1917, when delegates from Britain, France and Italy approved the development of an overland route with 'the object[s] of diminishing the length of communications by sea, which are at present seriously threatened by submarine attack, and reducing the [allies'] dependence on sea transport'.[34] The Italian minister of transport agreed to discuss the project with representatives from France and Britain, and the following day Lord Milner instructed Geddes to identify a suitable expert for the mission. The latter selected Guy Calthrop, the LNWR's general manager, who pursued the task with great energy in the weeks that followed.

Geddes's civil–military mission of the previous summer provided the blueprint for Calthrop's. Alongside the LNWR's superintendent, Calthrop was accompanied by the Royal Navy's Commodore Irwin and army officers from the QMG's department and the directorates of equipment and ordnance stores, medical services and railway traffic.[35] The party left Charing Cross on 14 January, just a week after the allies had agreed to develop the line, and completed its work within three weeks. Calthrop's final report was submitted on 7 February, exactly one month after the inter-allied conference in Rome had authorized the examination.[36] In it, he looked favourably upon the overland route but acknowledged significant obstacles to the scheme's realization. The available facilities at Cherbourg, the only port to which the cross-Channel voyage was short and that was not already in use 'to its full extent for the BEF in France', were limited. The French navy occupied the port's dockyard, the pier had been pressed into constant action to replace berthing accommodation used by the BEF at other French ports and the railway connections from the small commercial port that remained 'needed much alteration'.[37] Given these deficiencies, Calthrop estimated the capacity of Cherbourg to be between 1,200 and 1,400 tons per day. However, as with Francis Dent's projections of the Bassin Loubet's capacity two years earlier, Calthrop's figures proved over ambitious. As Henniker noted after the war, 'in actual practice only about 600 tons per day was attained' from the Bassin du Commerce at Cherbourg in 1917–18.[38]

If Cherbourg provided the northern terminus of the overland route by default, several options for the Mediterranean terminus were considered

[34] TNA, CAB 28/2, papers I.C. 13–32, Conclusions of a conference, 5–7 Jan. 1917, p. 1.
[35] TNA, CAB 24/7/11, proposed overland route to Salonica, Calthrop to Derby, 7 Feb. 1917, p. 1.
[36] TNA, CAB 24/7/11, proposed overland route to Salonica, Diary of the work of the War Office mission.
[37] TNA, CAB 24/7/11, proposed overland route to Salonica, Calthrop to Derby, 7 Feb. 1917, pp. 3–7; Henniker, *History of the Great War*, p. 289.
[38] Henniker, *History of the Great War*, p. 289 n. 2.

before Taranto was selected. The commercial port at Taranto could only be reached through 'a very congested station', and Calthrop was 'convinced that a very serious delay would take place' if traffic for the allied forces at Salonika ran through Taranto to the commercial quay.[39] To avoid the city the Italian naval commander at Taranto recommended the development of a location to the east, on the south side of the Mar Piccolo. An entirely new port had to be constructed on the site and a multinational force under the command of Lieutenant-Colonel Charles Langbridge Morgan, chief engineer of the LBSCR before 1914, commenced work that summer. By January 1918 they had created what Colonel Rhys Williams described as 'a remarkable achievement':

> A bare hill-side in July, 1917, had been transformed into the site of a camp capable of containing 15,000 men, most of whom are housed in stone or Nissen huts. One hundred and eighty-eight Nissen huts have already been erected. Hospitals containing 520 beds have been built. A stone-built quay with six wooden jetties, each of sufficient depth to load three barges at a time, is in working order. Alongside the warehouses six sidings each 700 yards in length have been completed, and the warehouses are connected with the quay by a Decauville line. The Triage is in working order with seven lines 1,000 yards long.[40]

The fact that sufficient materials to create the new port were redirected to Taranto underscores the importance attached to the development of the overland route during 1917, which the war policy committee hoped in June would eventually account for 36,000 of the 51,000 tons per month required by the British forces in Salonika and Egypt.[41] The committee's aspirations were never met. It took until early July for the labour and materials required to construct the terminus to be despatched to southern Italy and, although a passenger service commenced operations on 28 June, the first consignment of goods for Taranto did not leave Cherbourg until 8 August. A regular service of two trains per day from the Channel to the Mediterranean did not begin until the final week of October 1917.

At that point the war intervened to stymie the development of the Cherbourg–Taranto route. On 24 October, German and Austrian forces broke through in the upper Isonzo valley and sent the Italian army into a desperate retreat. Within a week Italian troops had fallen back as far as their

---

[39] TNA, CAB 24/7/11, proposed overland route to Salonica, Calthrop to Derby, 7 Feb. 1917, pp. 7–8.

[40] 'The Mediterranean line of communication', *Railway Gazette: Special War Transportation Number*, 21 Sept. 1920, pp. 101–7, at p. 103.

[41] UWMRC, Granet papers, MSS. 191/3/4/145, shipping allotted to overseas expeditions outside France. Interim report by General Smuts, 23 June 1917, p. 1.

rearmost defensive positions on the River Tagliamento, some forty-five miles away from the Isonzo, and the French and British governments had resolved to bolster their ally with reinforcements drawn from the western front. The goods service between Cherbourg and Taranto was suspended on 30 October, and the first British troops entrained for Mantua a week later.[42] Supply trains for Salonika and Egypt did not recommence until January 1918 and, following another suspension in response to the German spring offensives in March and April, the Cherbourg–Taranto line recorded its highest traffic figures in June 1918.[43] However, even the 725 tons per day carried on the line in the summer of 1918 fell well short of Calthrop's initial estimates.

The root causes for these relatively low returns lay with the allies rather than the central powers. While the construction of the facilities required to transfer goods from shore to ship demonstrated that the allies could work collaboratively to solve the supply challenges raised by the First World War, the provision of railway equipment to operate between Cherbourg and Taranto highlighted the technical constraints that existed between sovereign states lacking in standardized infrastructure. In his February 1917 report Calthrop produced various calculations for the route's locomotive, rolling stock and engine crew requirements (see Table 8.1). The Italians confirmed their ability to provide the locomotives required for four marches on their own soil, but warned that the engines for any additional marches had to be provided by the British. The French were similarly accommodating, and committed to supply the passenger coaches and locomotives for four marches on French soil. However, as Sir Guy Granet noted, this agreement stood 'only on condition that they are immediately replaced by British coaches and locomotives of equal capacity'. The French minister of war accepted that French rolling stock should be employed along the entire route, but demanded that the British shipped replacement wagons to run on the French main lines as soon as the Cherbourg–Taranto service commenced. Granet acknowledged that the demand for 6,000 wagons was 'a large number', but believed that 'the British railways might manage it at a pinch'.[44]

The outlook for motive power was far less positive. Before the war the British railways possessed approximately 23,000 locomotives. In peacetime roughly 600 of those locomotives were withdrawn from service and replaced with new stock each year. However, the outbreak of war had dislocated the locomotive building and overhauling processes in Britain, as materials and

[42] Stevenson, *1917*, pp. 227–31.
[43] Henniker, *History of the Great War*, pp. 294–5.
[44] TNA, CAB 24/7/11, proposed overland route to Salonika, Calthrop to Derby, 7 Feb. 1917, pp. 9–10; memorandum by Sir Guy Granet, 22 Feb. 1917, p. 3.

Table 8.1. The equipment and personnel required for operation
of the overland route to Salonika, February 1917.

	1 daily passenger train	1 daily passenger train and 3 daily goods trains (700–800 tons)	1 daily passenger train and 5 daily goods trains (1,100–1,300 tons) †	1 daily passenger train and 9 daily goods trains (2,000–2,400 tons) ‡
Passenger coaches	400	400	400	400
Wagons (10-ton capacity)	0	2,000	3,300	6,000
Locomotives	22	88	182	370
Drivers	22	88	132	220
Firemen	22	88	132	220

Notes: The figure of 2,000 wagons was arrived at as follows: each train was limited to 40 wagons at 3 trains per day. Fifteen days were allowed for the round trip, and a 10 per cent allowance was made for spares.

† = Projected capacity of Cherbourg without additional harbour works.

‡ = Ultimate capacity considered.

Source: TNA, CAB 24/7/11, proposed overland route to Salonica via Cherbourg and Taranto. Brief summary of Mr. Calthrop's report, memorandum on proposed overland route to Salonica by Sir Guy Granet, 22 Feb. 1917, p. 3.

manpower were redirected into the fulfilment of the armed forces' various requirements. The SECR built just two locomotives and reconstructed one engine during the war, while across the country as a whole just 803 new engines were put into traffic between August 1914 and April 1917. At the same time the British railway companies had despatched 420 locomotives overseas and received 'urgent requests' for another 150 engines. 'Owing to a want of men and materials', the REC advised the War Cabinet, Britain's railways were 'short of no less than 1,600 locomotives' in May 1917. The REC complained that 'when an undertaking was given to send … 380 locomotives to France it was distinctly understood that the Companies would be put in a position to replace them at once and details of all materials required for the purpose were sent to the Ministry of Munitions'. The ministry, the committee claimed, had failed to supply the items requested by the railway companies or ensure that engines under manufacture with private locomotive builders were requisitioned and placed at the railway companies' disposal.[45]

---

[45] TNA, CAB 24/14/83, memorandum by the REC. Shortage of materials for repairs and renewal of permanent way, locomotives, carriages and wagons, 24 May 1917, pp. 3–4.

Britain's transport experts were asked to do as much as required with as little as possible during the war, in conditions which exerted great strain on the domestic railway network. 'The movement of traffic over the Railways at the moment', the REC highlighted, 'is greatly in excess of what it was in 1913 which was the busiest year the Companies had experienced prior to the War. This greatly increased traffic is being operated with less locomotives, less wagons and a greatly reduced Staff of efficient Railwaymen'. Unless steps were taken to allocate materials to the railways for new construction, the committee warned, 'there is the possibility and even the probability of the position in this Country becoming from the traffic working point of view as bad as it had been in the North of France'.[46] Consequently, the REC rejected requests to supply the projected 370 locomotives required to fulfil Calthrop's highest estimate for the Cherbourg–Taranto line, as it was 'impossible for the Railways to find these engines and at the same time to deal with the traffic in this country'.[47]

However, the existing stocks of the British railway companies represented the only immediately available source of locomotives for the new route to the eastern Mediterranean. Therefore, in the wake of the REC's memorandum, a gathering of locomotive engineers, goods managers and railway company superintendents were tasked to revisit the decisions made in December 1916 and identify further restrictions to domestic traffic that could free up locomotives for service overseas. Holiday traffic was subjected to further controls, express services were withdrawn, lightly loaded goods trains run for government departments were curtailed, passenger services were cancelled and the committee restated its request for 'all the material [the railway companies] may require for the repair of their locomotives, wagons and permanent way'. By early July agreement had been reached for the despatch of 175 locomotives to the European theatres by November, but by the end of the year only 155 engines had actually departed British shores – well short of the number required to operate five goods trains per day between Cherbourg and Taranto.[48] Yet although the overland route did not achieve the ambitious targets set for it by Calthrop in February 1917, it did 'justify its existence' by permitting the redirection of shipping tonnage to other duties and offering respite for soldiers deployed in inhospitable theatres. By the end of the war it had removed roughly 500,000 shipping tons and 350,000 troops from the submarine-infested waters of the Mediterranean, and provided

---

[46] TNA, CAB 24/14/83, memorandum by the REC, 24 May 1917, pp. 6–7.

[47] TNA, CAB 24/14/83, memorandum by the REC, 24 May 1917, pp. 4–5.

[48] E. A. Pratt, *British Railways and the Great War: Organisation, Efforts, Difficulties and Achievements* (2 vols., London, 1921), ii. 656–60.

British troops in the unhealthy Macedonian theatre with opportunities to undertake periods of home leave.[49]

Britain's transport experts and the limits of inter-allied cooperation

The difficulties the allies experienced in providing the resources necessary to operate the Cherbourg–Taranto route indicated the diminishing quality and quantity of materials available to the belligerents as the war ground on. Like those of France, Germany and the other belligerents engaged in the material-intensive combat of the First World War, British resources were limited and required careful management. Britain's relative distance from the destructive effects of the fighting did not isolate it from the gradual processes of erosion, wear and eventual breakdown that afflicted the machinery of its partners and enemies over the course of the conflict. As David Stevenson has noted, 'governments and public alike resigned themselves to fighting into 1919 or even 1920' during the spring and summer of 1917 – unable to accurately predict an end date for the hostilities, both sides were forced to balance fulfilling the material requirements of the front line with the continued maintenance of an operable transportation network.[50]

The nature of the support that had hitherto been supplied to the British by its hosts lay at the heart of a bitter dispute between the coalition's senior partners shortly after Geddes's arrival in France. In mid November 1916 a letter from Joffre to the French mission at GHQ stated that the number of wagons allotted to British traffic had risen from between five and six thousand in January 1915 to 19,350 at the conclusion of the Somme offensive.[51] Consequently, the French demanded that the British provide 19,350 wagons as soon as possible to cover the BEF's transport requirements and asked that the British be ready to supply a total of 54,000 wagons by the time the allies reached the German frontier.[52] Following a meeting between British and French transport authorities a few days later, in which Geddes was adjudged to be 'very liberal in the way he proposed to meet the French difficulties in their transport', the British military attaché described the French approach in a letter to Lloyd George:

---

[49] Henniker, *History of the Great War*, pp. 296–7; 'Mediterranean line of communication', p. 107.

[50] D. Stevenson, *1914–1918: the History of the First World War* (London, 2004), p. 298.

[51] PA, Lloyd George papers, LG/E/1/5/16, translated copy of a letter from Joffre to the French mission at GHQ, including handwritten notes from Geddes dated 19 Nov. 1916.

[52] PA, Lloyd George papers, LG/E/1/5/17, locomotives and rolling stock for the British armies in France and Belgium, 24 Nov. 1916, p. 2.

They showed a huckstering spirit and I do not think had any intention of trying to help us in any way. I gave [Albert] Claveille a bit of my mind today, and told him quite frankly that if the French authorities did not show a more conciliatory spirit to us, than they had shown at the conference yesterday, it would no doubt be necessary to reduce the size of our Army in France.[53]

Geddes was not prepared to take the estimates of his coalition partners at face value,[54] and expressed doubts as to the motives behind Joffre's request. He wrote to Lloyd George that the French demands were 'excessive', and expressed his conviction that 'they neither expect to have them met in full, nor believe that it is possible to meet them in full'.[55] On 13 December Claveille and Geddes reached a preliminary agreement for the British to provide 29,000 wagons over the following year, and to be prepared to supply 42,000 wagons in the event of a general advance.[56]

The winter weather in 1916–17 exacerbated the poor condition of the French railways. A severe frost set in during January 1917, which increased congestion and reduced the circulation of rolling stock around the network. On 24 January the situation became so acute that the French authorities placed an embargo on all military traffic for the French army other than supplies, ammunition and railway material. On the same day Geddes 'extracted' a 'candid confession' from Brigadier-General Camille Ragueneau, director of the French rearward services, that the French could see 'no hope' of being able to deal with more than the 150,000 tons per week received by the BEF at that time.[57] Geddes's choice of language in this letter is indicative of the state of suspicion that existed between the two transport authorities. 'Unless we can get 200,000 tons carried from the ports weekly', Haig wrote in his diary on 28 January, 'we cannot carry out our offensive as early as we wish'.[58] The new French commander-in-chief, General Robert Nivelle, refused to be persuaded by his counterpart's position, and rather condescendingly advised the British Field Marshal that if it was

[53] PA, Lloyd George papers, LG/E/3/14/29, Le-Roy Lewis to Lloyd George, 22 Nov. 1916. Albert Claveille, an engineer and former director of the French State Railways, became under-secretary of state for transport in the French government on 14 Dec. 1916.

[54] The DGT's figures stated that approximately 12,000 of the wagons used for the BEF's supply needs in Nov. 1916 had been provided by the French. See PA, Lloyd George papers, LG/E/6/1/5(A), memorandum on the question of railway wagon supply for the British army in France, by W. Guy Granet, 7 Nov. 1916, p. 1.

[55] PA, Lloyd George papers, LG/E/1/5/16, Translated copy of a letter from Joffre; LG/E/1/5/17 Locomotives and rolling stock, p. 6.

[56] *Les armées françaises dans la grande guerre: la direction de l'arrière* (Paris, 1937), pp. 601–2; Henniker, *History of the Great War*, p. 247.

[57] UWMRC, Granet papers, MSS. 191/3/4/47, Geddes to Granet, 24 Jan. 1917.

[58] NLS, Haig papers, Acc. 3155/110, diary entry, 28 Jan. 1917.

impossible to collect all the personnel and material necessary to carry out the whole of the contemplated work [to improve the BEF's transport position], it is essential that the various works be carried out according to their urgency, and that those indispensable for the first operations be carried out first, that is to say, those which have reference to offensive operations near Arras.[59]

Four days later, Haig 'discussed the state of the French railways and the effect on our preparations for the offensive' with his deputy chief of staff, Major-General Richard Butler and Geddes. They concluded that an improvement in the traffic position was improbable even if a 'drastic curtailment of civil traffic' took place, which was unlikely, and agreed that the condition of the railway network ought to influence the decision to launch the BEF's offensive operations.[60] After hearing Geddes's views, the War Cabinet agreed with Haig's recommendation that French and British ministers, senior commanders and transport experts should convene to discuss the ongoing 'crisis of transportation'.[61] Consequently, a conference was arranged to take place in Calais on 26 and 27 February 1917.

Of all the inter-allied meetings to take place during the First World War, the Calais conference has proved the most controversial. The political machinations that led to Haig's subordination to Nivelle for the duration of the spring campaign – described by William Philpott as 'probably the most unfortunate episode' in the coalition's history – have become synonymous with the events that took place in Calais, and have dominated the published accounts of those who were present.[62] The conference's agenda contained six items for discussion, all of which related to the operation and development of the transport network rather than the existing command structure on the western front.[63] However, Sir William Robertson, who had made Haig aware of his reservations about the presence of French and British government minsters at the conference, dedicated just one sentence of his autobiography to the transportation crisis.[64] Lloyd George's *War Memoirs* did acknowledge the 'long delays over questions of transport and coordination' that had determined the need for a meeting of allied political and military leaders.

[59] NLS, Haig papers, Acc. 3155/110, diary entry, 9 Feb. 1917.

[60] NLS, Haig papers, Acc. 3155/110, diary entry, 13 Feb. 1917.

[61] LHCMA, Robertson papers, 7/7/5, Robertson to Haig, 13 Feb. 1917; NLS, Haig papers, Acc. 3155/110, Robertson to Haig, 14 Feb. 1917; diary entry, 15 Feb. 1917; note on the present transportation situation, 16 Feb. 1917.

[62] W. Philpott, 'Haig and Britain's European allies', in *Haig: a Reappraisal 80 Years On*, ed. B. Bond and N. Cave (Barnsley, 2009), pp. 128–44, at p. 136.

[63] TNA, WO 158/41, transportation: agenda and notes for the Calais conference, 26 Feb. 1917.

[64] LHCMA, Robertson papers, 7/7/7, Robertson to Haig, 14 Feb. 1917; W. R. Robertson, *From Private to Field-Marshal* (London, 1921), p. 307.

However, his account of those discussions – which he claimed had 'occupied much of our time' – comprises little more than an attempt to portray Haig as an obstructive figure who created 'difficulties' that contributed to the failure of Nivelle's offensive.[65] In his analysis of the prime minister's account, Andrew Suttie comprehensively demolished Lloyd George's version of events, arguing that the latter distorted the facts and omitted important material from his recollections of the conference.[66] Lloyd George's exploitation of the conference to subordinate Haig to Nivelle was an 'ambush', which led to 'very little progress' being made with respect to 'the railway question' in Calais.[67]

The internal power struggle between Britain's political and military leaders that played out in the Hotel Terminus was mirrored by a vituperative inter-allied disagreement between France's and Britain's transport experts. The French, as with their coalition partner, had centralized their transport services in the latter part of 1916 under a civilian – Albert Claveille.[68] At Calais, Claveille and Geddes continued their discussions in an adversarial rather than conciliatory tone. The latter began by pressing the French for a date upon which the British could expect the railways to be able to handle 200,000 tons per week on the BEF's behalf, the figure he considered necessary to service the force's demands in preparation for an offensive. Ragueneau stated merely that he 'hoped to reach' a figure of 194,000 tons per week by the end of March, to which Lloyd George responded by explicitly linking the capacity of the transport network to the likely success of the BEF's operations. If the tonnage demanded by the BEF could not be provided, the prime minister warned, 'either Sir Douglas Haig must make his attack insufficiently provided, or else he must postpone it'. He was, he said, 'very anxious' that Claveille and Ragueneau understood the connection between the provision of sufficient transport and the BEF's participation in Nivelle's campaign.[69]

The French representatives did not allow their guests to dictate proceedings. Claveille responded to British accusations that they had provided insufficient transport by directing the prime minister's attention to the locomotives the British had promised to despatch to assuage the transportation crisis. Claveille argued that British deliveries of valuable

[65] D. Lloyd George, *War Memoirs of David Lloyd George* (2 vols., London, 1938), i. 891–3.

[66] A. Suttie, *Rewriting the First World War: Lloyd George, Politics and Strategy, 1914–1918* (Basingstoke, 2005), pp. 116–19.

[67] Stevenson, *1917*, p. 126; LHCMA, Robertson papers, 7/7/8, Robertson to Haig, 28 Feb. 1917; NLS, Haig papers, Acc. 3155/110, Haig to George V, 28 Feb. 1917.

[68] The reasons behind this decision, which revolved around the need to maintain a balance between military traffic in the battle zone and civilian traffic across the rest of France, are summarized in *Direction de l'arrière*, pp. 513–14.

[69] UWMRC, Granet papers, MSS. 191/3/3/183, notes of an Anglo-French conference held at the Hotel Terminus, Calais, 26–27 Feb. 1917, pp. 2–5.

locomotives and rolling stock had fallen 'behindhand' during the first two months of 1917. Furthermore, Ragueneau added that the BEF used an unnecessarily large amount of rolling stock to service its requirements; while the French army requisitioned 2,800 wagons per day to supply its forces on the western front, the British demanded 8,000 per day for half the number of men.[70] Geddes, as Robertson noted ruefully after the conference, did not choose to question the accuracy of Ragueneau's figures or explain the reason why such a wide discrepancy between the two forces' needs existed.[71] When Lloyd George asked Geddes to respond to Ragueneau's charge, the DGT instead launched into a further attack on what he considered to be the French failure to properly manage the railways. Before considering the BEF's demands for rolling stock he

> stated that he first wished to reach an agreement about the question of tonnage and trains to railheads. He pointed out that the figure of 200,000 tons required in the ports has already been reduced from 250,000 by an abatement in Sir Douglas Haig's demands made in consequence of the shortcomings of the railways. Having obtained the railway facilities to serve the ports, as well as the local traffic, which consisted of such matters as stone for metalling the roads, and timber, which was just as essential to military operations as ammunition, the next step was to get it to the front. He had arranged that 200 trains a day should proceed to railhead. It was absolutely necessary to have these 200 trains. In this connection he reminded the conference that a good deal of railway traffic was required for the maintenance of stocks, and until you reached your total of 200 trains your forward dumps could not be realised. Today, however, we were only able to run 80 trains a day to the front.[72]

Following further disagreement between the French and British representatives – over the BEF's requirements during the preparatory period before the offensive, during the phase of military operations and in the prospective advance following any success on the battlefield – Lloyd George moved to segregate the technical and strategic components of the conference. He observed that 'the discussion might continue forever on these lines. The experts did not appear to agree on a single figure'. Therefore, the specialists were invited to 'retire and discuss the question among themselves'.[73] The debate on transportation, for which the conference had been scheduled,

---

[70] E. Greenhalgh, *Victory through Coalition: Britain and France during the First World War* (Cambridge, 2005), p. 141.

[71] LHCMA, Robertson papers, 7/7/8, Robertson to Haig, 28 Feb. 1917.

[72] UWMRC, Granet papers, MSS. 191/3/3/183, notes of an Anglo-French conference, pp. 5–6.

[73] UWMRC, Granet papers, MSS. 191/3/3/183, notes of an Anglo-French conference, p. 8.

lasted less than two hours and achieved nothing more than an ill-tempered airing of grievances.[74]

Like Francis Dent and Gerald Holland before him, Geddes's freedom of action in France and Flanders was constrained by the attitude and priorities of the coalition's leading partner. The railway network behind the western front was a core component of the French army and state's supply system as well as the BEF's logistics chain. Therefore, the successful development, management and operation of this strategically vital artery required a constant process of negotiation, renegotiation, collaboration and compromise to ensure the sustenance of both nations' war efforts. However, Geddes's immediate response to events at Calais was to exclaim that the BEF was 'practically being "rationed" in the matter of trains by the French' and to question the utility of his remaining in France 'if this state of affairs [was] to last'.[75]

The man dubbed 'Napoleon' and 'Sir Hindenburg Geddes' by some observers, because of his perceived absolute command over British transportation in early 1917, struggled to adapt to the requirements of diplomacy and conciliation upon which coalition warfare depended.[76] Claveille's observation that the British had failed to deliver the agreed quantity of locomotives and wagons was accurate. By the time of the Calais conference the British had fallen over 30 per cent behind on the monthly schedule of wagon deliveries, while only fifty of the 100 locomotives Geddes had 'hoped' to despatch in the programme's first month had arrived in France.[77] Yet less than a week after the Calais conference, Geddes produced a letter for Haig that detailed 'the history of the whole transaction with the French'. The document, in Haig's words, showed 'clearly' how the French had 'failed to keep their agreements' – a response that suggests the British transport expert had either neglected or refused to countenance the possibility that Ragueneau and Claveille had raised legitimate concerns.[78]

---

[74] TNA, CAB 28/2, papers I.C. 13–32, notes of an Anglo-French conference held at the Hotel Terminus, Calais, 26–27 Feb. 1917, pp. 1, 4. The first session, at which transportation was the central focus, began at 3:30 p.m. on the 26th. The railway experts withdrew, and the conference adjourned 'for a short time', before the second session commenced at 5:30 p.m.

[75] NLS, Haig papers, Acc. 3155/III, diary entry, 3 March 1917; I. M. Brown, *British Logistics on the Western Front, 1914–1919* (London, 1998), pp. 159–60.

[76] P. K. Cline, 'Eric Geddes and the "experiment" with businessmen in government, 1915–22', in *Essays in Anti-Labour History*, ed. K. D. Brown (London, 1974), pp. 74–104, at p. 75; K. Grieves, 'Improvising the British war effort: Eric Geddes and Lloyd George, 1915–18', *War & Society*, vii (1989), 40–55, at p. 47.

[77] TNA, CAB 24/7/11, proposed overland route to Salonica, memorandum by Granet, pp. 3–4. On 17 March 1917, by which time the schedule called for the British to have despatched 9,500 wagons to France, only 4,500 had arrived. See *Direction de l'arrière*, p. 602.

[78] NLS, Haig papers, Acc. 3155/III, diary entry, 3 March 1917.

Geddes's attitude towards the French remained truculent. At the next gathering of French and British leaders on 12 March – at which neither Claveille nor Ragueneau was present – the French minister of finance, Alexandre Ribot, acknowledged that the Chemins de Fer du Nord 'was in a terrible condition', warned that the civil population served by the railway had raised complaints over the shortage of rail traffic, and emphasized that the will to provide for the BEF's requirements was not matched by the possibility to do so. In the spirit of cooperation, the French minister of war, General Hubert Lyautey, proposed 'the establishment of a permanent Anglo-French bureau' to provide 'a continuous reciprocal examination' of the railway question 'both from the point of view of our requirements and the means of execution'.[79] Geddes 'stated that … nothing would be gained by the establishment' of such a bureau and reiterated his demand that the BEF be provided with 200 trains per day. For the French, Geddes's intransigence proved exasperating. Albert Thomas explained in response that, even if the French were able to meet their ally's request, 'no doubt from time to time they would have to desist for days from supplying the full number of trains in order to meet particular emergencies'. The creation of a permanent Franco-British organization for the exploration of transportation issues – as opposed to discussion at 'intermittent conferences' – offered a forum through which the traffic demands of the military and civilian users of the Chemins de Fer du Nord could be regularly examined, prioritised, and delivered according to the network's capacity and the needs of the military situation. However, Geddes demurred. 'So long as the French had the management of the Chemins de Fer du Nord', he stated, 'it was impossible for [the British] to share it or be responsible in any way for it'.[80] The opportunity to create an inter-allied forum for the allocation of finite transport resources thereby lapsed until the arrival of American troops on French soil imposed further pressures on the heavily burdened railway network.[81]

The preservation of harmonious relations with an equal, and in many ways a senior, partner proved difficult for Geddes. His recognized gifts of intuition, rapid decision and force, which proved crucial in the development of the Ministry of Munitions and the creation of the directorate-general of transportation, could not be exercised so liberally when the needs of the French military and civil population had to be considered alongside the BEF's requirements. The coalition environment effectively gloved the free

---

[79] TNA, CAB 28/2, papers I.C. 13–32, notes of an Anglo-French conference held at 10 Downing Street, 12 and 13 March 1917, pp. 3–4.

[80] TNA, CAB 28/2, papers I.C. 13–32, notes of an Anglo-French conference, 12 and 13 March 1917, p. 5.

[81] Greenhalgh, *Victory through Coalition*, p. 241.

hand Geddes employed behind British lines with Haig's and Lloyd George's support, and may have contributed to Geddes's withdrawal from direct involvement on the western front in May 1917. Lloyd George's decision to transfer Geddes to the Admiralty has been presented in previous accounts as the logical appointment of 'an organiser to carry out for the Admiralty the functions which the Ministry of Munitions had long operated for the army'. Certainly, the navy benefited from Geddes's introduction of managerial methods that were 'alien' to the senior service's 'badly co-ordinated existing administrative practices'.[82] Yet the railwayman lacked specialist knowledge of the shipping industry, while the move further exposed his self-acknowledged weaknesses as a political operator. 'I am a Political Chief among Naval Experts', Geddes wrote to Lloyd George in December 1917: 'I am essentially an executive man now employed in a non-executive job ... I am very conscious of the honour of being First Lord [of the Admiralty], but I am not a shipbuilder, Naval strategist, Speaker, or politician. I am a Transportation man, and I feel I can do my best work where my previous experience justifies my position'.[83] However, as the scope of transportation by that point covered 'the Allied world' rather than the coordination of resources for Britain's war effort alone, such a role required a collegiate approach. Lloyd George possibly suspected that Geddes, who less than a month earlier had told Lord Derby that the French were 'quite hopeless at running their own railways', was temperamentally unsuited to a task that demanded more conciliatory methods.[84]

Geddes's dismissive attitude towards his French counterparts was by no means unique among the senior figures in Britain's war effort. Sir Sam Fay reflected in his autobiography that 'the French railway organization throughout the war' had been 'anything but good',[85] while William Philpott has outlined how the relationship between Haig and his allies was 'beset by suspicion, antagonism and double-dealing'. Geddes displayed a similar jingoism to the commander-in-chief, with whom he had struck up an immediate and lasting friendship. Both perceived that their citizenship of the greatest empire on the planet possessed them with 'an innate sense of superiority over the foreigner', which contributed to the 'indifferent management of the joint campaign' on the western front.[86]

---

[82] Grieves, 'Improvising the British war effort', pp. 47–9. On Geddes's experiences at the Admiralty, see K. Grieves, *Sir Eric Geddes: Business and Government in War and Peace* (Manchester, 1989), pp. 40–68.

[83] PA, Lloyd George papers, LG/F/17/6/19, Geddes to Lloyd George, 20 Dec. 1917.

[84] PA, Lloyd George papers, LG/F/14/4/78, Derby to Lloyd George, 24 Nov. 1917. Geddes's assessment of Italy's railway officials was similarly derisive.

[85] S. Fay, *The War Office at War* (London, 1937), p. 104.

[86] Philpott, 'Haig and Britain's European allies', pp. 129–30.

Yet Britain's transport experts were not merely exposed to the challenges of coalition warfare in France and Flanders during 1917. In the first half of the year, as Geddes became exasperated by the divergence of priorities between the British and French war efforts on the western front, a succession of civilians travelled east to undertake missions in conjunction with Britain's allies in Russia and Romania. The efficient use of the transport infrastructure in eastern Europe was as critical to the sustenance of operations there as the French railway network was to the fighting on the western front. Yet Russia lacked both a dense system of railways and the option to thoroughly exploit alternative forms of transport. Three-quarters of Russian lines were single-track, its roads were primitive and its inland waterways were mostly impassable during the winter months as they froze over. Therefore, the effective operation of the comparatively sparse Russian railway network was fundamental both to the continuance of the eastern war effort and the maintenance of the domestic Russian economy.[87]

The profound differences between the Russian transport infrastructure and those in western Europe influenced the choice of expert despatched to investigate conditions in the east. When asked to suggest a suitable man to accompany Lord Milner's mission to Russia in early 1917, Sir Sam Fay put forward the name of a Canadian. George Bury had worked for the Canadian Pacific Railway throughout his career, and had attained the position of vice president of the line prior to the outbreak of war. 'When making this recommendation', Fay recalled in his autobiography:

> I had in mind the long stretches of single line in Russia for which there is no counterpart in [Britain]. The Canadian Pacific Railway on the other hand was mainly single line, reaching 3,000 miles from seaboard to seaboard. A Canadian railway man would, therefore, be more able to appreciate the position than an English railway manager. I knew Bury, and had seen him at work in Winnipeg, where he controlled all lines west of the city. His energy and cheery optimism had impressed me.[88]

Lloyd George acted upon Fay's advice immediately. By 1 February, Bury had received the prime minister's instructions, which were to 'obtain all the information that you can and render every possible assistance, in regard to the working of the Russian railway system'.[89]

---

[87] Stevenson, *1917*, pp. 93–4.
[88] Fay, *The War Office at War*, p. 29.
[89] Lloyd George to Bury, 1 Feb. 1917, quoted in T. Murray Hunter, 'Sir George Bury and the Russian Revolution', *The Canadian Historical Association: Report of the Annual Meeting*, cdxli (1965), 58–70, at pp. 60–1.

Within three weeks, Bury had submitted a memorandum to the cabinet in London, which outlined the parlous state of communications in Russia. The first sentence of his report, dated 20 February, stated baldly that 'Russia has not sufficient railway lines and those she has are not equipped adequately with waggons and locomotives'.[90] He estimated that the Russians had a backlog of locomotives and wagons awaiting and under repair that was 1,500 and 15,000 units respectively higher than it should have been given the circumstances – a consequence of inefficient working procedures, inadequate labour supplies and a lack of materials. These deficiencies were exacerbated by the retention of large numbers of wagons in the battle area for use as storehouses, living quarters and offices by the army (Bury gave a figure of 'more than eight thousand units' in his report). In addition, he asserted that 'very unnecessary delays' took place at the ports due to the continuance of lengthy customs practices and the inefficient deployment of labour tasked to discharge ships, and observed that the Russian railway officers lacked 'the organizing genius to be found in some other countries'.[91]

Alongside his diagnosis of the problems that afflicted the Russian network, Bury's memorandum contained a series of recommendations for their alleviation. His solutions to the transport problem included: the direction of 'all the waggons and locomotives that can be delivered during this year' into Russia; the introduction of operating practices designed to maximize the haulage capacity of individual trains and improve the circulation of rolling stock; the amendment of government regulations to permit higher levels of tyre wear; the establishment of a central authority to oversee the distribution of locomotives and rolling stock across Russia's various districts; the provision both of additional passing loops and terminal facilities on existing lines; and the construction of a greatly improved line between the port at Murmansk and the interior. However, he stopped short of recommending that British and French officials be – like Geddes had been on the western front – parachuted into key positions in the Russian railway administration. Russia was a sovereign nation upon which its western allies could apply 'a certain leverage' but nothing more. As Bury acknowledged in his report, 'to attempt to place British or French officers in charge of the more important positions on the Russian railways' was a diplomatic impossibility. Instead, his recommendations centred upon 'securing for [the Russians] all the waggons and locomotives it is possible to obtain even by

[90] Memorandum regarding transportation, prepared for the British war cabinet, 20 Feb. 1917, p. 1. My thanks to Anthony Heywood for providing me with access to this document, and for his guidance on the complexities of Russia's railways during the war more generally.
[91] Memorandum regarding transportation, 20 Feb. 1917, pp. 2, 4–5, 6–7.

paying premiums for prompt delivery ... This is something the [British] Government should take in hand at once and vigorously'.[92]

As in France prior to the Somme offensive, a lack of rolling stock to remove supplies from the ports had created congestion at the docks – a situation compounded by the quantity of wagons retained in the battle area for use as storehouses and living quarters.[93] Bury claimed that the accumulated stocks at Vladivostok in early 1917 had reached the port's annual capacity, and projected that some nine months would be required to clear the existing material inland at the prevailing rate of movement. Therefore, to relieve congestion at the port, Bury recommended the increased use of the Japanese-controlled port of Dalniy (now the Chinese city of Dalian) on the South Manchurian Railway and advised that Vladivostok should be used solely for the import of railway materials into Russia.[94] Even after such changes Bury warned that the paucity of rolling stock and the vast distances between the far-eastern ports and the eastern front meant that Russia would – in the continued absence of a southern sea route through the Dardanelles – have to rely upon its northern ports 'for at least a year'.[95]

Conditions at the northern ports provided scant cause for cheery optimism. Archangel possessed fifty-two berths, each capable of discharging approximately 300 tons per day, with a further twenty berths scheduled to become operational over the summer of 1917. However, the port could only be used with any degree of certainty from June until November and the Archangel–Vologda railway line responsible for moving goods inland only carried three freight trains per day at that time.[96] While the Russian authorities promised Bury that they could move 127,000 tons via Murmansk – the only ice-free port available in northern Russia – before May 1917, the Canadian expert advised the cabinet in London that 'the best we should figure on ... is four-fifths of the Russian expectations'.[97] In a complaint familiar to investigators of the Channel ports the previous summer, Bury observed that the labour at Murmansk was 'inefficient and inadequate' and that additional storage was 'urgently required ... as there is almost certain to be congestion, and materials stacked in the snow on the shore are bound

---

[92] Memorandum regarding transportation, 20 Feb. 1917, pp. 7–8.

[93] Memorandum regarding transportation, 20 Feb. 1917, p. 4.

[94] Memorandum regarding transportation, 20 Feb. 1917, pp. 1–2; TNA, CAB 28/2, papers I.C. 13–32, Allied conference at Petrograd, January–February 1917, Report on mission to Russia, by Major David Davies, 10 March 1917, p. 5.

[95] Memorandum regarding transportation, 20 Feb. 1917, p. 5.

[96] TNA, CAB 28/2, papers I.C. 13–32, Allied conference at Petrograd, Report by Davies, p. 5; P. Gatrell, *Russia's First World War: a Social and Economic History* (Harlow, 2005), p. 25.

[97] Memorandum regarding transportation, 20 Feb. 1917, pp. 5–6.

to deteriorate'.[98] Yet whereas on the western front Geddes had stressed the 'moral and physical impossibility' of scaling back on munitions imports for the BEF, Bury advocated that all shipments of munitions to Murmansk be suspended until congestion at the port had been cleared. 'Otherwise', he warned, 'the place will become a second Vladivostock [sic] and munitions urgently needed elsewhere will be left to deteriorate in the snow'.[99]

Bury's experience of operations on a single-track railway that stretched over long distances through extreme territory informed his recommendations for the improvement of the Russian railway infrastructure. He recognized that the line from Murmansk to the interior represented the 'main artery for traffic' to Russia during the winter months if Vladivostok was closed to imports, and argued it was essential that:

> [t]he line between Port Murman and Kem be closed to traffic and every effort be put forth so that it may be in the best of condition by next winter. It means lifting sags, strengthening bridges, ballasting, building passing tracks every six miles, improving terminals, installing signals, etc., and the work is of such magnitude and so essential to the Russian Empire that it must be laid out and prosecuted on a very large scale indeed. The line from Kem to the Junction needs lesser improvements but the work on additional wharves, sheds and tracks at Port Murman must proceed day and night.[100]

The maximum capacity for each train on the extant Murman Railway between Kola and Petrograd (see Figure 8.1) was between 140 and 160 tons. However, Bury believed that the line was capable of handling trains of between 400 and 600 tons if it were 'properly constructed' and 'the support of the [Russian] Government and especially the Minister of Ways and Communications' for the scheme was enlisted.[101] The latter, E. B. Kriger-Voinovskii, was lauded by Bury as 'undoubtedly the best administrator' to have held the position during the war, and he instructed that it 'should be the duty of [the British] Government to make every effort to have his hands strengthened and … [ensure that] he be given the freest rein [sic] to bring about changes that he will have to make if his administration is to be effective'.[102]

---

[98] Memorandum regarding transportation, 20 Feb. 1917, pp. 6–7; TNA, CAB 28/2, papers I.C. 13–32, Allied conference at Petrograd, Report by Davies, pp. 5–6.
[99] UWMRC, Granet papers, MSS. 191/3/3/4, Geddes to Lloyd George, p. 9. Bury's remarks reported in TNA, CAB 28/2, papers I.C. 13–32, Allied conference at Petrograd, Report by Davies, p. 6.
[100] Memorandum regarding transportation, 20 Feb. 1917, p. 6.
[101] Memorandum regarding transportation, 20 Feb. 1917, p. 8.
[102] Memorandum regarding transportation, 20 Feb. 1917, pp. 8–9.

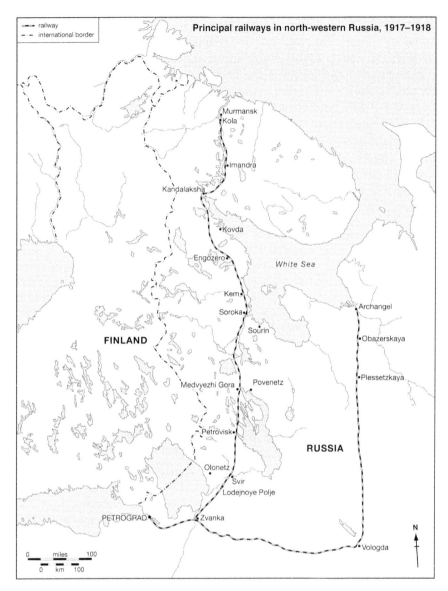

Figure 8.1. Principal railways in north-western Russia, 1917–18.

Source: Military Monograph Subsection M.I.2, Military Intelligence Division, General Staff, *Russia, Route Zone A: Murman Railway and Kola Peninsula: Information and Route Notes, Murmansk to Petrograd* (Washington, DC, 1918).. Map drawn by Cath D'Alton.

However, Kriger-Voinovskii's experience in 1917 underscored one of the two major limitations of Bury's mission. A fortnight after the latter's report was submitted the February Revolution broke out, and the Romanov dynasty – which had gratefully provided its Canadian visitor with a daily bottle of champagne to assist him in the production of his report – was overthrown.[103] Bury's desire that Kriger-Voinovskii's hands be strengthened was eclipsed by the domestic concerns of the Provisional government in Petrograd. Kriger-Voinovskii was replaced at the Ministry of Ways and Communications by Nikolai Nekrasov, who commenced a 'flurry of new railway appointments' chosen for their political desirability rather than their professional qualifications during the spring and summer.[104] While some improvements to the Murman Railway were made during 1917 – and orders for locomotives, rolling stock and other railway equipment were added to those placed with foreign suppliers earlier in the war – Bury's recommendations for the development of the Russian railways were not pursued in full by the new authorities.[105]

Yet the outbreak of the revolution and subsequent turmoil within the Russian railway industry was not the sole reason why Bury's report generated little activity behind the eastern front. The overall tone of his memorandum to the War Cabinet had been one of cautious positivity; Bury wrote with a belief that the Russian railways could be made effective through a combination of organizational changes, infrastructure improvements and material support. However, his diagnosis presented an overly optimistic image of Russia's transport position after two-and-a-half years of war, and was built upon a far from comprehensive understanding of the situation across Russia. Many of the organizational changes Bury advocated had been in place throughout the war, or had been introduced as the situation demanded. Furthermore, his recommendations for the improvement of the line from Murmansk to the interior – while unquestionably desirable for the development of Russian railway capacity – were wholly impracticable given the prevailing conditions. The line was built in appalling conditions by a labour force that mostly comprised German prisoners of war, for whom

[103] Murray Hunter, 'Sir George Bury and the Russian Revolution', pp. 61–2.

[104] On the 'chaos' within the Russian railway administration during the February Revolution, see A. Heywood, *Engineer of Revolutionary Russia: Iurii V. Lomonosov (1876–1952) and the Railways* (Farnham, 2011), pp. 151–8.

[105] Military Monograph Subsection M.I.2, Military Intelligence Division, General Staff, *Russia, Route Zone A: Murman Railway and Kola Peninsula: Information and Route Notes, Murmansk to Petrograd* (Washington, DC, 1918), p. 26; A. J. Heywood, 'Russia's foreign supply policy in World War I: imports of railway equipment', *Jour. European Econ. Hist.*, xxxii (2003), 77–108, at pp. 82–5; Heywood, *Engineer of Revolutionary Russia*, p. 157.

the provision of building materials and food were incredibly difficult. As Anthony Heywood has demonstrated, in 1916 the Russian railways had carried over 20 per cent more freight than had been transported in 1913.[106] The demand for freight transport remained incessant over the challenging winter of 1916–17, and shortages of civilian supplies existed across the country as military traffic took precedence.

These difficulties were not sufficiently appreciated by British observers, who lacked thorough knowledge of the situation outside Petrograd. The military attaché, Alfred Knox, recorded on 7 February that 'it had been suddenly discovered that many railways had only two to five days' supply of coal'. Consequently, he added, the 'railways were ordered to carry nothing but coal for a week. The week was extended till March 14th'.[107] By early March the condition of the network was adjudged by Sir Henry Wilson to be 'deplorable ... Coal cannot be carried from the pit-mouths to the railways and the manufactories; food cannot be distributed to the towns nor collected from the countryside; troops and *materiél* cannot be carried from one place to another'.[108] Yet over the same period, as Heywood has illustrated, the work performed on the Russian railways almost matched that achieved during the equivalent period of the previous year.[109]

Bury's memorandum betrayed similar ignorance of wider Russian problems. The Canadian railway expert failed to appreciate the complexity of the traffic problems that existed in Russia during the First World War, particularly in the vast area of the country east of the Urals. Consequently, he advocated solutions that would have been profoundly difficult to implement even had the Russian political situation been stable in 1917. Like his political and military contemporaries, Bury struggled to come to terms with what Catherine Merridale has termed Russia's 'extraordinary foreignness'.[110] The result was a report, researched and submitted in a period of little more than a fortnight, which failed to adequately appreciate the challenges that Tsarist Russia had faced in the sustenance of industrial war since August 1914.

[106] A. Heywood, 'Spark of revolution? Railway disorganisation, freight traffic and Tsarist Russia's war effort, July 1914–March 1917', *Europe-Asia Studies*, lxv (2013), 753–72, at p. 765.

[107] A. W. F. Knox, *With the Russian Army, 1914–1917* (2 vols., London, 1921), ii. 525–6. Knox's observations are almost certainly an exaggeration, as Russian authorities had been aware of the worsening fuel situation since the previous autumn.

[108] TNA, CAB 28/2, papers I.C. 13–32, Allied conference at Petrograd, Report by Lieutenant-General Sir H. H. Wilson, 3 March 1917, p. 2.

[109] Heywood, 'Spark of revolution?'

[110] C. Merridale, *Lenin on the Train* (London, 2017), p. 33.

For Lawson Billinton, who only departed Britain for Romania via Russia on 18 March, the effects of the revolution greatly impeded his attempts to sustain the war effort in the east. Prior to Billinton's arrival, Wilson had observed that:

> Partly owing to difference of gauge [between the Russian and Romanian railways], partly to incompetent administration, partly to German activities and sympathies, the railway situation [in Romania] is very serious indeed, and competent judges think there will be a famine in four to six weeks from now, and that the Russian troops [manning most of the Romanian front] will starve or have to fall back.
>
> The outlook on this front is bad ... it may be necessary to retire the whole line in order to save the armies from destruction.[111]

After a three-day voyage and sixteen days in Petrograd, Billinton's party made its way south to Iaşi – the Romanian government's temporary home following the fall of Bucharest. He immediately noted the disorganized condition of the railways and, when his attempts to find breakfast on his first morning in Iaşi proved fruitless, how short of food the Romanians were. It took him very little time to diagnose the problems:

> Very few trains were running and these were taking days instead of hours over a journey. The average speed from start to finish was in many cases as low as six or seven miles per hour. The locomotive side ... was in a very bad state. Over 60% of the locomotive stock was awaiting or under repair and the remainder was in a very indifferent condition. The output from the shops was at an especially low figure, and generally there was every indication of a complete paralysis of the railroad unless material alterations took place.[112]

Rather than merely report and advise, as Bury had done in Russia, Billinton exchanged his uniform for overalls and transformed 'from a military man into a jack-of-all-trades boilermaker, fitter, erector, etc., in order to carry out the work of reorganization'. He travelled across what remained of Romanian territory, visited all the available running sheds and repair depots, and gained a thorough knowledge of the 'contour of the road and how the locomotives were operated' by Romanian drivers. He remarked after the war that 'by very careful application there was very great improvement effected, not only in the number of trains running but also in punctuality' across the Romanian network.[113]

---

[111] TNA, CAB 28/2, papers I.C. 13–32, Allied conference at Petrograd, Report by Wilson, pp. 3–4.

[112] Marx, *Lawson Billinton*, pp. 80–1.

[113] Marx, *Lawson Billinton*, p. 81.

However, Billinton's efforts had little impact on the Romanian war effort. Raymond de Candolle, the general manager of the Buenos Aires Great Southern Railway who led the mission to Romania, reported pessimistically to Sir William Robertson in May 1917 that 'the working of Roumanian locomotive traffic departments still leaves much to be desired and is engaging our special attention although owing to [the] innate reluctance of [the] Roumanians to push through comprehensive programmes it will not be easy to bring about rapid improvement'.[114] By the summer of 1917, although the French still believed that a Russo-Romanian offensive on the eastern front was indispensable to the allied campaign, the British government had lost faith in Romania's ability to successfully conduct an offensive against the central powers. Arthur Balfour, the foreign secretary, suggested that the Romanians had been 'incompetent to the verge of a crime', while Lloyd George categorized allied obligations to Romania as of the lowest importance in Britain's evolving strategic calculations.[115] Unable to provide further support to the Romanians, Billinton embarked for the Caucasus in October 1917 to assist the Russian forces gathered at Rostov-on-Don under General Aleksei Kaledin. From there he embarked on the remarkable tour of the region that opened this chapter, and which ended the following summer with his return to Brighton.[116]

The wider military and political environments in which George Bury and Lawson Billinton operated worked against them in 1917. Events beyond their control overwhelmed their endeavours. Coordination of the Russian railway network largely disintegrated after the Russian state collapsed. The Provisional government's removal of undesirable elements, and the Bolsheviks' purge of 'large numbers of experts and administrators' from the railway industry, threw the work of governance into 'near total dysfunction' by the end of the year.[117] The Romanian decision to accept the armistice of Focşani on 9 December 1917, and the Russian descent into civil war, decreased the eastern front's prominence in British strategic plans and reduced its desire to divert precious human and material resources from the western front. However, as one set of allies in

[114] TNA, CAB 24/14/38, Roumanian communication. Tel. from Sir G. Barclay, dated 18.5.17, conveying General de Candolles's report, p. 3.
[115] G. E. Torrey, 'Romania in the First World War: the years of engagement, 1916–1918', *International History Review*, xiv (1992), 462–79, at pp. 466–7.
[116] Much of Billinton's own account of this lively period is reproduced in Marx, *Lawson Billinton*, pp. 87–111.
[117] E. Lohr and J. Sanborn, '1917: revolution as demobilization and state collapse', *Slavic Review*, lxxvi (2017), 703–9, at p. 706.

eastern Europe fell away, the constellation of nations assembled in France demanded reconsideration. The introduction of another sovereign army to the western front provided further challenges and opportunities for Britain's transport experts in the final year of the war.

The Supreme War Council and the Inter-allied Transportation Council

The directorate-general of transportation, first under Sir Eric Geddes and then Philip Nash, became an integral, civilian-led component of Sir Douglas Haig's military force. The organization created and fostered by successive DGTs provided the platform for talented transport administrators to apply their skills to the delivery of operational and infrastructural improvements in France. Their contributions raised the capacity and efficiency of the distribution network behind the BEF in the war's most important theatre of operations in 1917. Furthermore, as larger numbers of British personnel, locomotives and other transport-related equipment arrived on the western front during the year, British units became increasingly concerned with the maintenance of the shared logistics systems behind the allied forces.[118] In January 1917 the French handed over responsibility for the repair and maintenance of the British-built ambulance trains that had gradually entered service in France over the previous two years, while the chief mechanical engineer took over 'from the State Railway some partly-finished shops at St Etienne, near Rouen' that were subsequently completed and equipped for the repair of locomotives.[119]

The British presence outside the repair shops of France also expanded. ROD crews, having previously been largely restricted to the running of trains behind the Ypres salient, were given instruction in French signalling methods and gradually permitted to operate services on the French main lines. Between March and November 1917 the number of trains driven by the ROD over French tracks grew from ten per day to 341 – not all of which were dedicated to the BEF's maintenance.[120] General François Anthoine's French First Army, which arrived on the BEF's northern flank in Flanders ahead of the third battle of Ypres, was served exclusively by the ROD throughout the offensive. Liaisons between Anthoine's force and its British supply service were described as 'smooth and efficient' in the QMG's post-war report, in stark contrast to the fractious relationship that

---

[118] TNA, WO 107/69, work of the QMG's branch, p. 2.

[119] Simpson, 'Railway operating in France', pp. 709–10. The chief mechanical engineer, Colonel George Tertius Glover, was another of Geddes's civilian appointments. Glover had entered the North-Eastern Railway as a draughtsman in 1894, and was the Great Northern Railway of Ireland's locomotive engineer from 1912 onwards.

[120] Pritchard, *History of the Royal Engineers*, v. 622–3.

existed between Geddes and those in charge of the French transportation services earlier that year.[121]

Lloyd George's clumsy attempts to subordinate the BEF at Calais, Geddes's reluctance to acknowledge French concerns over the efficient operation of the allies' shared transport infrastructure and General Nivelle's failure to realize his grandiose plans retarded rather than reversed the trend towards greater allied cooperation at a political and strategic level in the second half of the war. Events both on land and at sea drew the allies closer together in search of answers to the war's evolving logistical and organizational complexities. At sea, the German decision to pursue unrestricted submarine warfare engendered a supranational consideration of shipping priorities, which resulted in the eventual formation of the Allied Maritime Council in November 1917 to 'make the most economical use of tonnage under the control of all the Allies, to allot that tonnage … in such a way as to add most to the general war effort, and to adjust the programmes of requirements of the different Allies in such a way as to bring them within the scope of the possible carrying power of the tonnage available'.[122] On land, the United States' entry into the war as an associated power also 'forced a degree of allied cooperation' to meet the challenge of inserting another army into the crowded space behind the western front.[123]

America's decision to send an expeditionary force to Europe reinforced the value of efficient railway use as the war intensified. In recognition of the railways' importance, an American military railway mission comprising civilian and military railway experts was despatched to Europe on 14 May. After arriving in Britain the party conferred with Geddes and Sir Guy Granet, who provided their new partners with an insight into the creation of the directorates-general of transportation and military railways. The Americans then crossed to France, where they observed the French army's lines of communications in action and met with Nash at GHQ.[124] The mission confirmed to William John Wilgus, a retired civil engineer who served with the AEF's transportation services for the duration of the war, the virtue of 'building up a new arm of the Service, headed by men from civil life who

---

[121] TNA, WO 107/69, work of the QMG's branch, p. 17.

[122] J. A. Salter, *Allied Shipping Control: an Experiment in International Administration* (Oxford, 1921), pp. 144–55. On the work of the Allied Maritime Council, see M. McCrae, *Coalition Strategy and the End of the First World War: the Supreme War Council and War Planning, 1917–1918* (Cambridge, 2019), pp. 187–236; Greenhalgh, *Victory through Coalition*, pp. 129–32.

[123] E. Greenhalgh, *Foch in Command: the Forging of a First World War General* (Cambridge, 2011), p. 393.

[124] W. J. Wilgus, *Transporting the A.E.F. in Western Europe, 1917–1919* (New York, 1931), pp. 3–7.

had been trained in the art of transportation and, therefore, could most quickly glean the required knowledge and construct and operate a machine that would function with efficiency'.[125] General Pershing concurred with the idea of replicating the British approach; the American commander-in-chief's personal sympathy towards the unification of transport within the American force under a civilian expert was confirmed following a visit to Nash's headquarters at Monthuis in July. William Wallace Atterbury, vice president of the Pennsylvania Railroad, was appointed the AEF's DGT in October 1917 and held the position for the duration of the conflict.[126]

The addition of an American army with its own requirements for port space, rolling stock, roads and vehicles introduced another thread to the patchwork of transport issues that occupied allied leaders. Franco-American discussions on the operation of the AEF's lines of communications; Franco-British conversations about the supply of rolling stock and personnel from the British empire to augment French resources; negotiations between French, British and Italian experts over the operation of overland communications to the eastern Mediterranean; and deliberations between British, French and Belgian representatives about the use of much-needed – but hitherto idle – Belgian locomotives were all ongoing when Pershing arrived in France. Furthermore, each topic was treated independently until July 1917, when a meeting of French, British, Italian, Belgian and American transport authorities took place and agreed to the organization of periodic conferences to discuss the common use of rolling stock, railway materials and technical labour. These conferences, at which Granet or a deputy from the War Office represented Britain's interests, met on a dozen occasions before the end of the year and dealt with 'numerous questions' prior to the formation of the Supreme War Council (SWC) in November.[127]

Created when the Italian disaster at Caporetto made it 'evident that a very much closer cooperation was necessary for a successful prosecution of the war by the Allied nations', the SWC provided the catalyst for British transport expertise to become engaged on work at a supranational level.[128] On 1 December the council recommended that a suitable expert be

[125] Wilgus, *Transporting the A.E.F.*, p. 550.

[126] J. G. Harbord, *The American Army in France, 1917–1919* (Boston, Mass., 1936), p. 116.

[127] TNA, CAB 25/110, inter-allied transportation council: organisation and functions, Nash to Storr, 4 May 1918; Henniker, *History of the Great War*, pp. 197–8. Unfortunately, Henniker does not provide any examples of the discussions that took place at the conferences held during this period.

[128] TNA, CAB 25/127, historical record of the SWC of the allied and associated nations from its inception on 7 Nov. 1917 to 12 Nov. 1918, the day after the signature of the armistice with Germany, together with a note as to its role and work subsequent to that date, p. 2.

appointed to examine and report on the allies' railway arrangements across the European theatre. Both the British and non-British representatives on the SWC identified Geddes as a suitable candidate for the task.[129] However, as noted above, Lloyd George refused to release his First Lord of the Admiralty for transportation duties. Consequently, Philip Nash was entrusted to conduct the investigation in January 1918.

The instructions issued to Nash, when compared to the parameters of Geddes's transportation mission to GHQ in August 1916, emphasize the increased scope of Nash's remit.[130] Alongside investigating the allies' transport resources and the capacity of the railways across France and Italy, Nash was tasked with the development of a framework for a formal, inter-allied authority capable of studying the implications for transportation of the SWC's strategic designs. The report he produced in February 1918 underlined the interconnectivity of allied operations and demonstrated that the compartmentalization of transport questions within regional or even national boundaries clearly limited the efficiency with which the extant infrastructure was exploited.

The existing provision of railway facilities for both civilian and military traffic was evaluated within Nash's report. On the former, he concluded that any further reductions in France – the volume of which Geddes had condemned bitterly at Calais twelve months earlier – on 'a scale likely to affect the position materially', were 'impossible'. By the end of 1917 passenger traffic on the French rail network was 65 per cent lower than it had been in 1913, and Nash was confident that the suppression of unnecessary goods traffic had been 'thoroughly dealt with by the French government'.[131] On the latter, Nash's investigations revealed the parlous state of the allies' mobility within the war's principal theatre of operations, and confirmed the extra layer of complexity that American involvement in the fighting had created for the coalition's transport administrators. Predominantly based upon the ports on France's Atlantic coast (see Figure 8.2), the AEF was served by 'a limited number of railways lines of communications of great length'. Provided the American forces remained deployed within the Champagne or Verdun regions, Nash wrote, the existing Franco-American arrangements for the improvement of lines and stations, the erection of storage depots and the provision of personnel and materials were likely to be sufficient. However, Nash warned:

[129] PA, Lloyd George papers, LG/F/17/6/19, Geddes to Lloyd George, 20 Dec. 1917.
[130] Nash's instructions are replicated in Appendix II.
[131] TNA, CAB 24/43/19, report on general transportation situation on the western front, 20 Feb. 1918, p. 6.

If it were found expedient to move any considerable American force to operate on a more northerly front – say in prolongation of the British line southwards – the American line of communications would have to be altered, and all the traffic from American bases for the maintenance of this force would have to pass through the neighbourhood of Paris, or, to avoid this neighbourhood by a wide detour over lines of limited capacity. This would involve passing a considerable traffic over the Ceinture Railways, a system which is already much congested by military and other traffic, and it seems extremely doubtful whether any considerable addition to this traffic would be practicable.[132]

The accumulation of American manpower in France was, Nash demonstrated, both an aid to and an added complication for the country's allies. Whereas the Germans could shift troops from east to west without the need to consult its allies,[133] any prospective redeployment of the AEF in response to the anticipated German offensive on the western front in the spring required the identification of traffic priorities between the forces of three independent national railway organizations: the British directorate-general of transportation, the American directorate-general of transportation and the French ministry of public works (which exercised its control through the *direction des transports militaires aux armées* at *GQG*).[134] As the Franco-British discussions at Calais in February 1917 had illustrated, there was little guarantee that agreement on those priorities was likely to be arrived at between the coalition partners easily.

Yet if strategic movements within France were likely to demand the curtailment of industrial traffic, the dislocation of services to and from Paris and disruption to the supply trains that fed and maintained the armies at the front, transport between France and Italy presented 'much more serious' difficulties for the allies because of the relative paucity of connections between the two nations.[135] Only two railway routes were available, one via the coast and the other through the Mont Cenis tunnel. Both presented considerable obstacles to the swift movement of men and goods; on the former route a section between Ventimiglia and Savona was single- rather than double-track, while on the latter a portion of the line between Modena and Bussoleno had been electrified. Consequently, the maximum capacity of the two lines was around forty trains per day, each restricted to a maximum

---

[132] TNA, CAB 24/43/19, report on general transportation situation, p. 7.

[133] On the scale and significance of Germany's movement of divisions to the western front during this period, see G. Fong, 'The movement of German divisions to the western front, winter, 1917–1918', *War in History*, vii (2000), 225–35.

[134] TNA, CAB 24/43/19, report on general transportation situation, pp. 18–19.

[135] TNA, CAB 24/43/19, report on general transportation situation, p. 8.

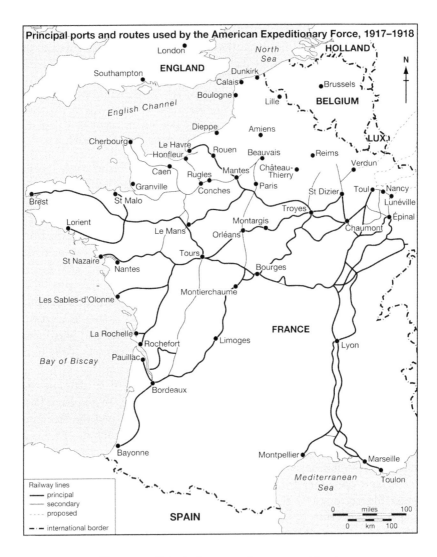

Figure 8.2. Principal ports and routes used by the
American Expeditionary Force, 1917–18.

Source: W. J. Wilgus, Transporting the A.E.F. in Western Europe, 1917–1919 (New York, 1931).
Map drawn by Cath D'Alton.

of 260 tons per train.[136]

The limitations of the land connections between France and Italy were recognized by the SWC even before Nash's report had been completed. On 21 January 1918 the council's twelfth joint note stated that Italy was 'safe', but that 'the power of rapid rail transport' both within Italy and between Italy and France had to be increased in order to 'secure strategic unity of action over the two theatres'.[137] In the wake of Caporetto, troops had been sent to Italy at a rate of forty-two trains per day, but the achievement of such figures had necessitated the suspension of all other traffic on the lines. Alongside the Cherbourg–Taranto service the Mont Cenis route carried a mail and passenger train, a train of supplies for the French forces in Macedonia and fourteen trainloads of coal for Italy each day. By February 1918, Nash reported, the nationwide reserves of coal within Italy – a nation that produced no coal of its own – had dropped to an average of twelve days' supply.[138] Any curtailment of coal imports to facilitate the movement of troops risked the further erosion of stocks upon which the continuance of Italy's war effort depended. Therefore, any decision to move troops from France to Italy in 1918 could only be made after careful consideration of the relative priorities of the military situation in France and Italy, the available stocks of raw materials within Italy and the reserves of supplies accessible to the allied forces engaged in Macedonia and further afield.

Nash recognized that such decisions could not be taken unilaterally. 'Speaking broadly', he stated, 'each operating agency is, within its own province, carrying out the work allotted to it in an efficient manner. These investigations have, however, convinced me that by freer interchange of ideas and experiences drawn from the different agencies concerned, highly important results in the direction of expediting movement might be obtained'.[139] To encourage the circulation and consideration of transportation issues across the coalition, Nash recommended the formation of an Inter-allied Transportation Council (IATC), comprising representatives from the four principal allied nations and under the direction of the SWC rather than any national body. The IATC, Nash proposed, should advise the SWC of the transportation implications 'of all plans of campaign' under consideration on the western front; negotiate with the allied governments over the provision of extra railway facilities necessary to 'give effect to any accepted plan of campaign'; prepare schemes for the movement of large bodies of troops when ordered to do so by the SWC, and liaise with the independent

[136] Henniker, *History of the Great War*, p. 298.
[137] TNA, CAB 25/120/2, nos. 1–150, joint note no. 12, 21 Jan. 1918, pp. 1–2.
[138] TNA, CAB 24/43/19, report on general transportation situation, p. 17.
[139] TNA, CAB 24/43/19, report on general transportation situation, p. 19.

allied governments concerned with the prosecution of such movements; 'study the enemy position regarding transportation facilities of every kind' and update the SWC on the enemy's logistical capabilities; prepare schemes for the development of railway lines with the explicit purpose of relieving seaborne communications; and oversee the 'performance of the different agencies operating on the lines of communication … bringing to the notice of the Governments or armies concerned cases in which the fullest use does not appear to be made of available resources, and suggesting remedies'.[140]

There was little need for discussion of Nash's report in London. Even before the document had been submitted to the War Office the concept of an inter-allied committee working with the SWC at Versailles had been accepted by the allied prime ministers, while his proposed framework had received the backing of the French minister of public works, Albert Claveille, the Italian ministers of transport and war, Riccardo Bianchi and General Vittorio Alfieri respectively, and the American commander-in-chief, General Pershing. Consequently, Nash was duly appointed as the British representative on the IATC by Lloyd George in mid March 1918.[141]

Superficially, upon his appointment to the IATC, Nash attained a position of higher authority over transportation in the First World War than that acquired by Geddes when the latter became DGT in October 1916. The IATC provided Nash and his colleagues with an administrative framework and the necessary international contacts to construct a wider understanding of allied transportation than had hitherto been possible. Geddes's organizational changes had provided GHQ and the British government with the tools to consider the transport requirements of Britain's global war effort, but in 1917 suspicions, disputes and disagreements took place at the intersections of British and non-British (principally French) authority. The IATC provided the allies with the bureaucratic machinery necessary to work collaboratively rather than in competition with one another, and permitted the SWC to treat the European theatre from Flanders to the Adriatic as one continuous front. As Meighan McCrae has noted, the allies' subscription to the IATC – and the appointments of senior transportation figures such as Nash and Claveille to sit upon it – illustrates the importance the coalition placed on improving communications between the French and Italian fronts in the winter of 1917–18.[142]

---

[140] TNA, CAB 24/43/19, report on general transportation situation, p. 20. Nash later advocated for a Belgian representative to join the council, to sit alongside representatives from France, Italy, Britain and the United States.

[141] TNA, CAB 24/43/19, report on general transportation situation, p. 21; Henniker, *History of the Great War*, p. 199.

[142] McCrae, *Coalition Strategy and the End of the First World War*, pp. 104–5.

However, the council's lack of executive authority hampered its ability to make fundamental changes to the allies' transportation facilities during the war's final year. The IATC set to work immediately, and between March and November 1918 it considered topics ranging from the most effective distribution of personnel, tools and materials among the allied armies to how best to transport onions from Egypt to the western front.[143]

The first task it was set provides an insight both into the organization's strengths and limitations and the instability of the circumstances in which the IATC operated as the war entered a more dynamic phase. On 23 March the SWC directed the newly formed council to undertake an examination of railway movements, which bore striking similarities to the exercises issued by the War Office to the ERSC in the latter part of the nineteenth century. The SWC responded to a warning from the Italian commander-in-chief, General Diaz – which stated that 'in the event of a powerful attack on [the Italian] front, he would demand the support of' eight divisions – with a request that the IATC 'undertake immediately a study of the betterment to be made in the transport facilities between France and Italy' to facilitate such a movement. Nash and his colleagues were tasked to examine whether the number of troop trains handled daily over the lines between the two countries could be more than doubled, and to identify what volumes of vital commodities such as coal had to be stockpiled in Italy to permit the temporary suspension of supply trains 'for some days during the critical period' when the troops were moved.[144] The council's report, submitted on 10 April, provided the allies' strategic planners with a response that took account of the difficulties involved in the supply of coal to Italy that could not be alleviated by further use of shipping; the impracticality and lengthy nature of all possible infrastructure improvements to the existing routes between France and Italy; the ongoing shortages of rolling stock across the western front; and the sustained, unavoidable demands for traffic 'of high military importance' from the allied forces located at Salonika and beyond. The IATC provided the SWC with a possible solution to the transport challenge – dependent upon the suppression of certain supply traffic, advantageous weather conditions and the availability of shipping tonnage – that required the exploitation of seven separate lines of communications.[145]

Nash was quick to emphasize that the scheme contained within the report was 'merely a study ... [that could not] be put into effect without

---

[143] Henniker, *History of the Great War*, p. 200.

[144] TNA, CAB 25/110, IATC, explanation of motives, 23 March 1918, p. 1; Joint questionnaire number 1, 27 March 1918, pp. 1–2.

[145] TNA, CAB 25/53, transportation of troops between France and Italy, Nash to Storr, 10 Apr. 1918.

considerable preparation'. The council's recommendations involved the exploitation of routes that comprised railways only, those that combined rail and road transport and those that used a combination of rail and sea transport to execute the required troop movements. Yet the council's responsibilities did not extend to the identification of suitable entrainment and detraining points, to the collection of suitable rolling stock and locomotives, to the provision of additional engine crews and other personnel required to facilitate the moves, or even to the 'clearing of the roads of snow'. Such questions remained the exclusive responsibility of the national governments and military authorities concerned with transport matters – the council merely acted as 'an extremely useful organ of information and coordination', which illustrated the decisions that needed to be made before a swift, successful movement of troops could take place.[146] Consequently, the ultimate decision over whether or not to act upon the IATC's recommendations in the British case had to be taken by the War Office rather than by Nash. Over a month after the SWC issued their joint note number twenty-two, which recorded the IATC's recommendations as 'certain precautionary measures' deemed necessary to ensure the efficient transfer of manpower to Italy,[147] Nash broached the subject with London. In a somewhat pathetic letter he begged 'that I may be permitted to know if the necessary arrangements' – which included the strengthening of the British army's supply organization in Italy, the identification and allotment of tonnage for the transport of troops from Marseille and the accumulation of sufficient stockpiles both of supplies for the soldiers to be moved to Italy and of coal for the Italian economy – had been made.[148] They had not.

In the spring and summer of 1918 British priorities lay elsewhere. Under the stresses engendered by the German advances in France the diversion of precious shipping tonnage, construction personnel and materials, alongside the arrangement of hypothetical programmes for the rapid movement of troops away from the western front, proved to be a low priority for those at GHQ and in London. The British government declined to approve joint note number twenty-two, much to the distress of the country's allies, as 'present conditions' made it impossible to spare the supplies required to build up stocks in Italy.[149] As the surviving documents demonstrate, what

---

[146] TNA, CAB 25/53, transportation of troops, Nash to Storr, 11 Apr. 1918; Draft for collective note number 2, 15 Apr. 1918, p. 2. The French were also fully alive to the implications of the IATC's absence of executive authority. See McCrae, *Coalition Strategy and the End of the First World War*, p. 105.

[147] TNA, CAB 25/53, transportation of troops, joint note number 22, 19 Apr. 1918, pp. 1–2.

[148] TNA, CAB 25/53, transportation of troops, Nash to Storr, 29 May 1918.

[149] TNA, CAB 25/53, transportation of troops, telegram: War Office to Britcil, 6 June 1918.

followed was a prolonged and sustained multilingual correspondence between the allies in response to the flow of events in France and Italy. The IATC continued to produce reports and investigate the material requirements of infrastructure improvements across the two nations for the remainder of the war. The council facilitated discussions between the Italian government and the British Ministry of Shipping, which led to an agreement on the establishment of a strategic coal reserve in Italy and the provision of extra cargo tonnage to reduce pressure on the cross-border railways. The French also agreed to provide 10,000 wagons to the Italians, which were replenished by additional imports of rolling stock from Britain for use on the French lines.[150]

Yet these successes were the result of agreements reached by the national governments of each allied partner rather than the outcomes of IATC-led discussions. The IATC's memoranda and reports provided the foundations for inter-allied negotiations, in which the internal priorities and national interests of the individual members remained of foremost importance to the delegates, rather than the blueprint upon which truly collaborative actions were decided upon and implemented. Nash, the recognized transport expert with over twenty years' experience in the railway industry, contributed little to the realization of projects that did not align with Britain's strategic priorities as the war entered its final phase. As the tide turned on the western front and the allies launched what ultimately proved to be the war's final offensive campaign, attitudes towards the provision of resources to Italy hardened among the British, French and Americans. While the Italians attempted to use the IATC's reports to press for additional support from their allies, the rest of the coalition 'increasingly expect[ed] the Italians to solve their own problems'.[151] Nash, far from being considered an asset to the conduct of a war that could only be conducted in collaboration, was seen to be acting in a manner that did not complement British priorities. By mid October 1918, Lord Milner, the secretary of state for war since April, had resolved to despatch Nash to the eastern theatres to report on the railway situation. 'His object', recalled Sir Sam Fay, 'was to get Nash away from Paris where, [Milner] said, he was a nuisance'.[152] The armistice came into effect before Milner could implement any changes to the IATC's composition.

---

[150] McCrae, *Coalition Strategy and the End of the First World War*, p. 117.
[151] McCrae, *Coalition Strategy and the End of the First World War*, pp. 122–3.
[152] Fay, *The War Office at War*, pp. 205–6.

Conclusion

The British government's refusal to automatically comply with the IATC's recommendations in 1918 exposes the limitations faced by an advisory body that wished to coordinate the war efforts of multiple sovereign nations. Sir Philip Nash and his colleagues from France, Italy, the United States and Belgium were provided with the bureaucratic machinery to identify and examine the transport implications of the evolving war effort, but their investigations were unable to break from the shackles of national insularity. Each of the coalition partners maintained their domestic priorities and interests in the second half of the war, which superseded the creation of a truly unified command structure even when the work of Britain's transport experts demonstrated the interconnectivity and scale of the allies' war efforts.

The move towards increasing cooperation between the allies came about in response to events rather than any innate desire for closer collaboration. The twin battles of Verdun and the Somme greatly accelerated the degradation of the French transport infrastructure in 1916, which recalibrated the relationship between host and ally on the western front. As discussed in the previous chapter, the abandonment of the pre-war agreement provided the catalyst for British locomotives, wagons and other transport equipment to be despatched to France in hitherto unanticipated numbers. At the same time, Britain's transport experts were tasked with ensuring that the resource requirements of the nation's domestic industry and other expeditionary forces were secure – a challenge complicated by the war's voracious appetite for finite raw materials and the continued absence of a realistic appraisal for when victory would be achieved.

British transport experts became an increasingly important component of the allied war effort within this constantly changing environment. In France, Italy, Egypt, Russia, Romania, Macedonia and at home, British civilians were liberally despatched to support and advise Britain's coalition partners and to report upon developments across the nation's own multiple theatres of operations. As George Bury's examination of the Russian network and Lawson Billinton's attempts to rehabilitate Romania's railways demonstrate, they were not always successful, but in many instances British investigators were warmly received. Guy Calthrop recorded that it afforded him 'the greatest pleasure to state that the French and Italian Authorities, from Ministers downwards, were most helpful' to the mission charged with examining the feasibility of the overland route to the Mediterranean.[153]

---

[153] TNA, CAB 24/7/11, proposed overland route to Salonica, Calthrop to Derby, 7 Feb. 1917, p. 15.

However, mistrust continued to dog the allies' attempts to expel the Germans from occupied territory in the war's principal theatre of operations. British materials were not sent to France and Flanders unquestioningly, while French demands for material and operational support were not accompanied by a willingness to relinquish control over the supply lines upon which allied troops depended in the second half of the war. Sir Eric Geddes's adversarial approach to coalition warfare exacerbated rather than smoothed tensions between the French and British armies over the winter of 1916–17. The bullish attitude that dominated the North-Eastern Railway proved less applicable to the inter-allied conference room. In an arena where a collegiate, diplomatic approach was required, Geddes's refusal to seek out a compromise with his coalition partners stymied the evolution of an inter-allied forum for the discussion of transportation issues until early 1918. Geddes was kept away from a body that ultimately provided the allies with the technical advice required to consider the transportation requirements of movement on a global scale. The IATC's deliberations and reports largely remained hypothetical exercises. Yet, on the western front, Geddes's organizational and operational changes to the BEF's transportation system were subjected to the ultimate practical test in the second half of the war.

9. The road to victory: transportation in the British Expeditionary Force, 1917–18

As DGT on the western front, Sir Eric Geddes was responsible for ensuring that the necessary transport arrangements were in place to support the BEF's future military operations. In 1917 Haig's force undertook four offensives – at Arras, Messines, Ypres and Cambrai – each of which dwarfed the battle of the Somme in 1916 in terms of the volume of ammunition fired in support of the infantry. As Ian M. Brown observed, in the second half of the conflict BEF ammunition expenditure was constrained by the working lives of the guns rather than the absence of sufficient reserves of shells.[1] In 1918 the dislocation caused to the BEF's transport infrastructure by the German spring offensives was substantial, but insufficient to critically impair the fighting abilities of the British troops. Over the summer months a vast allied transportation effort provided the conduit through which the counter stroke was delivered and sustained through much – but not all – of the war's final campaign.

The results of Geddes's reorganization of transport, maintained following his departure for the Admiralty in May 1917, were played out during the so-called hundred days offensive. In the eight-day bombardment prior to the Somme, the BEF had fired 1,732,873 rounds. Eight weeks later the transport network behind the front line came perilously close to collapse under the pressure of sustaining an ammunition expenditure of 28,000 tons per week. In contrast, when British troops assaulted the Hindenburg Line eight weeks after the battle of Amiens they were supported by an artillery that was able to fire a staggering 943,847 rounds in just twenty-four hours. In the week ending 29 September 1918 the BEF expended 3,383,700 rounds, a total of 83,170 tons. In the decisive period between 8 August and 11 November 1918 the BEF poured a colossal 621,289 tons of artillery ammunition into the retreating Germans' defences.[2] The war was won on the western front, and the BEF played a key role in its delivery.

---

[1] I. M. Brown, *British Logistics on the Western Front, 1914–1919* (London, 1998), p. 174.

[2] The statistics in this passage are derived from J. Terraine, *The Smoke and the Fire: Myths and Anti-Myths of War, 1861–1945* (London, 1980), pp. 118–19; D. Stevenson, *With Our Backs to the Wall: Victory and Defeat in 1918* (London, 2011), p. 379; B. Bond, *British Military Policy Between the Two World Wars* (Oxford, 1980), p. 5.

The importance of those munitions being in a position from which they could be fired into the German lines cannot be overstated. Yet the role of the transportation units responsible for the movement of voluminous quantities of material during the summer and autumn of 1918 has been largely overlooked in histories of the First World War. Numerous authors have charted the technological and tactical improvements made by the BEF between the nadir of 1 July 1916 and the end of the war, but few have documented the modernizing or learning process within the context of the allies' ability to apply those modifications effectively. The allies possessed superior access to both human and material resources than the central powers, especially following the United States' entry into the war in April 1917. However, 'many of those resources were in the wrong place: far away in overseas empires or the US'.[3] Colossal quantities of shells were of no use to artillery commanders at the front if the Channel ports could not process them from the ships, nor the trains, lorries, wagons or barges move them inland to the guns from which they were despatched on their final journey across no man's land. The successes of the BEF's supply services, in conjunction with myriad long- and short-term issues that combined to reduce the German army's effectiveness, greatly contributed to General Ludendorff's decision to seek an armistice in 1918.[4]

This chapter examines the BEF's operations in the second half of the war, and tracks the evolution of the DGT's role in France and Flanders. It underscores how important the transport factor was to the operational tempo of an army engaged in material-intensive warfare, and emphasizes the complexities experienced by those – soldier and civilian – tasked with the maintenance of an industrial army in both static and mobile environments. In 1917 the organizational structure conceived and implemented by Geddes delivered unprecedented volumes of firepower to the front, but proved incapable of responding swiftly to the consequences of its success. The devastation created in large part by the artillery supplied by Geddes's light railways proved an insurmountable obstacle in the mud around Passchendaele. When mobility did return to the battlefield the following year, the prevailing conditions engendered passionate discussion over the subordination of the civilian-created directorate to the QMG's department within the military hierarchy. Yet as the need for civilian input of a managerial, directorial nature declined in the latter stages of the war, the divide between recognizably civilian and military tasks became

[3] Stevenson, *With Our Backs to the Wall*, p. 223.
[4] D. Stevenson, '1918 revisited', *Journal of Strategic Studies*, xxviii (2005), 107–39, at pp. 113–19.

322

indistinguishable. The transport factor – exemplified in the universal tasks of road building, railway construction and the delivery of goods from distribution centres to consumers – exerted a powerful influence over the BEF's war-making capacity until the very end of the war.

Making a new world: transporting the war of material in 1917

Negotiations over the BEF's fair share of the allies' finite transport resources, discussed in the previous chapter, were not the only claim on Geddes's attentions during his tenure as DGT. He was also responsible for making the necessary transport arrangements to ensure that the BEF was adequately supported in its military operations. However, Geddes had no influence over the location of the BEF's 1917 offensives or upon the nature of Britain's contribution to those campaigns. On 15 November 1916 the allies' military leaders met at Chantilly and agreed to continue the strategy of simultaneous offensives against the central powers on all fronts in 1917. Haig and Joffre confirmed the location for operations on the western front two weeks later. Haig, although desiring an offensive in Flanders to clear the Belgian coast, accepted Joffre's plan for a British attack – to take place between Vimy and Bapaume – concurrent with a French assault launched between the Somme and Oise rivers.[5] Therefore, Geddes's concerns over the directorate-general of transportation's first winter in existence were dominated by the need to provide a transport infrastructure capable of safeguarding Haig's troops against the problems experienced during the Somme.

The effects of Geddes's reorganization of British transportation on the western front became apparent almost immediately. The tonnage carried by the hitherto underutilized IWT fleet increased by a third in the month after Geddes's appointment, as 94,073 tons were carried by canal and rivercraft in November 1916. The following month, as 'every effort continued to be made to relieve the railways and ports by utilising as fully as possible the carrying powers' developed by Gerald Holland over the previous years, IWT provided carriage for 108,000 tons of material that otherwise would have taken up space on the overburdened railway network.[6] Barges were increasingly loaded direct from ships in port, a practice that allowed the BEF to avoid crippling shortages when the port of Boulogne was closed for a month following the accidental sinking of the SS *Araby* in late December.[7]

[5] W. Philpott, *Anglo-French Relations and Strategy on the Western Front, 1914–18* (Basingstoke, 1996), p. 129.

[6] TNA, WO 95/56, director of IWT, war diary, memorandum 4, p. 1.

[7] NLS, Haig papers, Acc. 3155/109, diary entries, 22 and 26 Dec. 1916; Brown, *British Logistics*, pp. 156–8.

However, as on the railways, the winter weather of 1916–17 caused significant reductions in the circulation of waterborne traffic. 'Towards the end of January', Holland noted in his final memorandum as director of IWT, 'the frost was so intense that vigorous measures had to be adopted to keep open the navigation sections of the Northern Waterways system'. On 27 January, 'in spite of all efforts to keep the ice broken up, the canals became frozen to such an extent as to bring transport operations in many places to a standstill'. Activity continued near the coast at Calais and on the River Somme – where the strong current prevented the water from freezing entirely – but IWT was unable to continue its expansion until the thaw commenced on 15 February. By March 1917 the carriage figures achieved by IWT had not only bounced back to the levels recorded in December 1916 but had surpassed them. 150,000 tons were carried by IWT in the month before the battle of Arras, and at both Calais and Dunkirk between 37 and 39 per cent of imports were loaded direct from ship to barge – reducing the strain on the limited railway facilities at both ports.[8]

The cross-Channel barge service also contributed to the alleviation of rail and port congestion. In January 1917 cross-Channel barges carried 1,635 tons per week to France. By September that figure had risen to 14,460 tons per week, and from May 1917 the service was entrusted with ammunition deliveries. The Royal Navy's alleged resistance to, and eventual adoption of, the convoy system has dominated historical analyses of Britain's response to the German policy of unrestricted submarine warfare in 1917.[9] Yet the humble barge also made a telling contribution to the maintenance of the allies' cross-Channel communications, and helped facilitate materially intensive forms of warfare. Between January 1917 and the armistice 574,127 tons of ammunition passed between the IWT depots at Richborough and Zeneghem; not a single barge was lost to enemy action in the English Channel during the war. Alongside ammunition, the service carried the IWT directorates' stores and supplies, railway and engineering materials and even aircraft to the western front.[10]

The increased deployment of IWT exemplified the advantages that accrued from the centralization of transport. Geddes's authority to choose the most suitable mode of transport for the various classes of supply demanded at the front replaced the self-interested actions of 1916 – when

[8]   TNA, WO 95/56, director of IWT war diary, memorandum 4, pp. 2–4; A. M. Henniker, *History of the Great War: Transportation on the Western Front, 1914–1918* (London, 1937), p. 218.

[9]   D. Stevenson, *1917: War, Peace, and Revolution* (Oxford, 2017), pp. 67–87.

[10]   TNA, WO 158/851, history of IWT, p. 58; E. A. Pratt, *British Railways and the Great War: Organisation, Efforts, Difficulties and Achievements* (2 vols., London, 1921), ii. 1107–9.

individual departments and services had competed with each other for the limited rail capacity available – with an organization that allocated transport according to a broader consideration of the BEF's overall priorities. Goods for which there was a steady, largely predictable demand from the front could be sent forward by the comparatively slow barges, which freed up significant numbers of railway wagons for the carriage of more urgently required stores.[11] The exploitation of canal transport permitted both a larger and more responsive war effort from 1917 onwards, with clear implications for the intensity of the operations the BEF was able to conduct.

However, the BEF's offensive preparations in early 1917 demonstrated both the key role to be played by transportation in the conduct of modern warfare and the subordinate position still occupied by those responsible for the transport infrastructure's maintenance within the BEF's hierarchy. As on the Somme the previous year, the transport facilities in the area designated for what became known as the battle of Arras required significant improvement ahead of zero hour. The town of Arras was served by just two standard-gauge lines, both of which had to be doubled and provided with new railheads to satisfy the BEF's needs. An additional sixty-five miles of track and sidings were laid in preparation for the offensive using rails lifted directly from the LNWR's lines in England, and by mid February Haig appeared satisfied with Geddes's progress. The former recorded in his diary that the 'energy and knowledge' demonstrated by 'Geddes and his men' deserved 'very great credit'. The British railway outlook ahead of the 1917 campaigning season, he stated, was 'promising'.[12]

The winter frost, the German retirement to the Hindenburg Line, the sluggish arrival of the materials ordered by Geddes and Haig's initial decision to concentrate British construction work around the Somme in anticipation of offensive operations in that area all retarded the development of the transport system around Arras.[13] Yet for Cyril Falls, writing in the official history of the battle, the most significant challenge for the units charged with improving the transport infrastructure behind the front lay in the BEF's continued prioritization of the supply of ammunition over engineering materials ahead of Arras. Artillery was, as Keith Jeffery acknowledged, 'central to victory' in the First World War and the cause

[11] M. G. Taylor, 'Land transportation in the late war', *Royal United Services Institution. Journal*, lxvi (1921), 699–722, at p. 710. Adrian Hodgkin recorded that three fully loaded barges carried the equivalent of 100 wagons of 10-tons' capacity. See IWM, Hodgkin papers, diary entry, 28 June 1916.

[12] *History of the Corps of Royal Engineers*, ed. H. L. Pritchard (11 vols., Chatham, 1952), v. 630; NLS, Haig papers, Acc. 3155/110, diary entry, 17 Feb. 1917.

[13] TNA, WO 158/852, history of light railways, pp. 5–6.

of around two-thirds of the conflict's casualties.[14] With this in mind, Ian M. Brown's pioneering work on British logistics considered available ammunition stocks to be a 'useful gauge of the BEF's supply state' during the war and a valuable tool for understanding the force's logistical operations.[15] Between 17 March and the opening of the battle on 9 April, all but one of the ammunition trains requested by BEF commanders were received at railheads. 'When going round the batteries in action', Geddes told Falls after the war, 'artillery officers begged him to stop bringing up ammunition because they had all they wanted and their men were too tired to offload more'. 'This was indeed a happy state of affairs', Falls wrote, 'but it is certain that no engineer officer said as much respecting engineer stores'.[16] During the same period only 125 out of 206 stone trains and thirty-five of the fifty-six requested trains of engineering stores ran.[17]

The corollary of the BEF's focus upon the maintenance of ammunition deliveries over other supply traffic was that Geddes's light railway policy – which had originally been conceived to relieve road transport, assist the advance across shell-torn ground and to forward stone and equipment for road repairs – had yet to bear fruit before the battle of Arras began.[18] The Third Army had requested fifty miles of light railway for new construction, and a further fifty miles for repairs and extensions to connect the British system to captured German lines in November 1916. In addition, a light railway system inherited from the French by XVII Corps, in an area north of Arras 'where roads to the front line were almost non-existent, had been developed into a 'complete supply system' comprising around thirty-two miles by April 1917.[19] However, the slow arrival of rails and engines rendered the light railway operating companies unable to provide significant respite to the overburdened road network during the battle. Across the Third Army's zone of operations, teams of mules and drivers from the Royal Field Artillery provided the motive power required to haul trains of six wagons over lines caked with mud, which were highly susceptible to derailments and consequent delays. Such a poorly equipped light railway system could only contribute a very small amount to the Third Army's supply needs at

---

[14] K. Jeffery, *1916: a Global History* (London, 2015), pp. 257–8.

[15] I. M. Brown, 'The evolution of the British army's logistical and administrative infrastructure and its influence on GHQ's operational and strategic decision-making on the western front, 1914–1918' (unpublished University of London PhD thesis, 1996), p. 218 n. 13.

[16] C. Falls, *History of the Great War. Military Operations, France and Belgium, 1917* (3 vols., London, 1940), i. 546.

[17] Falls, *Military Operations, France and Belgium, 1917*, i. 191 n. 2.

[18] TNA, WO 107/69, work of the QMG's branch, pp. 20–1.

[19] Pritchard, *History of the Royal Engineers*, v. 296.

Arras, with the inevitable consequence that the inadequately supported road network suffered greatly in the inclement weather that accompanied the offensive.[20] Falls was left in no doubt as to the consequences of the BEF's prioritization of firepower over transportation before Arras:

> Had the roads behind the Third Army's front been in a better condition at the start and had the supply of stone and timber been more plentiful, it is probable that the long pause of nine days between the attacks of the 14th and 23rd April could have been considerably diminished. And every hour of this period was valuable to the enemy in his feverish preparation of new positions of defence.[21]

Sufficient stocks of artillery were undoubtedly a key component of the war-winning, combined-arms strategy deployed by the BEF in the latter stages of the First World War. However, the subordination of supplies for the upkeep and extension of the BEF's transport infrastructure drastically impaired the force's ability to maintain its operational tempo and advance in strength before the Germans could regroup at Arras.

The light railway network's rapid expansion over the spring and summer of 1917 permitted the British guns' unprecedented consumption of ammunition, which contributed greatly to the BEF's initial successes in Flanders. The transport infrastructure in the area tasked with supplying the operations that became known as Messines and Third Ypres contrasted favourably with the road and rail facilities available around the Somme the previous year. A double line from Hazebrouck to Steenwerck had been supplemented by the doubling of the line between Hazebrouck and Ypres in 1915, and the ROD had been responsible for traffic on the latter from November of that year. A further line between Bergues and Proven was laid in 1916 and doubled in 1917, which gave the BEF a total capacity of 180 trains per day in the battle area. A highly developed forward delivery system, which utilized both light railway and road transport in roughly equal volumes, provided the Second Army with facilities capable of removing around 300,000 tons of material from the railheads in the month preceding Messines.[22] For the first time in the war light railways were deployed to carry ammunition right up to the heavy batteries before the assault on Messines, which allowed General Herbert Plumer's force to accumulate a stockpile

[20] W. J. K. Davies, *Light Railways of the First World War: a History of Tactical Rail Communications on the British Fronts, 1914–18* (Newton Abbot, 1967), pp. 61–2; Pritchard, *History of the Royal Engineers*, v, pp. 293–4.

[21] Falls, *Military Operations, France and Belgium, 1917*, i. 547.

[22] J. E. Edmonds, *History of the Great War. Military Operations, France and Belgium, 1917* (3 vols., London, 1948), ii. 39–40; Davies, *Light Railways of the First World War*, pp. 67–8.

of 144,000 tons of shells prior to the battle.[23] Geddes may have departed for the Admiralty by the time the attack took place, but he bequeathed to his successor as DGT, Sir Philip Nash, an organization able to supply the Second Army's artillery with each of the 3,561,530 rounds it fired into the German defences between 26 May and 6 June 1917.[24] Designed from the outset as a limited operation to clear the ridge, and described by Nick Lloyd as 'perhaps the finest example of a "bite-and-hold" operation ever conducted',[25] all of Haig's objectives for Messines were in British hands within a week and the battle was closed down as attentions shifted north towards Ypres.

As the third battle of Ypres unfolded over the summer and into autumn, the limited applicability of Geddes's industrial supply system to the conditions of the industrial battlefield – and its considerable effects upon the landscape of the western front – were gradually and graphically exposed. Events at Messines had demonstrated what was possible with a 'massive accumulation of guns and shells, the unrelenting preliminary bombardment of trenches and strongpoints', and the employment of new techniques designed to protect the advancing infantry and suppress the German artillery.[26] The brief duration and comparatively restrained aims of the offensive concealed the outcome of such colossal expenditures of firepower. At Third Ypres the BEF assembled an even larger collection of artillery pieces to support an offensive with far more ambitious objectives, both in strategic and geographical terms. Haig had extended the distance of Plumer's prospective advance at Messines to between three and four thousand yards, but he demanded far more from the troops employed at Ypres.[27] The commander-in-chief's sights were set on the capture of Roulers, twelve miles beyond the front line on 31 July, and he emphasized on an early plan for the operation that 'there must be no halt on reaching Roulers'.[28] To prepare the way for the prospective advance, the BEF's artillery fired 4,283,550 rounds during the preliminary bombardment – an increase of 147 per cent over the number that had been hurled towards enemy lines ahead of the Somme.[29]

However, the early gains at Third Ypres – eighteen square miles of enemy territory were captured on the first day, against 3.5 square kilometres on 1 July

---

[23] R. Prior and T. Wilson, *Passchendaele: the Untold Story* (New Haven, Conn., 1996), p. 59.
[24] Edmonds, *Military Operations, France and Belgium, 1917*, ii. 49.
[25] N. Lloyd, *Passchendaele: a New History* (London, 2017), pp. 56–7.
[26] Prior and Wilson, *Passchendaele*, p. 65.
[27] Prior and Wilson, *Passchendaele*, p. 58.
[28] Edmonds, *Military Operations, France and Belgium, 1917*, ii. 18.
[29] Edmonds, *Military Operations, France and Belgium, 1917*, ii. 138.

1916 – were superficial. As had occurred in Picardy, the exertions in Flanders failed to definitively break the German lines and the battle descended into a series of piecemeal assaults undertaken in appalling weather.[30] On the lines of communications towards the rear the rain had little influence. On 7 August, Nash reported to Haig that 'everything has gone extremely well. The number of trains to rail-head has increased to an excessive number: 30 more per diem than we had calculated for, and told the French we would want!'[31] However, beyond the railheads the weather conditions proved a considerable handicap to the BEF's advance. Haig complained as early as 1 August about 'a terrible day of rain' that had left the ground 'like a bog in this low lying [sic] country!' The light railways and roads necessary for the sustenance of the troops were 'steadily … pushed forward' despite the 'terrible wet',[32] but within a week troops could only reach the front line 'via a thin network of duckboard tracks that had been laid across the sodden landscape'.[33] Conditions around the 18th Division's artillery made the movement of guns even more arduous. The ooze and slime were so thick that it took the men of one battery over six hours to move a single 18-pounder gun just 250 yards.[34] The German gunners, whose arsenal had been increased from 389 to 1,162 pieces before the battle, targeted the main roads and added to the devastation.[35] On 1 September X Corps's area commandant complained that 'there [was] considerable difficulty in getting the traffic through owing to the constant shelling and bombing and the Labour Corps are always repairing shell holes in the roads which block the traffic'.[36]

With the road network increasingly congested and impassable, the time had come for light railways to fulfil the role Geddes had conceived for them during his transportation mission the previous summer. As the battle continued, an increasing quantity of the goods unloaded from the 220 trains that arrived at railheads each day were sent forward on the ever-expanding light railway network. By September 1917 the BEF's light railway system comprised 623 miles of track, almost double the mileage available during the operations at Messines (see Table 9.1). The tonnage conveyed by light

[30] J. Hussey, 'The Flanders battleground and the weather in 1917', in *Passchendaele in Perspective: the Third Battle of Ypres*, ed. P. H. Liddle (London, 1997), pp. 140–58.

[31] NLS, Haig papers, Acc. 3155/116, diary entry, 7 Aug. 1917.

[32] NLS, Haig papers, Acc. 3155/116, diary entry, 1 Aug. 1917.

[33] Lloyd, *Passchendaele*, p. 123.

[34] Prior and Wilson, *Passchendaele*, p. 98.

[35] Stevenson, *1917*, p. 198.

[36] BLSC, Liddle collection, papers of Brigadier-General W. R. Ludlow, LIDDLE/WW1/GS0984, diary entry, 1 Sept. 1918.

Table 9.1. Selected weekly averages on the light
railway network for typical months, 1917.

	January	March (pre-Arras)	June (Messines)	September (Third Ypres)	December
Locomotives in traffic	Unknown	126	342	546	513
Tractors in traffic	Unknown	68	230	335	434
Wagons in traffic	Unknown	1,395	2,756	4,332	4,797
Miles operated	97	164	314	623	717
Tons conveyed	10,325	25,315	95,180	210,808	165,530

Source: W. J. K. Davies, *Light Railways of the First World War: a History of Tactical Rail Communications on the British Fronts, 1914–18* (Newton Abbot, 1967), p. 74.

railways increased by an even larger proportion over the same period, rising from an average of 95,180 tons per week in June to 210,808 tons per week in September. Light railways carried over 227,000 tons in the final week of September, which included 47,724 tons of ammunition both in support of the successful attack at Polygon Wood and in anticipation of the assault on Broodseinde.[37]

The operations at Polygon Wood and Broodseinde represented the second and third steps in a sequence of assaults designed by Plumer to replace the more ambitious and costly approach that had been followed by Gough's Fifth Army in August and September.[38] Plumer's designs drew upon his experiences at Messines in that they called for a succession of assaults, each of roughly 1,500 yards, which aimed for strictly limited objectives until the high ground of the Gheluvelt Plateau lay in British hands. Each offensive in the sequence was accompanied by concentrations of artillery fire that far exceeded even that which had accompanied Gough's initial attack on 31 July.[39] The principal result of the focused, intense bombardments that took place in late September and early October 1917 was the annihilation

[37] TNA, WO 158/852, history of light railways, p. 7; Davies, *Light Railways of the First World War*, pp. 74–5.
[38] For a discussion of Gough's attacks in the opening phase of the battle, see Prior and Wilson, *Passchendaele*, pp. 98–110.
[39] Prior and Wilson, *Passchendaele*, pp. 115, 128.

of the Flanders landscape. The shallow nature of Plumer's attacks meant that British troops always advanced over shattered ground, the undulating nature of which made light railway construction both increasingly difficult and material intensive. More than one-fifth of all traffic on the light railway network comprised materials for the repair and maintenance of the railways, and between fifteen and twenty men per mile were required to ensure that the system remained operable.[40]

The light railway network developed and operated by Britain's transport experts was, through its facilitation of the limited and sequential advances of September and October 1917, the architect of its own breakdown. The shells delivered to the guns by light railway created ground that was wholly unsuitable for the forward projection of railway lines as the Ypres offensive crept forward. In his memoirs Gough recorded a vivid account of the prospect that faced the British troops in Flanders in late 1917:

> Imagine [the] countryside battered, beaten, and torn by a torrent of shell and explosive – a torrent which had lasted without intermission for nearly three years. And then, following this merciless scourging, this same earth was blasted by a storm of steel such as no land in the world had yet witnessed – the soil shaken and reshaken, fields tossed into new and fantastic shapes, roads blotted out from the landscape, houses and hamlets pounded into dust so thoroughly that no man could point to where they had stood, and the intensive and essential drainage system utterly and irretrievably destroyed. This alone presents a battle-ground of tremendous difficulty. But then came the incessant rain. The broken earth became a fluid clay; the little brooks and tiny canals became formidable obstacles, and every shell-hole a dismal pond; hills and valleys alike were but waves and troughs of a gigantic sea of mud. Still the guns churned this treacherous slime. Every day conditions grew worse. What had once been difficult now became impossible.[41]

At Third Ypres the transport system behind the front line collapsed. Nash reported to Haig on 13 October that 'he [had] light engines on the 60-centimetre railways sunk halfway up the boilers in the mud! Track has disappeared!'[42] The conservative nature of Plumer's objectives meant that the German guns lay beyond the reach of the attacking infantry and remained free to direct their fire upon the BEF's forward communications,

[40] R. Thompson, 'Mud, blood and wood: BEF operational combat and logistico-engineering during the battle of Third Ypres, 1917', in *Fields of Battle: Terrain in Military History*, ed. P. Doyle and M. R. Bennett (Dordrecht, 2002), pp. 237–55, at pp. 141–2; Davies, *Light Railways of the First World War*, pp. 72–3; TNA, WO 107/69, work of the QMG's branch, p. 21.

[41] H. Gough, *The Fifth Army* (London, 1931), pp. 214–15.

[42] NLS, Haig papers, Acc. 3155/118, diary entry, 13 Oct. 1917.

which exacerbated the difficulties faced by those tasked with getting urgently needed resources to the fighting troops. Cuts to the line became increasingly frequent, and 'in certain sections it appeared to be the aim of the German command to paralyse our advance by denying the supplies requisite to enable it to be continued rather than by overwhelming the troops themselves'.[43] The severance of British rail lines rendered any trains on them at the time sitting ducks – unable to move forward to deliver their goods nor backwards to safety. 'Many a train load of ammunition or other vital supply', recalled one commentator, 'was ... shot to pieces before the lines could be repaired'.[44] The BEF's extensive light railway network soon became 'utterly unable to cope with the immense volume of ammunition, engineering stores, and supplies required by the troops' at the front.[45]

Light railways were incompatible with the mud and German artillery tactics, while the lack of construction materials sent forward during the battle left the BEF reliant on a road network that was wholly inadequate for the task. At Poelcappelle on 9 October the advancing British troops were supported not by the 1,295 guns that had been assembled ahead of the Menin Road assault on 20 September but by a paltry twenty-five guns. Most of the artillery was 'stuck uselessly on the blocked single-track roads further to the rear'.[46] Brigadier-General Ludlow of X Corps rode through Ypres the following day, where he observed 'an enormous amount of traffic' along the Menin Road, to Hell Fire Corner – a 'scene of utter devastation and appalling mud'.[47] Under such conditions the achievement of an operational tempo of suitably rapidity to permanently destabilize the German army was impossible. In the words of the Royal Engineers' history of the war:

> The battle of Messines, and indeed the whole preceding three years' experience of fighting around Ypres, had shown the need for a continually increasing effort to open up roads and tracks in order to maintain the supply of guns and ammunition. It became true to say that the more shells fired in the Ypres salient the more work had to be done to restore the ground over which the assaulting infantry must pass. Thus, the R. E. problems became principally a matter of labour and materials for roads and tracks ... The roads had to be planned for the service of the batteries and ammunition dumps; the infantry had to use narrow duck-board tracks, and the pack animals the so-called dry-weather tracks, maintained by a continuous filling of shell-holes by rubble from the

[43] Taylor, 'Land transportation', p. 707.
[44] Taylor, 'Land transportation', p. 707.
[45] TNA, WO 107/296, report of British armies, p. 24.
[46] Edmonds, *Military Operations, France and Belgium, 1917*, ii. 238; Thompson, 'Mud, blood and wood', p. 246.
[47] BLSC, Ludlow papers, LIDDLE/WW1/GS0984, diary entry, 10 Oct. 1917.

shattered farmsteads. The task of the Chief Engineers was probably at its most difficult peak during the operations of 'Third Ypres'.[48]

The industrial supply system devised, installed and operated by Britain's transport experts played a key role in creating the horrific battlefield conditions experienced by those who participated in the third battle of Ypres. The transport network's increasing efficiency may have permitted the BEF's senior commanders to prosecute a 'rich man's war' from 1917 onwards.[49] However, both Gough and Plumer chose to repeat the mistakes made the previous year: unprecedented volumes of firepower took priority over the provision of adequate supplies of human and material resources for the maintenance and development of the forward communications required to sustain the advance.[50]

The BEF's obsession with firepower created a logistical imbalance on the western front that generated particularly appalling connotations in the last month of the battle. The haunting images of men, machines and animals submerged in the mud and slime around Passchendaele provide a far more poignant demonstration of the BEF's failure to adequately comprehend the implications of artillery-intensive warfare than the eighteen-mile-long line of trains outside Amiens the previous summer. Yet both were symbolic of the emergence of a challenge hitherto underappreciated by those charged with directing the British war effort. The improved quantity, if not quality, of the Ministry of Munitions' output in 1916 had exposed the inadequacy of the BEF's transport infrastructure and engendered Geddes's reorganizations. The quicker throughput of goods at the ports, the increased number of trains arriving at railheads and the greatly expanded capacity of the transport network between the railheads and the front had each aided the creation of a logistics system that was able to deliver unprecedented volumes of material to the terminal point of the DGT's responsibilities. When the location of the front line shifted – an inevitability if the allies were to forcibly expel the Germans from occupied territory – the implications of that reorganization were revealed.

Britain's transport experts were not responsible for the provision of forward communications on the battlefield. Their role was confined to the supply of materials demanded by the army to the required place at the desired time. Ahead of the so-called DGT line it fell to the chief engineers of corps and divisions to carry out the construction and maintenance work necessary

---

[48] Pritchard, *History of the Royal Engineers*, v. 312.
[49] G. Sheffield, *The Chief: Douglas Haig and the British Army* (London, 2011), pp. 101–2; Brown, *British Logistics*, pp. 173–4.
[50] Thompson, 'Mud, blood and wood', p. 250.

to move men, guns and supplies into position quickly enough to resume the offensive before the enemy recovered. Significantly, none of the senior engineers who participated in the battle of Third Ypres identified mud as the principal factor that influenced the outcome of the campaign. Instead, all were critical of the logistic, engineering, manpower and administrative components of the BEF's approach to warfare in late 1917. The most lucid observations came from the chief engineer of the Canadian Corps, Brigadier-General William Bethune Lindsay. The Canadians had taken over from II Anzac Corps on 17 October and Lindsay had immediately abandoned construction on light railways in order to focus on road and tramway building projects. The corps demanded that both the time and manpower to ensure sufficient communications were in place before it attacked, that adequate pauses were made between assaults to allow the road network to be advanced and repaired, and that the objectives selected for successive operations took account of the deteriorating conditions.[51] Between 26 October and 10 November the Canadians struggled through the morass and took the Passchendaele ridge at a cost of almost 16,000 casualties.[52] Their material requirements – in terms of ammunition, engineering stores, road stone, planks and pit props – employed 960 lorries and countless thousands of wagons and pack animals per day,[53] all of which had to travel forward over a landscape obliterated by the munitions furnished by Britain's transport experts.

The BEF's experience at Ypres in 1917 amplifies recent calls from historians of the First World War to 'look beyond the British army's struggles to master the technical battlefield', and to acknowledge the multiple practices of adaptation that occurred on both sides of no man's land during the conflict.[54] On the western front alone the belligerent armies applied new methods and technologies proactively, and were forced to respond to the innovations of their opponents. The BEF's material-intensive form of warfare depended heavily upon the maintenance of a dependable supply link to the front line. Light railways, a response to the problems of forward communications encountered during the Somme in 1916, represented a civilian-led attempt

[51] Thompson, 'Mud, blood and wood', p. 248. Thorough records of the work undertaken by the Canadian engineers in support of the Passchendaele offensives can be found in TNA, WO 95/1063, Canadian Corps. Chief engineer war diary, Aug. to Dec. 1917.

[52] Lloyd, *Passchendaele*, pp. 266–86.

[53] TNA, WO 95/1063, Canadian Corps. Headquarters branches and services: chief engineer, Methods of distribution employed for ammunition, 10 Nov. 1917, pp. 1–2.

[54] W. Philpott, 'Beyond the "learning curve": the British army's military transformation in the First World War', *RUSI*, 2009 <https://rusi.org/commentary/beyond-learning-curve-british-armys-military-transformation-first-world-war> [accessed 16 Dec. 2017].

to improve both the capacity and flexibility of the connection between the front line and the BEF's ever-expanding supply of munitions. However, the Germans' redirection of artillery fire towards the disruption of British supply operations, the British army's ongoing sacrifice of engineering and transport materials on the altar of firepower, and the horrendous weather conditions severely eroded the BEF's ability to advance in the autumn of 1917. As Robin Prior and Trevor Wilson noted, Third Ypres 'did not cease because it had reached some meaningful culmination. It simply came to a halt'.[55] When British troops began to move again the following year, they did so in the opposite direction.

A diminished role? Britain's transport experts and the return to mobility

The organization responsible for the supply of those troops underwent a number of personnel changes between the suspension of Third Ypres and the start of the German spring offensives. In London, Sir Sam Fay 'reluctantly' accepted the post of DGMR following Sir Guy Granet's decision to leave for America and take up the post of food controller. Fay was replaced as DOM by the soldier who had previously acted as his deputy, Colonel Herbert Delano Osborne.[56] In France, Sir Philip Nash was also replaced by a soldier rather than a civilian when he left the post of DGT to undertake his examination of allied transport resources on the western front. Major-General Sydney D'Aguilar Crookshank, a Royal Engineers officer whose pre-war experience had predominantly been acquired in India, became the BEF's first non-civilian DGT.

However, the promotion of these professional soldiers to roles previously held by civilians was not the result of a resurgent military voice within the British war effort. Instead, the turnover of personnel within the upper echelons of the BEF's transport services between the campaigning seasons of 1917 and 1918 was instigated by the prime minister and Sir Eric Geddes. Crookshank was first recommended as a suitable DGT by Geddes when the latter attempted to acquire Nash's services for the Admiralty in October 1917.[57] Haig was content to see Nash leave at the time on the proviso that the change was not made until after the planned operations at Cambrai had been concluded.[58] The German-Austrian breakthrough at Caporetto scuppered Haig and Geddes's arrangement. Nash took on the duty of

---

[55] Prior and Wilson, *Passchendaele*, p. 194.

[56] S. Fay, *The War Office at War* (London, 1937), pp. 141–4.

[57] NLS, Haig papers, Acc. 3155/118, diary entry, 5 Oct. 1917. Unfortunately, Haig's diary does not elaborate upon Geddes's justification for the recommendation.

[58] UWMRC, Granet papers, MSS. 191/3/4/160, Haig to Geddes, 21 Oct. 1917.

coordinating the emergency movement of troops from the western front to Italy and then, in January 1918, embarked upon his examination of communications between the English Channel and the Adriatic that led to the creation of the IATC.[59]

A further casualty of the 'increasingly irate' Lloyd George's rejuvenation of GHQ in the winter of 1917–18 had profound implications for the directorate-general of transportation.[60] The QMG, Sir Ronald Maxwell, left France at the end of the year, his poor health offered as an excuse to remove him from active duties. Following what Haig dubbed 'the decision of the Army Council to replace him with a younger man',[61] the forty-six-year-old Major-General Sir Travers Clarke took up Maxwell's post and acted as QMG throughout the final year of the war. His appointment was described later by Sir Frank Fox as 'a daring experiment on Lord Haig's part; for he was a comparative youngster to be put into a post which was then the most anxious and onerous in the Army'.[62] By the summer, Clarke's onerous responsibilities had expanded to include authority over the directorate-general of transportation.

Viewed through the surviving observations of those indirectly involved, the machinations that brought the directorate-general of transportation under Clarke's control give the appearance of a naked power grab by the new QMG. As early as January 1918 Brigadier-General John Charteris noted the existence of a 'permanent feud' between Crookshank and Clarke, while in April Fay recorded the appearance of 'a set against Crookshank' who wanted 'to get the transportation business under the Q.M.G. again, same as it was before Geddes's appointment'.[63] Haig became aware of the ill feeling between the two men during the same week, when a delegation of senior officers – with whom Clarke was in accord – openly questioned Crookshank's abilities and recommended his replacement. Haig thought it a 'serious matter to change a highly placed Administrative Officer' at such a critical period of the war, and attempted to diffuse the situation by reducing the direct personal contact between the DGT and QMG to weekly rather than daily conferences.[64]

---

[59] PA, Lloyd George papers, LG/F/17/6/14, Geddes to Derby, 17 Nov. 1917; Fay, *The War Office at War*, pp. 101, 107.

[60] For a brief overview of the prominent changes to Haig's senior staff in 1917–1918, see I. Beckett, T. Bowman and M. Connelly, *The British Army and the First World War* (Cambridge, 2016), pp. 346–8.

[61] NLS Haig papers, Acc. 3155/120, diary entry, 16 Dec. 1917. Brown, *British Logistics*, pp. 180–1 suggests that Sir John Cowans, the QMG at the War Office, orchestrated Maxwell's removal from GHQ.

[62] 'G.S.O.', *G.H.Q. (Montreuil-Sur-Mer)* (London, 1920), p. 226.

[63] J. Charteris, *At G.H.Q.* (London, 1931), pp. 282–3; Fay, *The War Office at War*, p. 160.

[64] NLS, Haig papers, Acc. 3155/126, diary entries, 21–23 Apr. 1918. Crookshank, Haig

Haig's careful man-management temporarily reduced tensions in France, but the question of Crookshank's position remained open in London. On his first day as secretary of state for war, Lord Milner told Fay that 'he had had conversations with General Staff G.H.Q. on [the] appointment of Crookshank, [and] thought he was not good enough as Director-General of Transportation'. Consequently, Fay was ordered not to confirm Crookshank's permanent appointment as DGT despite the latter having performed the role ever since Nash had departed for the IATC.[65] At a tense meeting in Milner's office on 25 May, attended by most of the Army Council, Clarke attempted to circumvent Haig's support for Crookshank and advocated the elimination of the post of DGT altogether. He stated that he 'wanted to split everything up into separate directorates' and return to the extant organization of 1914–16.[66]

There were sound operational reasons underpinning Clarke's suggestion, even though it appeared to augur a return to the watertight organizational structure that had retarded the development of a coherent transport policy before Geddes's arrival in France. Clarke argued that the return of a more mobile form of warfare 'made it clear ... that the separation of the Transportation services from the Q.M.G. was a serious defect of organization and a possible source of danger. The Q.M.G. was responsible for supplies but he was not in a position to co-ordinate and control all the means of supply'.[67] As the BEF retreated into a more circumscribed area under the pressure of successive German attacks, the uninterrupted daily provision to the fighting troops of 1,934 tons of supplies per mile of front became 'the whole question' of the war.[68] However, with Crookshank's status officially on a par with Clarke's, any divergence in policy between the two men required referral to the commander-in-chief for a decision. Any delays in the decision-making process courted disaster in the fluid conditions not experienced on the western front since 1914, and were made more likely by the changed character of battlefield supply. Based at GHQ, Clarke's direct access to the latest intelligence and the immediacy with which he received Haig's operational priorities contrasted favourably with the location of Crookshank's offices at the so-called 'Geddesburg' a few miles away – although the distance between GHQ at Montreuil-sur-Mer

---

discovered, resented what he perceived as Clarke's use of the daily interdepartmental meetings as an opportunity to criticize the work of the DGT's office.

[65] Fay, *The War Office at War*, pp. 189–90.

[66] Fay, *The War Office at War*, p. 193.

[67] TNA, WO 95/38, branches and services: quarter-master general, an explanatory review of the work of Apr. 1918, p. 2.

[68] Nash, quoted in Fay, *The War Office at War*, p. 190.

and the DGT's headquarters at Monthuis was far less pronounced than that between GHQ and the IGC's offices had been earlier in the war.

GHQ's comparative responsiveness had profound implications for the BEF's ability to meet the challenging mobile conditions of 1918. For Clarke the major lesson of Third Ypres had been the light railways' inadequacy to cope with heavy and accurate German artillery fire. Consequently, he observed, 'upon the roads and mechanical transport fell the bulk of the maintenance work of the armies'. Between the campaigning seasons of 1917 and 1918 Clarke implemented a complete reorganization of the BEF's mechanical transport, under the principle that lorries were for common use rather than to be dedicated to specialist units. As he explained in his post-war report:

> By this means considerable reductions in the number of mechanical transport vehicles allowed by existing War Establishments were made both in vehicles and man-power. The economy involved in this re-organization placed a large balance of vehicles at the disposal of the Quartermaster General. These were put into a general reserve, part of which was used to provide replacement vehicles for Mechanical Transport Companies, and the residue formed into G.H.Q. Reserve Mechanical Transport Companies.[69]

Clarke's new mechanical transport policy came into force on 13 March.[70] The German army's final offensive push began just over a week later.

The progress of the German operations over the following weeks and months appeared to confirm Clarke's beliefs both in the comparative advantage of road transport over light railways and of the desirability of centralized control over transportation from GHQ rather than the separation of duties between the QMG and a DGT. By 21 March – thanks to what David Stevenson has described as a 'masterpiece of staff work' – the Germans outnumbered their opponents at the junction of the British Third and Fifth armies by ratios of 2.6:1 in men and 2.5:1 in guns, and had accumulated 1,079 aircraft in the sector against the allies' 579.[71] The assembled German forces commenced a preliminary bombardment of unrivalled ferocity before dawn, which focused principally upon paralysing the BEF's rearward communications and silencing the British artillery.[72]

---

[69] TNA, WO 107/69, work of the QMG's branch, pp. 2–3.

[70] TNA, WO 95/37, branches and services: Quarter-master general, Circular to all armies, 28 Feb. 1918. The necessary work of reconditioning lorries for service in the GHQ reserve companies 'was perhaps sixty per cent' complete by 21 March. See Taylor, 'Land transportation', p. 708.

[71] Stevenson, *With Our Backs to the Wall*, p. 42.

[72] Stevenson, *With Our Backs to the Wall*, pp. 53–4.

Both road and rail networks suffered badly at the hands of intense German fire. 'Practically every road' was swept with 5.9-inch shrapnel and gas, while the area surrounding the railhead at Roisel, east of Péronne, received a combination of gas, high explosive and shrapnel shells at a rate of seven or eight rounds per minute. In the rear of the Fifth Army the bombardment destroyed the track, cut telephone lines and rendered the forward section of the light railway system unworkable.[73] By nightfall, the troops of the Third and Fifth armies had been driven back distances of up to eight miles.

The German advance continued at pace in the days that followed, and rapidly disorganized the network upon which Geddes's resupply system depended.[74] As light railway lines became untenable, British attentions swiftly turned to the evacuation of rolling stock or otherwise denial of its use to the enemy. Most of the equipment in the northern part of the battle zone was evacuated through Fosseux; over 300 locomotives and tractors were disabled by the removal of essential parts, and nearly 2,000 wagons were burnt by the BEF in the days after 21 March.[75] By the end of the following month the route mileage operated by light railways had been reduced from 920 to just under 360 miles, while the demands of the fighting had led to a rapid decrease in the personnel available for the network's operation and maintenance. From a peak figure of 262,000 tons per week just before the German spring offensives were launched, the tonnage carried by the BEF's light railways fell by over 50 per cent in the three months that followed.[76]

Further back, the standard-gauge railways also presented severe complications for the BEF's logisticians during the spring. By 29 March the Germans held Bapaume, Albert and Péronne, and within a week they were able to shell the railways around Amiens from positions on the outskirts of Villers-Bretonneux. Almost all of the Third and Fifth Army's railheads of a fortnight before the battle had been captured, along with many miles of valuable track.[77] As the German advance continued into April the surrender of important main lines, or their proximity to the enemy, made it increasingly difficult for Clarke and Crookshank to maintain fluidity throughout the rail network (see Figure 9.1). The loss of engine depots,

[73] R. H. Beadon, *The Royal Army Service Corps: a History of Transport and Supply in the British Army* (2 vols., Cambridge, 1931), ii. 132; Henniker, *History of the Great War*, p. 371.

[74] J. Boff, *Haig's Enemy: Crown Prince Rupprecht and Germany's War on the Western Front* (Oxford, 2018), pp. 211–14.

[75] Henniker, *History of the Great War*, p. 372; Davies, *Light Railways of the First World War*, pp. 90–1.

[76] TNA, WO 158/852, history of light railways, p. 22; Davies, *Light Railways of the First World War*, p. 92.

[77] Brown, *British Logistics*, p. 189.

the effects of German pressure on the crucial Amiens–Hazebrouck forward lateral line and the increased demands for railway traffic as troops from the Portuguese, French and American armies were rushed into the fray all increased the traffic on the remaining unmolested railways behind the British forces. During April alone the BEF ran as many ammunition trains as had been run in the entirety of the Somme campaign of July–November 1916, while the loss to the allies of the Amiens–Arras line reduced the available return routes for empty rolling stock and increased congestion across the network.[78] 'Good circulation', Clarke explained, 'is the essence of economical railway working; and a block at any point has an affect similar to that of an aneurism on a human artery'.[79] By May even the coastal lateral route had become seriously clogged with human and material traffic, and the continued forward progress of the German troops forced Clarke to utilize railheads far back from the fighting to sustain the BEF. However, the British forces continued to receive food, supplies and even a regular mail service despite the disruption.[80]

A coastal barge service between the northern waterways and the Seine was rapidly implemented as a temporary improvisation to aid the overburdened railways in the rear, but the road network provided the crucial, widely accessible connection between the railheads and troops engaged in fierce combat during the spring. Consequently, mechanical transport – the deployment of which lay outside the DGT's purview – became a progressively more valuable component of the BEF's umbilical cord as the German offensives unfolded. As Clarke emphasized in his review of May 1918, effective 'man power depend[ed], in the final result, on road power'.[81] The QMG-controlled reserve mechanical transport companies were allotted to formations according to the demands of the situation, 'kept untasked until the last possible moment', and withdrawn into reserve – or recalled for service elsewhere – as soon as their specific task had been completed.[82] The latter was far more common in 1918. In some cases, Clarke noted in his review of the QMG department's work in April 1918, 'M[echanical] T[ransport] drivers were on duty almost continuously for five days at a stretch' during the month. 'Under conditions of stress and

---

[78] TNA, WO 95/38, QMG war diary, explanatory review of the work of April 1918, p. 1; WO 107/296, report of British armies, pp. 8–9; Brown, *British Logistics*, p. 189; W. G. Lindsell, 'Administrative lessons of the Great War', *Royal United Services Institution. Journal*, lxxi (1926), 712–20, at pp. 715–16.

[79] TNA, WO 95/38, QMG war diary, explanatory review – May 1918, pp. 3–4.

[80] Stevenson, *With Our Backs to the Wall*, pp. 66–7.

[81] TNA, WO 95/38, QMG war diary, explanatory review – May 1918, p. 1.

[82] M. Young, *Army Service Corps, 1902–1918* (London, 2000), p. 121.

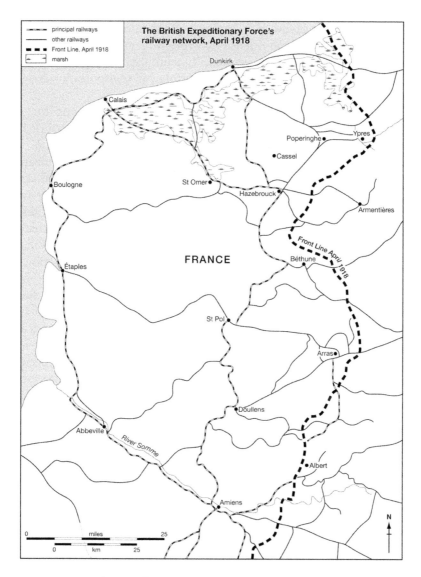

Figure 9.1. The British Expeditionary Force's rail network, April 1918.

Source: D. T. Zabecki, The German 1918 Offensives: a Case Study in the Operational Level of War (Abingdon, 2006), p. 87. Map drawn by Cath D'Alton.

danger', he continued, 'they had kept their vehicles on the road and assisted greatly to relieve the situation'.[83] 'Not only did the new organization prove its value' during the spring offensives, wrote Major Wilfrid Lindsell in a post-war article for the *RUSI Journal*, 'but unmistakable proof was also afforded of the necessity for centralised control of all transportation services under the Quartermaster-General'.[84]

The provision of separate but interlinked administrations for the operation and maintenance of an interconnected transport network had proven incompatible with the requirements of modern warfare in 1916. The directorate-general of transportation provided the BEF with an organization better equipped to service the material requirements of an industrial army, but the experiences of 1917 and early 1918 had confirmed that they were predominantly suited to the development and sustenance of a stationary army. The German spring offensives created conditions not experienced in France since the autumn of 1914, when the then QMG, Sir William Robertson, had accepted overall responsibility for the coordination of traffic on the western front. By mid April 1918 Sir John Cowans, the QMG of the forces, was explicit in his belief that the 1914 decision to subordinate the IGC had to be replicated. He wrote to Clarke that 'the introduction of the D.G.T. was all very fine when we were in a stationary period, but now either he or his representative ought to be in your office to take your orders'.[85] Fay, in the Army Council meeting held on 25 May, managed to convince his colleagues to retain the directorate-general of transportation as a distinct body, but was unable to protect the department's independence. The council advised Haig of their decision to subordinate Crookshank to Clarke in early June and by the end of the month the matter had been decisively settled in Clarke's favour.[86]

The status and identity of the directorate-general of transportation elicited strong reactions from Britain's transport experts. Fay predicted gloomily in April 1918 that the organizational changes advocated by Clarke and Cowans would 'certainly mean the importation of a great deal of feeling, with a probable loss of efficiency or interest' among the civilians embedded within the military machine.[87] He returned to the theme in Milner's office, where he urged the Army Council 'that great tact must be exercised or

[83] TNA, WO 95/38, QMG war diary, explanatory review of the work of April 1918, p. 6.
[84] Lindsell, 'Administrative lessons', p. 716. Lindsell later reinforced the message in an instruction manual on the subject of military administration published in 1933. See W. G. Lindsell, *A. and Q.: or Military Administration in War* (Aldershot, 1933), pp. 127–38.
[85] TNA, WO 107/16, IGC, general correspondence, Cowans to Clarke, 16 Apr. 1918.
[86] Brown, *British Logistics*, pp. 194–5.
[87] Fay, *The War Office at War*, p. 191.

the railway men who were volunteers in the organization of Geddes, both officers and men, might become mulish and difficult to handle'.[88] Fay's warning implied that the patriotic good will provided to the army by Britain's transport experts was far from an inexhaustible commodity, and suggests that civil–military relations within the British war effort remained fragile. Two civilian responses to the events of 25 May seemed to confirm Fay's assessment. Geddes, from his vantage point at the Admiralty, labelled the decision to subordinate Crookshank to Clarke a 'military conspiracy'. In France, Ralph Wedgwood's stance indicated that his patience with the army had worn thin by the summer of 1918. As confidence in the DGT's ability remained low among the professional soldiers at GHQ and in the War Office, Fay – who later told Milner that 'nothing would induce' him to take the post if it was under the direct command of a QMG rather than Haig – suggested Wedgwood as a potential successor to Crookshank.[89] Wedgwood, the director of docks, was eminently suitable for the role. He had considerable pre-war experience of handling freight traffic within the North-Eastern Railway's goods department, had acted as a railway transport officer at the beginning of the war and had overseen a colossal expansion in output at the BEF's docks since the directorate's establishment in early 1917. However, he had become embroiled in an increasingly acrimonious dispute with the Labour Corps over the employment of unskilled workers within the docks during 1918 and told Fay unequivocally that 'he would have nothing to do with [the position of DGT]. "The army got into a mess before, and were going to get into another now, let them get out of it in their own way." He was', Fay reflected, 'deaf to any argument'.[90]

Wedgwood's emphatic refusal to take on the role of DGT, founded upon a belief that another transportation 'mess' was imminent, illustrates that self-interest retained a powerful influence over men's actions throughout the conflict. His comments to Fay demonstrated that his motivations were born of a desire to avoid association with a project he anticipated would end in failure, rather than emerging from a sense of loyalty to the incumbent. Indeed, not all of Britain's transport experts were inclined to support Crookshank as DGT. Henry Maybury, the director of roads, complained to Fay that 'he could not get a definite decision out of Crookshank, and that things had never worked out well' during the latter's tenure as DGT. He implored Fay to 'send us a boss' in place of the vacillating soldier. George McLaren Brown was the 'boss' Fay identified, a man who had risen

---

[88] Fay, *The War Office at War*, p. 193.
[89] Fay, *The War Office at War*, p. 205.
[90] Fay, *The War Office at War*, pp. 193–4.

from ticket agent for the Canadian Pacific Railway to head the company's European operations before the outbreak of war. Control of the Canadian Pacific's extensive fleet of ocean-going vessels lay within Brown's remit, and he had been drawn into Fay's directorate of movements to provide 'general control of the movement of all war material and stores to France ... [and the] working of ports in the United Kingdom'.[91]

The friction that accompanied Brown's appearance in France to shadow Crookshank revealed the other strand of self-interest that attended personal relationships across the British war effort between 1914 and 1918. Clarke, echoing the attitude taken by his predecessor when faced by Geddes in 1916, claimed that 'some of the directors [in the QMG's department] would not work' alongside Brown. However, on this occasion the lack of cooperative spirit was not caused by tendencies towards self-preservation within the professional army. Indeed, the most vociferous critic of Brown's possible appointment as DGT was a civilian, Brigadier-General John Stewart. Like Brown, Stewart's background was in the railway industry and he had close associations with the Canadian Pacific. Unlike Brown, he had been heavily involved in the British army's global war effort for much of the conflict. He had been fundamental to the organization of the Canadian Overseas Railway Construction Corps and the rapid construction of the BEF's light railway network, had been entrusted with the investigation of railway facilities in Egypt, Palestine and Mesopotamia in 1917 and 1918 and had taken on the role of director of construction in the summer of 1918.[92] Stewart dismissed Brown as 'only a ticket agent' who 'knew nothing about railways'. Fay put the former's comments down to a personal grudge and an ambition to secure the post of DGT for himself.[93]

Whether a valid assessment or not, the entire episode demonstrated the continued delicacy of interpersonal relationships within the diverse, pressurised environment of the wartime British army. Britain's transport experts were human, and susceptible to the same inclinations towards self-preservation, ambition, jealousy, fatigue and obstinacy as their military counterparts. Ultimately, neither Brown nor Stewart succeeded the professional soldier as DGT. Circumstances on the battlefield rather

[91] W. Stewart Wallace, *The Macmillan Dictionary of Canadian Biography* (3rd edn., Toronto, ON, 1963), p. 85; Fay, *The War Office at War*, p. 47.

[92] 'General's death closes colourful saga of west', *Vancouver Sun*, 24 Sept. 1938, p. 3. A partial account of Stewart's wartime contributions is given in G. W. Taylor, *The Railway Contractors: the Story of John W. Stewart, His Enterprises and Associates* (Victoria, BC, 1988), pp. 106–20.

[93] Fay, *The War Office at War*, pp. 202–3.

than in council chambers or offices behind the lines ensured that the war was over before Crookshank could be replaced.

From Amiens to the armistice: the transport factor on paper and in practice, 1918

The town of Amiens has acquired a central position in considerations of the German and allied campaigns of 1918, one which stretches beyond the battle that bears its name. Like Ypres in 1914 and 1915, and Verdun in 1916, the protection of Amiens in 1918 took on a symbolic role for General Ferdinand Foch in the wake of his appointment as allied commander-in-chief in late March.[94] Yet the town possessed a far more significant, practical importance to the BEF that spring. Amiens and Hazebrouck comprised the two principal bottlenecks in the BEF's railway communications. Almost all of the traffic received by rail from the BEF's northern ports passed through Hazebrouck en route to the front, while most of the traffic despatched from the southern Channel ports went via Amiens. Around 80 per cent of the allies' north–south traffic – which in early 1918 averaged 140 trains per day, and was liable to comprise as many as 212 trains per day if a large-scale strategic movement of troops took place – either went through or skirted the town.[95] Construction of an avoiding line to improve rail capacity around Amiens had commenced in early March, but work on the nine-and-a-half-mile-long deviation had not been completed when the German offensive began.[96]

The depth and pace of the German advance forced the allied high command to acknowledge the threat to Amiens in the opening week of the campaign. The relative distances from the front line of Hazebrouck and Amiens had led Haig to concentrate his defensive efforts on the former, the latter being located 'at a greater depth than most … commanders would have believed to be vulnerable' following the events of 1914–17.[97] Within a few days of 21 March such beliefs had been overturned. In his diary entry for 26 March Haig recorded that the attendees of a meeting at Doullens – which included, among others, the French president and premier, Foch, and Pétain – unanimously 'decided that *Amiens must be covered at all costs*' to prevent its loss to the enemy.[98] The situation appeared precarious. Construction units

[94] Stevenson, *With Our Backs to the Wall*, p. 67; E. Greenhalgh, *Victory through Coalition: Britain and France during the First World War* (Cambridge, 2005), p. 198.
[95] D. T. Zabecki, *The German 1918 Offensives: a Case Study in the Operational Level of War* (Abingdon, 2006), pp. 85–6; Henniker, *History of the Great War*, pp. 398–9.
[96] Pritchard, *History of the Royal Engineers*, v. 636–37.
[97] Zabecki, *The German 1918 Offensives*, pp. 111–12.
[98] NLS, Haig papers, Acc. 3155/124, diary entry, 26 March 1918. Emphasis in original.

hitherto engaged on the avoiding line around the town were redeployed to construct defences under Stewart's command, while Clarke began to prepare for the removal of personnel, animals and stores from the area between Amiens, Abbeville, Blargies and Dieppe. 'Scheme X' was ready on 31 March, and was accompanied shortly after by 'scheme Y' – evacuation plans that dealt with the area surrounding Calais and Dunkirk as well as that between Abbeville, Abancourt and Dieppe.[99] Henry Rawlinson, when he replaced Hubert Gough as army commander in the area on 28 March, captured the town's significance in a letter to Henry Wilson. He wrote that there could 'be no question but that the Amiens area is the only one in which the enemy can hope to gain such a success as to force the Allies to discuss terms of peace'.[100] Consequently, he sent an urgent appeal to Foch for more troops. 'I feel anxious for the safety of Amiens', he wrote, confiding in his diary on the same day that 'if the Bosche attack heavily tomorrow I fear he will break our last line of defence in front of Amiens and the place will fall'.[101]

The town did not fall on 29 March. After a week in which the front line had shifted almost forty miles, the German army's advance slowed as the transport factor began to assert itself. By 24 March the German Eighteenth Army 'was starting to feel the effects of fatigue and stretched supply lines' as the gap between the assault troops and their railheads widened. By the time the Germans occupied Albert two days later they had left their railheads far behind, and the front-line units had been without fresh rations for two days.[102] Between the tiring attackers and their supplies lay ground shattered by the fighting of 1916 and obliterated by the Germans in their retirement during 1917. On 29 March General Georg von der Marwitz, commander of the German Second Army, described the landscape through which his troops had progressed in a letter to his wife:

> [T]he region in which we are engaged is appalling. It is the area of the earlier Somme Battle and is a giant desert. Villages are scarcely recognizable as such and topography resembles upland covered with brush and thicket. Our front lines reach to the edge of the *undestroyed* region, but it is not pretty there either, for the British have wasted no time in devastating everything. How they will ever make this land inhabitable again is anybody's guess.[103]

---

[99] TNA, WO 107/35, Amiens–Abancourt–Dieppe–Abbeville area: Measures to be taken in the event of the abandonment of, March to June 1918; Henniker, *History of the Great War*, p. 402; Pritchard, *History of the Royal Engineers*, v. 646. By 4 April Stewart was responsible for the work of 67 different units – a workforce of 22,400 men.

[100] Quoted in Zabecki, *The German 1918 Offensives*, p. 86.

[101] Quoted in R. Prior and T. Wilson, *Command on the Western Front: the Military Career of Sir Henry Rawlinson, 1914–18* (Oxford, 1991), at pp. 280, 282.

[102] Stevenson, *With Our Backs to the Wall*, pp. 224–5.

[103] Quoted in M. Pöhlmann, 'Return to the Somme 1918', in *Scorched Earth: the Germans*

The forward movement of heavy artillery, ammunition and food became increasingly difficult as the advance progressed. Under such circumstances the Germans were forced to follow a path familiar to the BEF from previous operations; the campaign paused so that guns, fresh troops and supplies could be dragged into position. By the time fourteen divisions were ready to attack on the morning of 4 April the allies had reinforced and fortified their defences ahead of the town. The German Second Army reached the outskirts of Villers-Bretonneux, around ten miles from the allies' vital railway centre, but the town itself remained in allied hands. Ludendorff terminated the offensive the following day and redirected his energies towards operations further north.[104]

The German high command may have won 'nothing of value' from operation Michael,[105] but their army's failure to take Amiens did not ease allied anxieties as to the situation astride the Somme. Foch, Haig and Rawlinson all recognized the town's psychological significance, but it was the task of civilian railway expert Philip Nash to articulate its practical importance to the allied transportation effort. On 2 April, two days before the final German push for the town began, the British section of the SWC asked Nash to examine the implications for the allies' traffic capabilities should Amiens fall:

> In the event of Amiens being no longer in Allied hands it is thought that considerable difficulty may be experienced in maintaining adequate communications between the Allied armies operating north and south of that place, while it is of the utmost importance that such communications should be fully maintained. A wedge driven in the Allied line with its point at Amiens would in fact result in all Allied communication north and south of the Somme having to be maintained through the comparatively narrow space between Amiens and St Valery which, taking into consideration the fact that a portion of this is certain to be under shell fire, is very limited indeed (some 60 kils.) and through this narrow *trouée* all movement would have to pass.[106]

Only three double-tracked routes for traffic across the Somme were available to the allies, two of which crossed the river around Amiens, while two single lines with a far lower capacity provided further options to the west.[107]

---

on the Somme 1914–1918, ed. I. Renz, G. Krumeich and G. Hirschfeld (Barnsley, 2009), pp. 179–201, at p. 192. Emphasis in original.

[104] Boff, *Haig's Enemy*, p. 218.

[105] A. Watson, *Ring of Steel: Germany and Austria-Hungary at War, 1914–1918* (London, 2015), p. 521.

[106] TNA, CAB 25/111, German possession of Amiens: effect on transportation, Storr to IATC secretary, 2 Apr. 1918.

[107] I. M. Brown, 'Feeding victory: the logistic imperative behind the hundred days', in *1918: Defining Victory*, ed. P. Dennis and J. Grey (Canberra, 1999), pp. 130–47, at pp. 134–5.

The report Nash submitted a week later illustrated just how restricted the allies' options were. If the crucial lines around Amiens were to be rendered unsafe for rail traffic, the remaining lateral routes between the town and the coast were ill-equipped to replace them. In March 1918 the allies had moved an average of 140 trains per day along the north–south routes over the Somme. However, Nash stated that the capacity of the only routes across the river to the west of Amiens – which ran Eu–Abbeville and Gamaches–Longpré – permitted just ninety train movements per day. Furthermore, as his projections did not include any contingency for bad weather or movements required for 'railway exigencies' such as the return of empty wagons, the actual capacity of the lines was likely to be at least 10 per cent lower.[108]

Nash's investigation emphasized how the movement of men and materials across the Somme would be constrained by the loss of Amiens. However, his report also warned of the 'much more serious situation' that would occur in the event of a German thrust towards Abancourt. The main line that headed west from Abancourt was the crucial link in Nash's higher estimate of allied rail capacity (see Figure 9.2). If use of the line were denied by enemy action, 'the possibility of through movement between North and South would be limited by the capacity of the Dieppe–Eu section' along the Channel coast. 'This is a single line section with heavy gradients and poor facilities', Nash explained, 'so that only eight daily train movements in each direction can be counted upon'. To make matters worse the country surrounding Serqueux, which fed Abancourt along a line that accommodated 100 movements per day, was 'so hilly and broken that the construction of any new connection to the main line north of the latter was 'impracticable'.[109] In short, Nash's examination highlighted that if the allies were unable to use Abancourt then they would lose access to almost their entire capacity for lateral rail movements across the River Somme – the coalition troops and French civilians located north of the river could not be supplied from the ports of Le Havre and Rouen, while the French munitions factories around Paris could not be fuelled by coal delivered direct from the Bruay-Béthune mines.

The precarious situation outlined in Nash's report was further complicated by events that began on the day it was submitted. The next phase of the German attack forced the allies to consider the parlous state of transportation across the western front, not just in the area immediately jeopardized by the fighting. Following the launch of operation Georgette on 9 April the Germans achieved more spectacular tactical successes, and by

---

[108] TNA, CAB 25/111, German possession of Amiens, Nash to Storr, 9 Apr. 1918, pp. 1–2.
[109] TNA, CAB 25/111, German possession of Amiens, Nash to Storr, 9 Apr. 1918, p. 6.

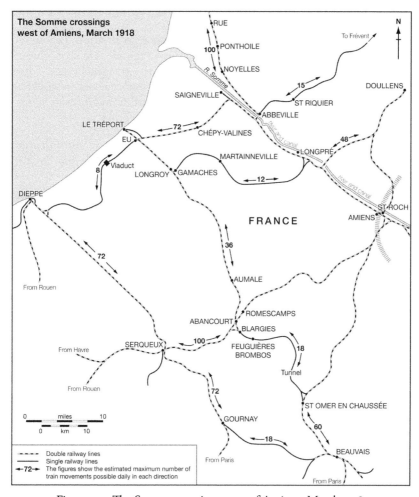

Figure 9.2. The Somme crossings west of Amiens, March 1918.

Source: A. M. Henniker, *History of the Great War: Transportation on the Western Front, 1914–1918* (London, 1937), p. 399. Map drawn by Cath D'Alton.

nightfall on 11 April the leading German units occupied territory around six miles from the outskirts of Hazebrouck. By the following day the Germans were within artillery range both of the key junction on the BEF's northern line of communications – which connected the ports of Calais, Boulogne and Dunkirk to the front line – and the Bruay-Béthune coalfield from which 70 per cent of the French munitions industry's coal was mined.[110] As the advance developed Nash informed the British section of the SWC that

> instructions ha[d] been given by the French Authorities to suspend all imports of coal and raw materials, which are not for use in the adjacent areas, through the ports of Dunkirk, Calais, Boulogne and Treport. Arrangements are in hand for diverting the supply of locomotive coal for British engines in the North to Dieppe or Rouen. The total tonnage thus diverted from North to South is about 50,000 tons per week.[111]

This traffic represented an additional pressure on the communications across the Somme at a time when the presence of active German artillery had severely curtailed movement in the vicinity of Amiens.

The depth of the German penetration in Flanders compelled the allies into action. Within the BEF, Clarke's hitherto localized plans for the evacuation of personnel and stores evolved into a sophisticated and detailed scheme, designated 'Z', for the complete abandonment of the area north of the Somme. As Ian M. Brown has noted, the BEF 'had spent the better part of three years building this line of communication, along with its attendant infrastructure – rail improvements, depots, bases, port improvements, road improvements, light railways, ammunition depots'.[112] All were scheduled to be removed, destroyed or otherwise rendered inoperable to the enemy. Scheme 'Z' ultimately provided the QMG with a timetable for the evacuation of 250,000 men, their attendant equipment and 600,000 tons of stores from northern France in just twenty-eight days.[113] While many of the preparations remained hypothetical, on 11 April the units of the locomotive repair shop at Borre near Hazebrouck – established in April 1917 with the capacity to undertake 200 heavy and 120 light repairs per year – were ordered to dismantle its machinery and remove the contents from the enemy's grasp. 'The lads worked so well', recalled Colonel L. S. Simpson, 'that in three days and nights they had got out, loaded up, and dispatched to Audruicq practically everything except the big wheel lathes and an engine or two that were not down on their wheels – this under shell fire most of the time – and I am glad to say three hundred of them

---

[110] Stevenson, *With Our Backs to the Wall*, pp. 71–2; Boff, *Haig's Enemy*, pp. 220–2; Watson, *Ring of Steel*, p. 521. Dunkirk itself was also under shell-fire, 'and in consequence could be little used'. See 'G.S.O.', *G.H.Q.*, p. 255.

[111] TNA, CAB 25/111, German possession of Amiens, Nash to Storr, 10 Apr. 1918, p. 1.

[112] Brown, 'Feeding victory', p. 137.

[113] TNA, WO 107/34, programme of railways: memorandum forecast of trains and railheads, 1918. For a detailed account of the wider administrative responsibilities demarcated in scheme Z, see Henniker, *History of the Great War*, pp. 402–11.

turned up at Audruicq safe and sound on the 13th, followed the next day by the remainder'. The evacuation did not halt work for long. Within a week the ROD had opened up a repair shop at Rang-du-Fliers, to the south of Étaples, which was 'turning out three to five engines a week, besides repairing metre gauge stock and making ballast ploughs, telegraph post slotting machines, and a hundred and one other things' for the remainder of the war.[114]

The German advance also created immediate concerns for the BEF's allies. On 12 April Haig received warning from the French president, Georges Clemenceau, that only five days' reserves of coal existed in France.[115] Without access to continuous supplies of coal the French munitions industry could not function. However, Nash was emphatic in his belief that it was 'absolutely necessary, from the purely military point of view, to immediately free the single North and South lateral in Allied hands from all but military traffic. This means that the whole of the Coal traffic from the Pas de Calais to the area South of the Somme must immediately be shut down'.[116]

The manner in which the allies solved this dilemma illustrates the extent to which the French and British war efforts had become intertwined by April 1918. On 19 April, when Nash was emphasizing to the British section of the SWC that traffic from the Bruay-Béthune coalfield had to be suspended, the French minister of armaments informed him that 'in order to make good the loss of the Pas de Calais mines' the French would require 600,000 tons of coal per month from England.[117] Consequently, an inter-allied meeting took place at the Ministry of Armaments in Paris on 23 April, during which experts from both nations – taking into account the higher calorific content of British coal in comparison to French coal – thrashed out an agreement for Britain to import 450,000 tons of coal per month into ports south of the Somme in the event that the Germans rendered the northern coalfields inoperable.[118]

[114] L. S. Simpson, 'Railway operating in France', *Journal of the Institution of Locomotive Engineers*, xii (1922), 697–728, at pp. 717–18.
[115] Stevenson, *With Our Backs to the Wall*, p. 72.
[116] TNA, CAB 25/111, German possession of Amiens, transport position in France. Memorandum by Major-General Sir P. A. M. Nash, 19 Apr. 1918, p. 1.
[117] TNA, CAB 25/111, German possession of Amiens, transport position in France, 19 Apr. 1918.
[118] TNA, CAB 25/111, German possession of Amiens, summary of a meeting held at the Ministry of Armaments, Paris, on 23rd Apr. 1918, to consider the requirements for the export of coal from the United Kingdom to France in certain eventualities. Sir Richard Redmayne, the chief inspector of mines and former chair of mining engineering at the University of Birmingham, led the British delegation.

Nash's experiences of coalition warfare during the German spring offensives provide a startling contrast both to the terse and suspicious atmosphere that surrounded Sir Eric Geddes's negotiations with the French in 1917 and Nash's own attempts to coordinate transportation matters on the Italian front in 1918. The speed and depth with which the German army advanced after 21 March compelled the French and British to subordinate their insular domestic objectives for the benefit of the alliance's survival as an effective fighting force. By late April, thanks to the energies of transport experts from both nations and a commitment to the maintenance of communications between them, the allies had managed to reduce north–south coal traffic by one-third and had taken steps to ensure the continued supply of fuel for the French economy should events at the front compel the abandonment of the Bruay-Béthune mines.[119]

However, the reduction of traffic flows across the Somme in spring 1918 was 'merely a palliative' and comprised only half of the parameters for Nash's examinations.[120] Alongside his instructions to identify the potential economies available on the north–south routes behind the allied front, the SWC requested that he assess the 'desirability and practicability' of railway improvements that 'would materially improve the general transportation situation' were Amiens to be captured.[121] By 9 April he was able to report that 'certain works' that could 'immediately give some relief' to the allies' transport constraints were already 'in hand', while further construction efforts to increase the routes available for northsouth traffic were being contemplated.[122]

A fortnight later that contemplation had resulted in an action plan agreed by both the British and French railway authorities.[123] As coalition troops continued to frustrate successive German assaults over the following months, allied engineers undertook an extensive scheme of railway construction that greatly increased the allies' capacity to move men and materials within the truncated space between the front line and the Channel coast. British engineers improved stations on the Gamaches–Longpré line to provide the Fourth Army with more railheads, doubled the thirteen-

---

[119] TNA, CAB 25/111, German possession of Amiens, Nash to Storr, 13 Apr. 1918; telegram: G.H.Q. to Britcil, 13 Apr. 1918; note on the organisation of supply of the Allied forces operating in the area north of the Somme, 15 Apr. 1918, p. 2; extract from War Cabinet, 395, dated 19th Apr., 1918; Belin to Sackville-West, 24 Apr. 1918.

[120] Henniker, *History of the Great War*, p. 400.

[121] TNA, CAB 25/111, German possession of Amiens, Storr to IATC secretary, 2 Apr. 1918.

[122] TNA, CAB 25/111, German possession of Amiens, Nash to Storr, 9 Apr. 1918, p. 2.

[123] TNA, CAB 25/111, German possession of Amiens, Nash to Storr, 27 Apr. 1918, pp. 2–3; Pritchard, *History of the Royal Engineers*, v. 645.

mile-long line between Longpré and Martainville and the twenty-seven miles of line between Abbeville and Frévent, contributed to the duplication of the line south from Étaples to Port-le-Grand, and engaged in a range of improvement works behind the Flanders front.[124] In May 1918 alone the British built or reconstructed 148.74 miles of broad-gauge track.[125] Yet their efforts were dwarfed by those of the French, whose principal achievement was to survey, prepare and construct an entirely new double-line some fifty-five miles long from a position south-east of Abancourt to a new connection across the Somme.[126] The line, which had only been set out on the ground on 30 April, was linked through from end to end by 15 July and ready for traffic a month later (see Figure 9.3). As Henniker noted admiringly in the British official history, 'a trunk line complete with engine sheds, water supplies, signalling, telephones, station buildings, etc., had been constructed in … 106 days from the date on which work started'.[127] It was, concurred the Royal Engineers' history of the war, a 'very remarkable feat'.[128] Upon completion the new line, combined with the other construction undertaken that summer, reduced the proportion of north–south traffic that had to pass through Amiens: three double-lines, with a capacity of 144 trains per day in each direction, provided the allies with more lateral rail options than they had possessed before the Germans had attacked. In short, by the late summer of 1918 the BEF's railway communications were in better shape than they had been before 21 March.[129]

Yet despite the comparative reduction in the town's importance to the allies' lines of communications, Amiens still represented the perfect location from which they could exploit their freshly installed transport options. Haig had already raised the idea of pushing the Germans beyond artillery range of the town before 18 July, when a counterattack on the Marne by troops of the French Tenth Army indicated that Ludendorff had lost the initiative on the western front.[130] The German troops who occupied the sector 'were

[124] Henniker, *History of the Great War*, pp. 400–2; J. H. F. Le Hénaff and H. Bornecque, *Les chemins de fer français et la guerre* (Paris, 1922), pp. 237–8; D. Lyell, 'The work done by railway troops in France during 1914–19', *Minutes of the Proceedings of the Institution of Civil Engineers*, ccx (1920), 94–147, at pp. 109–12.

[125] TNA, WO 95/40, QMG war diary, statistical summary, Oct. 1918, p. 5.

[126] Further details of the railway construction work completed in the Somme region during this period are given in TNA, CAB 25/111, German possession of Amiens, Le Hénaff to Nash, 18 July 1918; *Les armées françaises dans la grande guerre: la direction de l'arrière* (Paris, 1937), pp. 669–71.

[127] Henniker, *History of the Great War*, pp. 400–1.

[128] Pritchard, *History of the Royal Engineers*, v. 645.

[129] Brown, *British Logistics*, p. 194; Brown, 'Feeding victory', p. 139.

[130] W. Philpott, *Bloody Victory: the Sacrifice on the Somme* (London, 2009), pp. 517–18, 520.

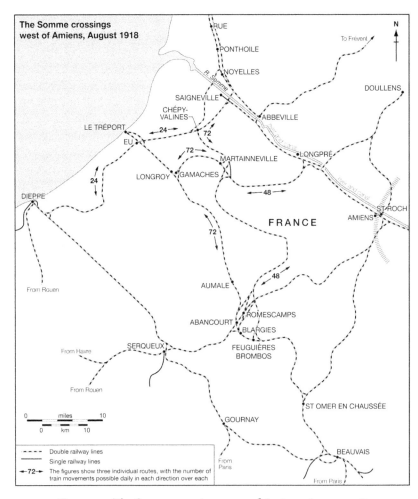

Figure 9.3. The Somme crossings west of Amiens, August 1918.

Source: A. M. Henniker, *History of the Great War: Transportation on the Western Front, 1914–1918* (London, 1937), p. 400. Map drawn by Cath D'Alton.

not the formidable fighters of the March offensive'; their defences were 'inadequate', their morale was poor, and the difficulties of supply across the devastated landscape were such that their rations were deemed 'very bad and scarce'.[131] Between March and the end of July the German army had suffered 977,555 casualties, which it had found increasingly difficult to replace.[132]

By contrast, the transport infrastructure behind the allies facilitated the accumulation of significant human and material resources in the sector ahead of the battle of Amiens. Between 27 July and 10 August Rawlinson's Fourth Army expanded greatly; the number of men attached to the army rose from 257,567 to 441,538, and the number of horses from 54,323 to 98,716.[133] The BEF drew together three cavalry and seventeen infantry divisions for the offensive, with each battalion of the latter equipped with thirty Lewis guns, eight light trench mortars and at least sixteen rifle-grenadiers. They were supported by 534 tanks, 800 aeroplanes and over 2,000 lavishly stocked guns – each 60-pounder possessed sufficient shells to fire four rounds per minute for four hours on zero day. Alongside them the French First Army comprised fifteen infantry divisions, over 1,000 aeroplanes and 1,624 light and heavy guns. Across no man's land sat eleven depleted German divisions, supported by fewer than 400 aircraft and inadequate numbers of artillery.[134] 'Given the relative strengths of the forces involved', argued Rob Thompson, 'it was clear that this assault would be victorious' for the allies.[135]

The scale and manner of the allied victory on 8 August has given the battle of Amiens a prominent position in the (especially Anglophile) revisionist history of the conflict.[136] The results of the first day have provided ample evidence to support 'the perception that the British applied sophisticated, fire-power-based combined arms methods to overcome German defences

---

[131] Prior and Wilson, *Command on the Western Front*, pp. 289–90.

[132] Watson, *Ring of Steel*, p. 524.

[133] J. P. Harris, *Amiens to the Armistice: the BEF in the Hundred Days' Campaign, 8 August–11 November 1918* (London, 1998), p. 73.

[134] Philpott, *Bloody Victory*, pp. 520–1; Prior and Wilson, *Command on the Western Front*, pp. 309–15.

[135] R. Thompson, '"Delivering the goods". Operation *Landovery Castle*: a logistical and administrative analysis of Canadian Corps preparations for the battle of Amiens, 8–11 August, 1918', in *Changing War: the British Army, the Hundred Days Campaign and the Birth of the Royal Air Force, 1918*, ed. G. Sheffield and P. Gray (London, 2013), pp. 37–54, at p. 40.

[136] For a selection of accounts, see Harris, *Amiens to the Armistice*, pp. 116–17; G. Sheffield, *Forgotten Victory. The First World War: Myths and Realities* (London, 2001), pp. 237–41; C. Messenger, *The Day We Won the War: Turning Point at Amiens, 8 August 1918* (London, 2008); N. Lloyd, *Hundred Days: the End of the Great War* (London, 2013), pp. 54–5.

and restore mobility to the battlefield'.[137] The BEF's leading units had driven the Germans back some eight miles by nightfall on the first day, captured over 12,000 prisoners and 374 guns and cleared any lingering threat to the railway communications around the town. When operations were suspended after 11 August the British and French between them had inflicted 48,000 casualties and asserted the allies' material, psychological and tactical superiority over their opponents – a predominance they did not relinquish for the remainder of the war.[138]

Those advantages were entirely underpinned by, and their maintenance dependent upon, transportation. The allies' material superiority over the Germans provided them with the means by which to prosecute a more intensive form of warfare in the second half of 1918. During the hundred days the BEF participated in a series of concentric operations, conceived by Foch as a 'sequence of offensives, each one *within the capabilities of a single army's fighting power and logistics*, ... engaged at a pace that would exhaust Germany's ability to sustain battle once and for all'.[139] To sustain pressure on the retiring German forces, the BEF and its allies had to ensure that the tempo of their operations over the late summer and autumn of 1918 were sufficient to prevent the enemy from preparing positions that could withstand the allied onslaught. Once the enemy had been compelled to fall back, a successful pursuit depended upon the timely movement of ammunition, supplies and materials across the battlefield in sufficient volume to render the Germans unable to prepare new defensive lines. Faced by a tenacious, determined opponent, the BEF could not abandon the industrial machinery upon which successful combined-arms operations relied. Therefore, as much as could be physically moved had to be shifted forward over the available roads, railway lines and waterways to maintain pressure on the enemy. Ultimately, the transport infrastructure in France and Flanders represented a fundamental component of the allies' weapons system during the advance to victory – one that had to be exploited with greater success than the Germans had achieved in the spring.

The altered situation on the western front engendered a re-evaluation of the BEF's transport organization. The conditions within which the BEF operated as a mobile fighting force differed significantly from those that had been in place when Sir Eric Geddes had created the directorate-general of

[137] J. Boff, 'Combined arms during the hundred days campaign, August–November 1918', *War in History*, xvii (2010), 459–78, at p. 461.

[138] Boff, *Haig's Enemy*, pp. 234–5; Harris, *Amiens to the Armistice*, pp. 103–4; Prior and Wilson, *Command on the Western Front*, p. 320.

[139] W. Philpott, *Attrition: Fighting the First World War* (London, 2014), pp. 327–8. Emphasis added.

transportation. Clarke was quick to perceive how the return of movement to the battlefield had changed his department's task, and maintained his insistence that a military-led organization was better suited to provide transport for a mobile army than a civilian one:

> A military organization to be efficient must at any time be prepared for either emergency – advance or retreat. This is a fact which there seemed a tendency to ignore sometimes in the past – the very great difference between the problems which the civilian expert has to meet under conditions of stability and those which confront the military leader under conditions of war.[140]

The QMG's transport policy, issued as guidelines to the BEF's individual armies on 27 August, reflected his attitude. He recommended that work on light railways, the medium most closely associated with Geddes's reforms of the BEF's transport system, was to cease unless conditions for their repair were 'specially favourable'.[141] A combination of mechanical and horse-drawn transport would 'take up most of the load from railheads' to the front line as the army pressed forward. In essence, Clarke recommended a return to the transport policy envisaged within the army's pre-war – pre-civilianized – organization.[142]

There was more to Clarke's new transport policy than his slight against the perceived blind spot of civilian specialists suggests. The QMG's decision to recommend that the main line railways and roads received priority attention underlined both the BEF's reliance on the former for the bulk movement of goods inland from the ports and the latter's advantages in the fluid conditions of a mobile war. Put simply, light railways were poorly equipped to deal with the rigours of a general advance over a substantial distance. As Henniker explained in the official history:

> The net train loads on the 60-cm. lines were very small – say 30 tons; each standard gauge train arriving at a railhead would need, say, 10 light railway trains to clear it; traffic on light railway trunk lines would therefore be ten times as intense as on the standard gauge; the line must therefore be solidly built and equipped like a first-class main line. To follow up a moving front and to maintain the lines behind it would require a continuous traffic of material, ballast, coal, etc.; beyond a limited distance a great part of the capacity of the line would be required for its own maintenance and extension. It was estimated

---

[140] TNA, WO 95/40, QMG war diary, explanatory review, Aug. 1918, p. 1.
[141] TNA, WO 95/40, QMG war diary, diary entry, 27 Aug. 1918.
[142] TNA, WO 95/40, QMG war diary, explanatory review, Aug. 1918, p. 1; G. R. Winton, 'The British army, mechanization and a new transport system, 1900–14', *Journal of the Society for Army Historical Research*, lxxviii (2000), 197–212; Henniker, *History of the Great War*, p. 451.

that it might be worth extending an existing light railway system for 12 to 15 miles, but that the maintenance by light railways alone of an army advancing over a greater distance, even if practicable at all, would require an enormous equipment of locomotives and rolling stock, and a number of skilled personnel to maintain and operate it far in excess of what had been provided to enable it to fulfil its role during stationary warfare.[143]

Furthermore, the light railway organization that had been steadily developed over the course of 1917 had been severely dislocated by the German spring offensives. Some 560 miles of light railway had been destroyed or captured by the Germans, while the redeployment of light railway troops into broad gauge construction units meant that by August 1918 the BEF's light railway directorate lacked the human and material resources that a comprehensive use of the medium required.[144]

However, despite the medium's limitations and the contents of Clarke's guidance, light railways were exploited whenever circumstances permitted during the hundred days. As has been previously demonstrated for the BEF's devolved approach to battle planning, GHQ provided army and corps commands with considerable latitude to exercise initiative in the realm of transport improvements.[145] In practice, Clarke requested that the BEF's individual armies obtain permission for light railway construction during the hundred days only when the proposed projects required substantial resource commitments, affected the principal west–east trunk routes or demanded inter-army coordination.[146] Early in the advance, when German systems were captured in good order or existing lines could be connected to the British network with relative ease, light railways provided valuable bulk distribution lines between the standard-gauge railheads and forward dumps. During August the BEF's advance was supported by a light railway network that conveyed a weekly average of 157,651 tons – significantly lower than the figures recorded during the previous year's operations at Third Ypres, but a substantial increase on the 109,172 tons carried each week

[143] Henniker, *History of the Great War*, pp. 450–1.
[144] Taylor, *Railway Contractors*, p. 118.
[145] The extent to which Haig and GHQ successfully delegated authority during the hundred days has been subject to sustained examination. For a range of views, see T. Travers, *How the War Was Won: Command and Technology in the British Army on the Western Front, 1917–1918* (London, 1992), pp. 175–82; Sheffield, *The Chief*, pp. 293–339; W. Reid, *Douglas Haig: Architect of Victory* (Edinburgh, 2006), pp. 449–86; Prior and Wilson, *Command on the Western Front*, pp. 397–8; J. Boff, *Winning and Losing on the Western Front: the British Third Army and the Defeat of Germany in 1918* (Cambridge, 2012), pp. 214–15; A. Simpson, *Directing Operations: British Corps Command on the Western Front, 1914–18* (Stroud, 2006), pp. 156–76.
[146] TNA, WO 95/39, QMG war diary, 12 Aug. 1918.

during May 1918. Over the course of the hundred days the mileage of light railways operated by the BEF increased by 35 per cent,[147] shaped into a series of 'long antennae' that ran west–east in pursuit of the advancing troops. The Third Army built up a particularly successful line, which crossed the St Quentin Canal and extended over thirty-five miles between Fosseux and Crèvecœur-sur-l'Escaut.[148]

Clarke's observations on the difference between civilian and military leadership in the field of army transportation fixated on a distinction that had become increasingly blurred by August 1918. The QMG's subordination of the directorate-general of transportation did not mean that Britain's transport experts lacked influence over the BEF's operations in the war's final campaign. Rather, their principal contribution was felt on a practical rather than managerial level. Between late 1916 and the summer of 1918 the directorate-general of transportation had been conceived, supplied with skilled personnel and industrial equipment and embedded within a military supply machine that stretched from the Channel coast to the front line; the organizational changes demanded by industrial warfare had been devised and implemented long before the battle of Amiens. After the battle, the directorate-general of transportation experienced its 'severest test' of the war:

> The organization for the advance of practically the entire British army over the devastated zone from 30 to 50 miles deep (Arras to Mons) along the whole front, which had been scientifically and systematically demolished by an enemy who understood demolition and devastation to a nicety, and carried it out with the systematic practice and painstaking detail for which he had a justly high reputation. For the transportation troops this meant ceaseless work under high pressure and under great difficulties and discomfort.[149]

That ceaseless work – rather than entailing the creation and establishment of new organizational structures and using pioneering managerial techniques to monitor the performance of a dispersed workforce – primarily comprised unloading, loading, driving, navigating, building, repairing, and operating the BEF's panoply of transport options in support of the fighting troops.

The predominantly civilian workforce within the directorate-general of transportation responded to the changed circumstances of the war with vigour. Their endeavours were prodigious across all areas of the BEF's

---

[147] TNA, WO 95/40, QMG war diary, statistical summary, Oct. 1918, p. 4.

[148] Davies, *Light Railways of the First World War*, pp. 98–102, 104; Boff, *Winning and Losing*, p. 88.

[149] S. D'A. Crookshank, 'Transportation with the B.E.F.', *Royal Engineers Journal*, xxxii (1920), 193–208, at p. 194.

transportation service in the final months of the war. At the six principal docks under the direction of Brigadier-General Ralph Wedgwood, the North-Eastern Railway's chief goods manager, average weekly imports rose from 150,300 to 173,270 tons between August and October 1918.[150] At an individual level, the tonnage handled per man per hour increased by 15 per cent over the same period. Improvements also took place within the IWT service commanded by Brigadier-General Cyril Luck – the ex-Royal Indian Marine successor to the LNWR's Gerald Holland. In October the BEF's fleet of tugs and barges conveyed an average of 66,368 tons per week across the navigable waterways of France and Flanders, an increase of 20 per cent over the figures recorded the previous June.

Substantial as these improvements were, the BEF's operational tempo during the hundred days primarily depended upon the speed with which the railways and roads behind the front line could be extended. The railway construction troops' initial progress was good. By 8 September the most easterly divisional railheads available to the Fourth Army were located around Bapaume, some thirty-two miles away from their position before the battle of Amiens, and the entire length of the Amiens–Albert–Arras line was handed over to the ROD for operation the following day.[151] From the nadir of May, when the German spring offensives had left the BEF in possession of just 220.7 miles of operable railways, by the end of September the British portion of the rail network behind the western front comprised 485.3 miles of track. In the first two months of the hundred days an average of 153 trains per day arrived into the BEF's railheads, loaded with ammunition and supplies for onward distribution to the advancing troops.[152]

The ROD, commanded by the Midland Railway's superintendent of the line, Colonel Cecil Paget, played a critical role in facilitating the British contribution to the battles of the Hindenburg Line in late September. The BEF had fired a daily average of 4,748 tons the previous June, which had required the provision of sixteen daily ammunition trains by Paget's division. Across September as a whole the ROD ran an average of twenty-four, and as many as thirty-three, ammunition trains per day as the force accumulated the firepower considered necessary to destroy the German army's last major defensive system. Across August and September as a whole the BEF hurled a daily average of more than 8,000 tons of munitions across no man's land, a period of artillery expenditure that peaked with the colossal bombardment

[150] Unless otherwise stated, all figures in this passage are taken from TNA, WO 95/40, QMG war diary, Statistical summary, Oct. 1918.

[151] TNA, WO 95/40, QMG war diary, sketch map illustrating how the railways followed up our advancing troops, Sept. 1918; Henniker, *History of the Great War*, p. 435.

[152] Henniker, *History of the Great War*, p. 456.

detailed at the beginning of this chapter.[153] In the opening phase of the hundred days campaign, which climaxed when the 46th Division crossed the St Quentin Canal and pierced the Hindenburg Line, the requirements of Foch's strategic plan rather than logistical constraints influenced the scale and timing of British military operations.[154]

However, the efforts and endeavours of Britain's transport experts could only ameliorate rather than eliminate the transport factor in the weeks that followed. The prodigious weight of fire employed by the BEF's artillery during August and September was not sustained in October, as the difficulties inherent in the supply of a large, moving army over the available railway and road networks steadily accumulated. First, the pace of railway construction behind the advancing force slowed considerably.[155] During August and September 86.04 miles of new track were laid and 518.19 miles reconstructed. In early September Sir Sam Fay had secured agreement with the Ministry of Munitions to raise the reserve levels of track in France to 500 miles by the end of October (an increase of 20 per cent). However, Clarke recorded on 5 October that shipments of track had not kept up with the programme and that consequently 'the stock of rails [was] very much reduced'.[156] Delay-action mines, lodged in railway embankments and bridges by retiring German engineers, added further complications for the railway construction troops – in the words of one commentator they 'caused more transportation difficulties than the whole of the previous four years had produced'.[157] On 19 October the QMG's diary recorded that a supply train had run into a crater, while progress in the Cambrai–St Quentin area was particularly affected by 'a large number of accidents and the explosion of delay-action mines' over the course of the following week.[158] When the Third Army successfully dislodged the Germans from defensive positions

[153] Henniker, *History of the Great War*, pp. 456–7. As Henniker noted, the traffic figures given above only included ammunition trains from the main ammunition depots. Additional traffic from the advanced depots in army areas often demanded the running of as many as 12 additional trains.

[154] Boff, *Winning and Losing*, pp. 31–2, 89–90.

[155] See TNA, WO 95/40, QMG war diary, sketch map, Sept. 1918. The distances between the most easterly railheads available to divisions on 8 Oct. lay on aggregate far closer to those available on 8 Sept. than those in use on 8 Aug.

[156] Fay, *The War Office at War*, p. 213; TNA, WO 95/40, QMG war diary, diary entry, 5 Oct. 1918. Fay's concern was wholly justified. Between Aug. and Oct. 1918 the BEF's railway construction troops laid 485 miles of new track. See Pritchard, *History of the Royal Engineers*, v. 659.

[157] J. C. Harding-Newman, *Modern Military Administration, Organization and Transportation* (Aldershot, 1933), p. 23.

[158] TNA, WO 95/40, QMG war diary entries, 19 and 26 Oct. 1918.

around the River Selle in a night attack on 20 October, it was 'unable to exploit its success' for three days due to the slow arrival of the force's ammunition trains.[159] Increased congestion on the rail network meant that a daily average of 133 trains ran into railheads during October, a decrease of twenty per day that contributed markedly to a decline of one-third in the BEF's artillery use in October 1918.[160]

The reduced capacity and reliability of the BEF's rail lines had profound implications for the rest of the force's transport infrastructure. A reduction in the deliveries of stone slowed the construction of new railheads and roads closer to the front, which increased the distances to be covered by mechanical transport at the same time as it decreased the amount of material available for the upkeep of the road surface. Furthermore, when an undetected device exploded on 3 November and destroyed a bridge on the Cambrai–Busigny line that had only been repaired a fortnight earlier, it forced the divisions of IV and V Corps to draw their supplies from railheads ten miles further to the rear.[161] At the same time the bulk of the Fourth Army's railheads were located between twenty and twenty-five miles behind the front. Over the next week Rawlinson's troops advanced a further twenty-five miles, but a combination of mines, destroyed bridges, and accidents ensured that few new railheads could be opened to traffic. 'At the date of the Armistice', Henniker recorded, 'the only reliable railheads for the Fourth Army were 50 miles behind the Armistice line; in the north even the most advanced railheads of the Fifth Army were 30 miles behind it'.[162]

The task of bridging the ever-widening gap fell chiefly upon the BEF's fleet of lorry drivers, whose bodies and vehicles were subjected to punishing workloads as the hundred days progressed. The lorries of the 14th GHQ Reserve Mechanical Transport Company were on the road for around one hundred hours over the course of five days in late September, while Henniker recorded cases of 'columns taking seventy-two hours to complete what should have been a daily round'. The combination of tired drivers, poor roads and long journeys – which reduced the time available for vehicles to be serviced – increased the number of accidents and breakdowns

---

[159] J. Boff, 'Logistics during the hundred days campaign, 1918: British Third Army', *Journal of the Society for Army Historical Research*, lxxxxiv (2011), 306–21, at p. 320. All classes of supply were affected by congestion on the railways in late Oct. The regular daily supply trains for 61st Division arrived a day late on four occasions, while the deliveries for 16 and 19 Oct. did not get through at all.

[160] TNA, WO 95/40, QMG war diary, statistical summary, Oct. 1918, pp. 4, 7.

[161] Boff, 'Logistics during the hundred days campaign', p. 314.

[162] Henniker, *History of the Great War*, pp. 460–61.

among the BEF's pool of lorries.[163] At the end of October Clarke reported that 'some 3,600 lorries were out of action (out of a total of about 20,000 working in Army areas) chiefly due to broken springs',[164] which further degraded the supply situation at the front. Eighty lorries were required to carry ammunition for VI Corps' heavy artillery. By 2 November only fifty-six were available to accumulate firepower ahead of the battle of the Sambre.[165]

The BEF's victory on the Sambre sent the German army into a general retreat. For the remainder of the war the 'British trudged on through cold and wet across the sabotaged and booby-trapped landscape that the Germans had left behind them'. The 'weather and logistical difficulties rather than the Germans were the main obstacle to the advance' in the final phase of the hundred days.[166] Since 8 August the BEF's spearhead formations, the Third and Fourth armies, had advanced approximately sixty and 100 miles respectively.[167] The repairs of mine craters in the roads could not keep pace with the advance, as congestion, delays, accidents, the demands of civilian populations in liberated territory and breakdowns across the transport network severely decreased the volume of material to reach the front. By the week ending 9 November the BEF's transportation network echoed the paralysis experienced on the Somme two years earlier. Ten accidents and sixteen mine explosions – mostly on the critical Cambrai-St Quentin line – reduced the speed with which trains arrived at railheads, were unloaded and returned to the bases for fresh loads. Consequently, a 'serious shortage of trucks' existed at almost all of Wedgwood's ports, which 'resulted in congestion of quays and very considerable delay in discharging ships'.[168] Beyond the railheads, the belt of country only passable by low-capacity animal transport rapidly widened. The roads directorate, under the road board's Brigadier-General Henry Maybury, was responsible for the construction, repair and maintenance of a network comprising 4,412 miles of road – he had, Fay reflected after the armistice, 'left his mark all over the roads in Northern France'. However, like the majority of the men employed on supply and transportation duties in November 1918, Maybury

[163] TNA, WO 95/454, headquarters branches and services. Deputy Director Supplies and Transport, the Fourth Army, notes on conference, 4 Oct. 1918; WO 95/40, QMG war diary, Explanatory review, Oct. 1918, p. 3; Henniker, *History of the Great War*, p. 461.
[164] TNA, WO 95/40, QMG war diary, explanatory review, Oct. 1918, p. 3.
[165] Boff, *Winning and Losing*, pp. 86–7.
[166] Stevenson, *With Our Backs to the Wall*, pp. 168–9.
[167] Boff, *Winning and Losing*, p. 36; Harris, *Amiens to the Armistice*, p. 295.
[168] TNA, WO 95/40, QMG war diary, diary entries, 2 and 9 Nov. 1918.

was worn out.[169] Men, mostly of advanced years and lower standards of physical fitness, had been tested to their physical and psychological limits; their lorries had been 'knocked about' and overworked.[170]

The demands of mobile warfare, waged by a mass army backed by the entire *impedimenta* that successful combined-arms operations required, could no longer be effectively met by the manpower and resources available for its supply. On the afternoon of 9 November the Third Army instructed IV, V and XVII Corps to consolidate their positions and echelon back in depth to reduce the army's transport burden. VI Corps took over responsibility for the whole of the Third Army's front, and was ordered to act as an advanced guard and keep in touch with the German retirement.[171] To the Third Army's right, responsibility for the Fourth Army's pursuit was devolved upon Major-General Hugh Keppel Bethell's 66th Division on the same morning. The Fourth Army had deployed over half-a-million men and animals and 2,000 guns at the battle of Amiens. Bethell's force comprised only one infantry brigade, one cavalry brigade, seven armoured cars, a field gun battery, two sections of 4.5-inch howitzers, an anti-aircraft section, three field companies of engineers, a pioneer battalion and the division's machine-gun battalion. It represented all that the available transport infrastructure could sustain just three months later.[172] The heavy artillery, tanks, light railways, repair workshops and sundry units and services that had contributed to the successful operations conducted by Rawlinson's forces throughout the course of the hundred days were abandoned. The transport factor made it impossible to feed a larger, hungrier force.[173] Consequently, Bethell's force was ordered to maintain contact with the Germans the following day, but to do no more.[174] At 11 a.m. the next morning the war on the western front came to an end.

---

[169] Fay called upon Maybury on 27 Dec., and recorded in his diary that the latter was 'not at all well—has nerves and cannot sleep'. See Fay, *The War Office at War*, p. 208.

[170] TNA, WO 95/454 Deputy Director Supplies and Transport, the Fourth Army, diary entries, 31 Oct. and 10 Nov. 1918; Crookshank, 'Transportation with the B.E.F.', p. 206.

[171] Boff, *Winning and Losing*, p. 35.

[172] A. Montgomery, *The Story of the Fourth Army in the Battle of the Hundred Days, August 8th to November 11th, 1918* (London, 1920), pp. 260–1; J. E. Edmonds and R. Maxwell-Hyslop, *History of the Great War. Military Operations, France and Belgium, 1918* (5 vols., London, 1947), v. 528.

[173] J. C. Darling, *20th Hussars in the Great War* (Lyndhurst, 1923), p. 127.

[174] Edmonds and Maxwell-Hyslop, *History of the Great War*, v. 533.

Conclusion

The BEF's transport problems did not end when the guns fell silent. The reduced demand from the front for materials directly related to the conduct of military operations, such as ammunition, was counteracted by requests for the construction materials necessary to reconnect the pre-war rail links between France, Belgium and Germany. 'There was', recorded Clarke in his review of November 1918:

> A tremendous extent of damage to be repaired all at once, our line having already advanced an abnormal distance beyond its railheads, while every mile of railway progress towards the German frontier meant an addition to the number of trucks required in order to maintain the supply service at the same number of trains per day. In practice it could not be done.[175]

The limited train capacity of the 'hastily reconstructed lines in the forward areas, devoid of all ordinary facilities for working, remained; reconstruction across the [devastated area] only increased the length of line over which traffic was precarious and intensified the shortage of rolling stock'.[176] The BEF could not advance to the German border in strength even when the military force opposing it had ceased to offer resistance.

However, the transport infrastructure that by November 1918 was inadequately equipped to move and supply an industrial army had proven itself good enough to facilitate the allied success on the western front over the previous two years. Britain's transport experts played a key role in that success. The integrated transport directorate established by Sir Eric Geddes, populated with civilians, and embedded within the military hierarchy over the winter of 1916–17 facilitated the prosecution of a material-intensive war on a scale beyond the BEF's capability during the battle of the Somme. The directorate-general of transportation created the circumstances in which British gunners could add significantly to the destruction wrought upon the French and Belgian landscape by the artillery of the First World War, not least at Third Ypres. What David Lloyd George referred to as 'the campaign of the mud' in his indictment of the BEF's senior commanders was – at least in part – the responsibility of the prime minister's desire to introduce civilian specialists to the administration of the British army.

The events of 1917 and 1918 demonstrate the inaccuracy of Lloyd George's statements regarding the imposition of supposedly superior civilian methods upon a backward-looking, reactionary army. Geddes's system was well suited to the task of supplying a stationary force, but it sank into

---

[175] TNA, WO 95/40 QMG war diary, explanatory review – Nov. 1918, p. 2.
[176] Henniker, *History of the Great War*, p. 467.

the quagmire it had created at Passchendaele and proved ill-equipped to service the requirements of a mobile force. The BEF's ultimately successful transportation effort during the final year of the war came about through the amalgamation of civilian and military expertise; recognisably non-military technologies, methods, and personnel were applied to the identifiably military problem of sustaining a relentless offensive pressure against a retreating opponent. By the summer of 1918 the line between civilian and military existed more in the minds of the personalities at work in the BEF's administrative hierarchy than it did in the directorate-general of transportation's practical accomplishments. Sir Douglas Haig 'could launch simultaneous offensives or sequential ones on widely separated fronts – something that had been unthinkable before 1918'.[177] Those offensives were a crucial factor in the German decision to seek an armistice.

Yet neither Haig, nor Geddes, Travers Clarke, Sydney Crookshank, or any of the other individuals responsible for the supply of the BEF during the hundred days could entirely eliminate the realities of mobile, material-intensive, modern warfare. The German spring offensives provided a graphic demonstration that the armies of the First World War could not tear loose from their railheads, penetrate deep into enemy territory, and maintain the intensity and tempo of their operations.[178] The BEF's advance could not have taken place without adequate, sustained access to supplies of food, ammunition, and myriad goods of both direct and indirect relationship to the conduct of military operations. Consequently, in the words of a staff officer who was responsible for XIII Corps' supply arrangements throughout the hundred days, the final campaign of the war was 'the most cumbrous steam-roller affair it [was] possible to conceive'.[179] Britain's transport experts had helped to keep it in motion for long enough.

[177] Brown, *British Logistics*, p. 179.

[178] M. Van Creveld, 'World War I and the revolution in logistics', in *Great War, Total War: Combat and Mobilization on the Western Front, 1914–18*, ed. R. Chickering and S. Förster (Cambridge, 2005), pp. 57–72, at p. 67.

[179] W. N. Nicholson, *Behind the Lines: an Account of Administrative Staffwork in the British Army, 1914–18* (London, 1939), p. 215.

10. Conclusion

Throughout the First World War, the British railway industry's trade press acknowledged the magnitude of the conflict and detailed the railways' ongoing support to the nation's armed forces. In the war's opening months, as men streamed into the recruiting stations, the numbers and proportions of each company's workforce to have answered Kitchener's call were recorded in frequently updated league tables of patriotic service.[1] As the war expanded in scale and scope, the activities and increased prominence of railwaymen like Sir Eric Geddes and Sir Sam Fay were reported on with familial pride. After the fighting had ceased, the *Railway Gazette* marked the industry's involvement with a special issue, which exclaimed in an editorial that although 'transport has always been an important factor in war ... never in the history of the world has it played such a great part as in the war now terminated'.[2] The fundamental requirements of modern, industrial warfare had presented unprecedented challenges for the transportation services behind all of the belligerent armies during the war: the accumulation and sustenance of millions of men; the transport of thousands of machines; and the provision of innumerable combinations of goods and services among others. Without vast bureaucratic organizations and complex, integrated supply systems the First World War could not have taken on the course or the character that it did. Transportation was central to the conduct of the war, and cannot be divorced from the discussion of the military campaigns that took place between 1914 and 1918. However, the prediction made by the *North-Eastern Railway Magazine* in 1916 – that 'when the history of the present war is written it would be found that our railways and our railwaymen had taken a very large share in operations' – has not proven to be the case.[3] This book has rectified this deficiency.

This volume opened with three questions, the answers to which challenge Lloyd George's principal assertion about the wartime British army and shed new light on the military's application of civilian specialists between 1914 and 1918. It has demonstrated that Britain's senior political and military

---

[1] 'Railwaymen and the war', *Railway Gazette*, 6 Nov. 1914, pp. 493–97.

[2] 'The organisation of war transportation', *Railway Gazette: Special War Transportation Number*, 21 Sept. 1920, p. 1.

[3] 'The romance of the railways', *North-Eastern Railway Magazine*, vi (1916), p. 38.

'Conclusion', in C. Phillips, *Civilian Specialists at War: Britain's Transport Experts and the First World War* (London, 2020), pp. 367–73. License: CC-BY-NC-ND 4.0.

figures acknowledged the value of industrial knowledge and technologies long before the outbreak of war in August 1914. The emergence and growth of the railway industry stimulated the development of an enduring professional association between the army, the government and Britain's transport experts. This peacetime relationship manifested itself both in organizational and practical terms. The formation of the ERSC in 1865 underlined the army's respect for the expertise of, and methods applied by, those tasked with the construction and management of a global transport and distribution network. The establishment of the REC in November 1912 represented a desire among both civilian and military figures to work harmoniously in the event of war. The army also benefited substantially from its pre-war interactions with Britain's transport experts outside the committee's deliberations. A new generation of officers received academic and vocational introductions to the operation of railways thanks to the contributions of British transport experts to the LSE's administrative staff course and the Midland Railway's engineering programme.

The importance of close collaboration between Britain's transport experts, the army and the government was demonstrated graphically in the opening days and weeks of the war. The production of the WF scheme was a joint endeavour. The impressment of horses, the routing and rerouting of thousands of specially provided trains, the identification of infrastructural improvements that were required at the French Channel ports, the allocation of sufficient locomotives and rolling stock and myriad other tasks necessary for the mobilization and movement of the BEF to the continent could not have been accomplished by the military alone. A modern, industrial army equipped with weapons, vehicles, aeroplanes and other *impedimenta* could not have been despatched from Britain's shores swiftly and effectively unless its transportation was properly coordinated. Henry Wilson's principal contributions to the WF scheme was that he recognized how crucial the involvement of civilian specialists were to the formulation of a reliable, executable mobilization plan, and that he ensured the army's preparations for war were supported by the companies that put them into practice.

The bonds forged before 1914 were thoroughly exploited over the years that followed. As the BEF grew exponentially, and the war's insatiable demands for manpower and materials stretched into every corner of the empire (and beyond), Britain's transport experts were called upon to play a multitude of roles in support of the military effort. They, and the private enterprises that employed them, were essential to the empire's response to the demands of an industrial war. Within the first twelve months of hostilities the largest transport companies had placed their human and material resources at the War Office's disposal, and they continued to do so throughout the war.

Men with British, imperial and wider experience of railway construction and operation became core components of the supply services that were established to support the army's various expeditionary forces. A plethora of experts examined, reported upon and enhanced the BEF's transportation services on the western front. The industrial capacity possessed by Britain's largest railway companies was redirected from its peacetime applications towards the manufacture of items as diverse as drinking cups and six-pounder Hotchkiss guns. Furthermore, when the demand for munitions necessitated the establishment of a dedicated organization in 1915, Britain's transport experts were among the first technocrats to populate Lloyd George's nascent Ministry of Munitions. Civilian specialists intensified British military power, and facilitated the colossal expenditures of ammunition that characterized the war on the western front.

The multiplicity of contributions recorded in this book – both collectively and, as in the case of Sir Francis Dent's numerous responsibilities, individually – have stressed the extent to which civilian skills and expertise were applied to the prosecution of war-related duties during the First World War. Manufacturing, engineering and labour, engaged in dispersed but interlinked activities, were all vital to the continued capacity of an army engaged in a war of attrition. As Major Wilfred Lindsell acknowledged in his post-war textbook on military administration:

> Modern wars … are no longer won by decisive battles, but by sustained and adequate maintenance arrangements. The army requires its fighting troops, its supply, transport, medical and repair organizations, etc., but behind all this military paraphernalia it requires the entire resources of the Empire, and it requires that these shall be organized to meet the needs of the Empire in arms.[4]

As senior managers within some of Britain's largest companies, the pre-war railway industry provided figures such as Geddes, Fay and Sir Guy Granet with experience that made some of the challenges – if not the global scale – of wartime transport organization recognizable ones.

The recalibration of the Franco-British coalition, which took place concurrent with Geddes's installation as DGT and DGMR, also increased the opportunities for Britain's transport experts to play a larger role in the prosecution of the First World War. France and Britain entered the conflict without an adequate managerial framework to ensure that national priorities were subordinated to the shared aim of expelling German forces from occupied territory. During 1914 and 1915 both Dent and Gerald Holland

---

[4] See W. G. Lindsell, *A. and Q.: or Military Administration in War* (Aldershot, 1933), p. 129.

found their designs to improve the BEF's logistical capabilities constrained by the French authorities. As hosts, senior partners and the suppliers of the vast majority of the machinery, personnel and infrastructure required to operate the shared transport network behind the allied forces, the French army and state's desire to retain overall control of the apparatus upon which its national defence rested overrode all other considerations. Even after the BEF grew to number more than a million men, and the colossal battles of 1916 had eroded France's ability to honour its pre-war agreements, the latter proved reluctant to combine demands for further British support with a willingness to surrender executive control to Britain's transport experts.

Geddes's first real exposure to the Franco-British alliance underlined the inherent complexities of coalition warfare. The transportation discussions that opened proceedings at the Calais conference on 26 February 1917 have been overshadowed by the political machinations that followed, but the disagreements between Geddes and Albert Claveille emphasize the need for further studies on the mechanics of the allies' partnership. This book has documented the accumulation of British manpower, machines and materials behind the western front that followed Geddes's appointments as DGT and DGMR; it has raised doubts about the businessman's temperamental suitability for a role that demanded a conciliatory and diplomatic approach; and it has illustrated how British transport expertise was disseminated throughout the global war effort with varying degrees of success after 1916.

However, further research is required to understand how the other belligerents' military and railway authorities responded to the unprecedented challenges of industrial warfare. Elizabeth Greenhalgh, in a reference to 'conflict' between Claveille and GQG, has provided a tantalising glimpse into the civil–military relations that shaped the French war effort.[5] Yet the view from the other side of the conference table and the other side of the hill remains partial. More work is needed in this direction to produce a comprehensive account of transportation on the western front and beyond, one which covers topics such as: how the different pre-war transport arrangements developed by Britain and the continental powers affected the establishment of efficient supply organizations behind the front lines; how effectively the Franco-British partnership exploited the available transport infrastructure and supported its weaker allies; how efficiently the German railway authorities were able to utilize the rail capacity available in occupied territory to support their troops in a multi-front war; and how the continued presence of domestic concerns and post-war strategic

---

[5] E. Greenhalgh, *Victory through Coalition: Britain and France during the First World War* (Cambridge, 2005), p. 241.

considerations manifested themselves in the evolving coalitions that fought the war. As Sir Philip Nash's attempts to secure a British contribution to improve the transport connections between France and Italy in 1918 demonstrate, internal calculations retained a powerful influence over individual belligerents' actions throughout the war.

In highlighting the breadth and diversity of Britain's transport experts' contributions to the war effort, including Nash's with the IATC, this book has demonstrated that Geddes's transportation mission to GHQ in August 1916 was far from unique in scale or scope. The foregoing discussion has argued that Geddes's contribution must be considered as part of a wider narrative of civil–military relations, one which permits a more nuanced understanding of how Britain's senior political and military leaders conceptualized the war as it unfolded. The full implications of industrial warfare's material and organizational requirements revealed themselves only gradually. In 1914 and 1915 the manpower and materials required to overburden the extant rail and road systems on the western front had yet to be accumulated in France and Flanders. The military inclination and political justification for a large-scale examination of the BEF's transportation services lay dormant until 1916, when the unprecedented effort of the battle of the Somme illuminated the weaknesses of the infrastructure and organization upon which any substantial allied advance depended. Under such circumstances the use of civilian specialists was confined; the talents of men such as Dent and Holland were applied selectively in response to comparatively small-scale conundrums, such as the increased throughput of goods at the Bassin Loubet or the development of an efficient IWT fleet. Prior to the Somme, when the true extent of the commitment required to defeat the Germans remained unclear, it was both militarily undesirable and politically impossible for large quantities of precious raw materials to be redirected from the ever-growing demand for munitions and weapons of destruction into the production of locomotives and rolling stock. Until the absolute necessity for substantial infrastructure improvements was made abundantly clear during the second half of 1916, there was insufficient compulsion for transportation to be considered a priority issue in the allocation of raw materials.[6] Geddes's most important contribution to the British war effort was that he produced the organizational systems, manpower and equipment required to conduct warfare on a hitherto unimaginable scale. That a civilian, rather than a professional soldier, was given the opportunity

---

[6] As Granet noted in a memorandum on the supply of wagons in late 1916, there already existed a 'great shortage of steel owing to the large demands for big gun ammunition'. See PA, Lloyd George papers, LG/E/6/1/5(A), memorandum on the question of railway wagon supply, p. 3.

and the support to do so, has tinged many of the military's histories of the conflict with a sense of resentment.

Yet by moving the consideration of Britain's transport experts in the First World War beyond Geddes, this book has presented a more balanced interpretation of the relationship between civilians and soldiers than that which emerged from the post-war period. Lloyd George's claims about the triumph of civilian ingenuity and innovation over a hidebound, conservative, obstructive military must be revised in light of this study's findings. The reluctance of officers such as Frederick Clayton, Ronald Maxwell and Richard Montagu Stuart-Wortley to support the transportation mission has provided the bedrock for historical accounts of the so-called civilianization process that took place after August 1916. The foregrounding of insular, individualistic officers with self-preserving tendencies has created an imbalance in representations of the civil–military relationship at play within the British war effort. Haig proved able to work constructively with successive DGTs, while officers such as Henry Mance and Albert Collard were clearly highly respected members of the hybrid civil–military team assembled in the directorate-general of military railways.

Furthermore, this study has highlighted that the civilian specialists drawn into the military machine were not immune to engaging in boundary disputes, nor did they entirely embrace the army's existing hierarchies. Ralph Wedgwood's adversarial approach to those who pointed out inefficient practices at the docks under his control, while not directly increasing inefficiency, did nothing to alleviate the problems either. Furthermore, John Stewart's designs on the role of DGT in the war's final months demonstrate that not all civilians were driven by purely altruistic motivations during the conflict. The interactions between civilians and soldiers – and, indeed, between civilians and civilians or soldiers and soldiers – within the crucible of the First World War cannot be reduced to simplistic stereotypes. The relationships that developed during the conflict, whether friendly or unfriendly, depended upon such variables as personality, circumstances and timing.

This book has studied the interface between the British army and the empire for whose protection it was responsible. It has shown how the army came to reflect the society from which it came, argued that it was an industrial machine forged from an industrial population, and emphasized that it was sustained by many of the same techniques, methods and procedures that drove a world-leading economy. Between 1914 and 1918, and particularly after 1916, civilian specialists were redirected from the pursuit of profits towards the production of military power on a colossal scale. Britain's transport experts, from within and without the army's organizational structure, permitted the empire to pursue a far more material-intensive form of warfare than had hitherto been possible.

The impact of the First World War on British society was profound and abiding. It has left a long shadow over the nation in the century since the guns fell silent. Yet the influence of British society over the conduct of the war was equally significant. From the very outset, and indeed for many years prior to the outbreak of war, Britain's transport experts and the army conspired to ameliorate the logistical challenges to be addressed in the prosecution of a modern conflict. Between them they planned Britain's response to the war, enlarged the scope and scale of the empire's contribution to the fighting and sustained the full implications of modern, combined-arms warfare until victory had been secured.

Appendix I: Information requested by the secretary of state for war from the transportation mission led by Sir Eric Geddes, August 1916[1]

Requirement statistics: the following information to be obtained in quantities per week for each month up to 30 June 1917, in respect of the details set out below.

Tonnage and numbers to be conveyed, and number of railway, road and canal vehicles or craft of various kinds required:

From point of origin to home ports and vice versa.

From French ports, and vice versa.

From ports in other theatres of war, and vice versa, for:

• Officers and men	• Sick, wounded and leave men	• Horses and mules
• Motor vehicles	• Horse-drawn vehicles	• Spare parts for vehicles and guns
• Numbers of guns and weights	• Gun ammunition	• Machine-guns
• Rifles	• Small-arms ammunition	• Bicycles
• Trench warfare ammunition (including gas cylinders)	• Salvage	• Food supply
• Clothing, boots and other equipment	• Harness	• Petrol
• Mails, parcels and private consignments	• General stores	• Railway material
• Building material	• Other RE stores	• Medical supplies

[1] TNA, WO 32/5164, Facilities and arrangements for Sir E. Geddes in conducting his investigation on transport arrangements in connection with the British Expeditionary Force at home and overseas, 9 Aug. 1916.

- Munitions and raw materials for French government
- Fuel
- Voluntary Aid Detachments
- Red Cross
- YMCA
- Blue Cross
- Church Army
- Any other large traffics

Units of requirement of each item, e.g., per Corps, or per Division, per 1,000 men etc. where possible.

Provisions for strategic reasons and to meet requirements about today's railhead:

Construction, repair etc. of:

- Railways
- Docks
- Canals or roads

Necessary in the event of an advance, for the movement of troops, ammunition, stores etc., or to feed civil population.

Provision of:

- Railway material
- Girders
- Dock equipment:
- Locomotives
- Road material
- Gates
- Carriages and wagons
- Road transport vehicles
- Power
- Barges
- Material for repairs of canals
- Cranes
- Labour (repair, maintenance, operating and workshops)
- Fuel
- Rails
- Stores
- Dredgers

Special memoranda required on:

1   Existing organisation in this country.

2   Existing organisation in France.

3   French organisation and arrangements for working BEF traffic, including relationship with French government authorities and railway, dock or canal officials.

4   Relation of British military traffic to French traffic (military and/or civilian).

5   Relations with Belgian government qua Railways and ports in the future.

6   Present position of Belgian railways rolling stock.

7   Repairing facilities for locomotives and rolling stock in France and Belgium, including supply of labour and material.

8   Proposals in hand or contemplated for provision of additional lines in France or arrangements with French railways.

9   Relations with REC, with any existing memorandum on the subject.

10 Relations and procedure with Admiralty in France, on the sea, in England, and in other theatres of war.

11 Relations with Admiralty, Army Medical Service, etc., as to the evacuation of sick and wounded.

12 Reports made or any special instructions issued during the period of the war:

a. Labour at home or abroad.

b. Dock facilities at home or abroad.

c. Rail facilities at home or abroad.

d. Canal facilities at home or abroad.

e. Road transport at home or abroad.

f. Evacuation of sick and wounded.

13 Position as regards:

a. Railways.

b. Sea Transport.

c. Docks.

d. Canals.

e. Roads in France.

With maps and plans where available. Memorandum to give details as to all difficulties which are being experienced: all probable tight places being specially marked on the maps and plans. Details of steps in progress or in contemplation to counteract the difficulties.

14 General flow of traffic at home and abroad, through various ports and by the different routes. Descriptions of traffic generally forwarded by rail, canal and road.

15 Storage depots in France and in this country so far as transport questions are affected.

16 Requirements of special capacity wagons and numbers available.

17 Armoured trains.

18 All special regulations as to despatch and storage or loading on railways of mixed cargoes, ammunition, guns, men. Any restrictions against bulk cargoes of any kind.

19 Memorandum with specimen forms of all traffic returns submitted to WO or IGC.

20   Statement of all railway, dock or canal works, rolling stock, craft accommodation and equipment generally provided by the British government in France.

21   Extent to which railway telegraphs and telephone circuits are used for the business of other departments.

Appendix II: Instructions issued to General Nash, 10 January 1918[1]

1. To investigate and report on the existing transportation facilities by railway of the Allies on the Western front as a whole, that is, between the North Sea and the Adriatic, involving enquiry as to: —

(a.) The capacity and use of Trans-Continental main railway systems.

(b.) The resources of the Allies on the Continent in locomotives, rolling stock, railway material generally, and railway personnel.

(c.) The extent to which existing facilities can deal with movements of troops from one point to another on the Western front, and the manner in which they can be improved.

2. To make recommendations as to the constitution of an Inter-allied Co-ordinating Authority to deal with questions of Military Transport by rail on the Western front, including the following functions: —

(a.) To advise on the transportation aspect of any strategical proposals which are under consideration.

(b.) To study in advance and prepare for the carrying out of movements which may be decided upon.

(c.) To formulate special shipping requirements involved in any contemplated policy.

3. To indicate the extent to which the existing British organization for Transportation in the Field may require to be modified or altered in the event of the constitution of an Inter-allied Transport Authority.

[1] TNA, CAB 24/43/19, Report on general transportation situation on the western front, 20 Feb. 1918, p. 4.

Bibliography

Unpublished primary sources

Coventry, University of Warwick Modern Records Centre
Papers of Sir William Guy Granet

Edinburgh, National Library of Scotland
Papers of Field Marshal Sir Douglas Haig

Henley-on-Thames, Greenlands Academic Resource Centre
Papers of Colonel Lyndall Fownes Urwick

Keele, Keele University Special Collections and Archives
Papers of A. D. Lindsay

Leeds, Brotherton Library Special Collections
Bamji collection, Hospital barges in France: correspondence from a nursing sister, with the British Expeditionary Force, during World War 1
Liddle collection, papers of Major L. W. Conibear
Liddle collection, papers of Sapper W. J. Hill
Liddle collection, papers of Brigadier-General W. R. Ludlow
Liddle collection, papers of Captain R. H. D. Tompson
Yorkshire Archaeological and Historical Society collection, Personal file of Captain Bryan Fairfax

London, Imperial War Museum
Papers of Brigadier A. E. Hodgkin
Papers of Brigadier R. Micklem
Papers of Sir Eric de Normann

Papers of Field Marshal Sir Henry Wilson

Papers of Brigadier-General C. R. Woodroffe

London, Institute of Civil Engineers

OC/4277, H. A. Ryott, The provision of personnel for military railways in the war of 1914–18

London, Liddell Hart Centre for Military Archives

Papers of Field Marshal Sir George Milne

Papers of Field Marshal Sir William Robertson

London, The National Archives of the United Kingdom

ADM 116, Admiralty: record office: cases

CAB 15, Committee of Imperial Defence, committee on the co-ordination of departmental action on the outbreak of war, and war cabinet, war priorities committee: minutes, papers and war books

CAB 2/2, nos. 83–119

CAB 22/1, minutes of meetings

CAB 24, war cabinet and cabinet: memoranda

CAB 25, Supreme War Council: British secretariat: papers and minutes

CAB 28, war cabinet and cabinet office: Allied war conferences and councils minutes and papers

CAB 45/205, Lieutenant-Colonel G. E. Holland

MT 23, Admiralty, transport department: correspondence and papers

MUN 9/35, 'Sir Eric Geddes'

PRO 30/66/9, correspondence and papers relating to the shipment of troops to Ireland and France, and the establishment and organization of the director-general of military railways

RAIL 491/815, train control office at Derby

RAIL 1053/258, railway dispute: conference between David Lloyd George, president of the Board of Trade, and representatives of the railway companies

WO 32, War Office and successors: registered files

WO 33, War Office: reports, memoranda and papers

WO 95, War Office: First World War and army of occupation war diaries

WO 106, War Office: directorate of military operations and military intelligence, and predecessors: correspondence and papers

WO 107, office of the commander-in-chief and War Office: Quartermaster-general's department: correspondence and papers

WO 114/114, territorial force: establishment and strengths, 1908–14

WO 158, War Office: military headquarters: correspondence and papers, First World War

ZLIB 6/88, Midland Railway train control issued by Midland Rly

ZLIB 10/11, Great Western Railway: war reports of the general manager to the board of directors 1914–19

ZLIB 29, British transport historical records office library: York library collection

ZPER 7/103, records of railway interests in the war

ZPER 39/41, *The Railway Magazine*

Parliamentary Archives

Papers of David Lloyd George

Oxford, Nuffield College Library

Papers of John Edward Bernard Seely, Lord Mottistone

Journals, newspapers and periodicals

Commercial Motor

Great Central Railway Journal

Great Western Railway Magazine

London and North-Western Railway Gazette

North-Eastern Railway Magazine

Railway Gazette

Railway Magazine

The Times

Vancouver Sun

Published sources

Printed primary sources

Acworth, W. M., 'English railway statistics', *Journal of the Royal Statistical Society*, lxv (1902), 613–64.

— 'Railway economics', *Econ. Jour.*, ii (1892), 392–8.

Addison, C., *British Workshops and the War* (London, 1917).

Aitken, M., *Politicians and the War, 1914–1916* (2 vols., London, 1928).

Aspinall, J. A. F., *Train Control Arrangements: a Survey of the Comprehensive Control System Operating on the Lancashire and Yorkshire Railway* (Manchester, 1915).

Asquith, H. H., *Memories and Reflections* (2 vols., London, 1928).

Baker Brown, W., *History of the Corps of Royal Engineers* (11 vols., Chatham, 1952), iv.

Bate, T. R. F., 'Horse mobilisation', *Royal United Services Institution. Journal*, lxvii (1922), 16–25.

Beadon, R. H., *The Royal Army Service Corps: a History of Transport and Supply in the British Army* (2 vols., Cambridge, 1931), ii.

Beharrell, J. G., 'The value of full and accurate statistics: as shown under emergency conditions in the transportation service in France', *Railway Gazette: Special War Transportation Number*, 21 Sept. 1920, pp. 37–9.

Boag, G. L., *Manual of Railway Statistics* (London, 1912).

Bonham-Smith, R., 'Railway transport arrangements in France', *Royal United Services Institution Journal*, lxi (1916), 47–62.

Buckland, R. U. H., 'Experiences at Fourth Army headquarters: organization and work of the R.E.', *Royal Engineers Journal*, xli (1927), 385–413.

Burtt, P., *Control on the Railways: a Study in Methods* (London, 1926).

Callwell, C. E., *Field Marshal Sir Henry Wilson Bart, G.C.B., D.S.O.: His Life and Diaries* (2 vols., London, 1927), i.

Chapman-Huston, D. and O. Rutter, *General Sir John Cowans, G.C.B., G.C.M.G.: the Quartermaster-General of the Great War* (2 vols., London, 1924).

Charteris, J., *At G.H.Q.* (London, 1931).

Crookshank, S. D'A., 'Transportation with the B.E.F.', *Royal Engineers Journal*, xxxii (1920), 193–208.

Darling, J. C., *20th Hussars in the Great War* (Lyndhurst, 1923).

De Tarlé, A., 'The British army and a continental war', trans. H. Wylly, *Royal United Services Institution Journal*, lvii (1913), 384–401.

'Detailed history of the railways in the South African War, 1899–1902', *Royal Engineers Journal*, i (1905), 133–35.

'Directorate of inland waterways and docks', *Royal Engineers Journal*, xxix (1919), 338–64.

'The directorate of inland waterways and docks', *Railway Gazette: Special War Transportation Number*, 21 Sept. 1920, pp. 141–51.

Edmonds, J. E., *History of the Great War. Military Operations, France and Belgium, 1914* (2 vols., London, 1928), i.

— *History of the Great War. Military Operations, France and Belgium, 1916* (2 vols., London, 1932), i.

— *History of the Great War. Military Operations, France and Belgium, 1917* (3 vols., London, 1948), ii.

Edmonds, J. E. and R. Maxwell-Hyslop, *History of the Great War. Military Operations, France and Belgium, 1918* (5 vols., London, 1947), v.

Edmonds, J. E. and G. C. Wynne, *History of the Great War. Military Operations, France and Belgium, 1915* (2 vols., London, 1927), i.

Falls, C., *History of the Great War. Military Operations, Macedonia* (2 vols., London, 1934), i.

— *History of the Great War. Military Operations, France and Belgium, 1917* (3 vols., London, 1940), i.

Fay, S., *The War Office at War* (London, 1937).

Fewtrell, A. C., 'The organisation of the transportation services of the British armies on the western front', *Minutes of Proceedings of the Engineering Association of New South Wales*, xxxiv (1919), 153–72.

Findlay, G., *The Working and Management of an English Railway* (London, 1889).

French, J. D. P., *1914* (London, 1919).

Geddes, A. C., *The Forging of a Family. A Family Story Studied in its Genetical, Cultural and Spiritual Aspects and a Testament of Personal Belief Founded Thereon* (London, 1952).

Girouard, É. P. C., *History of the Railways during the War in South Africa, 1899–1902* (London, 1903).

— 'Railways in war', *Royal Engineers Journal*, ii (1905), 16–27.

Gough, H., *The Fifth Army* (London, 1931).

Grey, E., *Twenty-Five Years, 1892–1916* (2 vols., London, 1925).

'G.S.O.', *G.H.Q. (Montreuil-Sur-Mer)* (London, 1920).

Gwynn, C. W., 'The administrative course at the London School of Economics', *Royal Engineers Journal*, vi (1907), 229–35.

Haig, D., *Sir Douglas Haig's Despatches (December 1915–April 1919)*, ed. J. H. Boraston (London, 1919).

Haldane, R. B., *Richard Burdon Haldane: an Autobiography* (London, 1929).

Hall, L. J., *The Inland Water Transport in Mesopotamia* (London, 1921).

Harbord, J. G., *The American Army in France, 1917–1919* (Boston, Mass., 1936).

Harding-Newman, J. C., *Modern Military Administration, Organization and Transportation* (Aldershot, 1933).

Henniker, A. M., *History of the Great War. Transportation on the Western Front, 1914–1918* (London, 1937).

History of the Ministry of Munitions: General Organization for Munitions Supply (12 vols., London, 1922), ii.

History of the Ministry of Munitions: Industrial Mobilisation, 1914–1915 (12 vols., London, 1922), i.

Hurd, A., *History of the Great War. The Merchant Navy* (3 vols., London, 1924), ii.

Jacqmin, F. P., *Les chemins de fer pendant la guerre de 1870–1871: leçons faites en 1872 à l'École des Ponts-et-Chaussées* (2nd edn., Paris, 1874).

Kemp, A. E., *Report of the Ministry, Overseas Military Forces of Canada, 1918* (London, 1919).

Knox, A. W. F., *With the Russian Army, 1914–1917* (2 vols., London, 1921), ii.

Le Hénaff, J. H .F. and H. Bornecque, *Les chemins de fer français et la guerre* (Paris, 1922).

Les armées françaises dans la grande guerre: la direction de l'arrière (Paris, 1937).

Lindsay, A. D., 'The organisation of labour in the army in France during the war and its lessons', *Econ. Jour.*, xxxiv (1924), 69–82.

Lindsell, W. G., 'Administrative lessons of the Great War', *Royal United Services Institution Journal*, lxxi (1926), 712–20.

— *A. and Q.: Or Military Administration in War* (Aldershot, 1933).

Lloyd George, D., *War Memoirs of David Lloyd George* (6 vols., London, 1933; 2 vols., London, 1938).

— *Through Terror to Triumph: Speeches and Pronouncements of the Right Hon. David Lloyd George, M.P., since the Beginning of the War*, ed. F. L. Stevenson (London, 1915).

Lyell, D., 'The work done by railway troops in France during 1914–19', *Minutes of the Proceedings of the Institution of Civil Engineers*, ccx (1920), 94–147.

Marcosson, I. F., *The Business of War* (New York, 1918).

'The Mediterranean line of communication', *Railway Gazette: Special War Transportation Number*, 21 Sept. 1920, pp. 101–7.

Military Monograph Subsection M.I.2, Military Intelligence Division, General Staff, *Russia, Route Zone A: Murman Railway and Kola Peninsula: Information and Route Notes, Murmansk to Petrograd* (Washington, DC, 1918).

Montgomery, A., *The Story of the Fourth Army in the Battle of the Hundred Days, August 8th to November 11th, 1918* (London, 1920).

Murray, V., 'Transportation in war', *Royal Engineers Journal*, lvi (1942), 202–32.

Neele, G. P., *Railway Reminiscences* (London, 1904).

Nicholson, W. N., *Behind the Lines: an Account of Administrative Staffwork in the British Army, 1914–18* (London, 1939).

'Organisation and work of the transportation directorate', *Railway Gazette: Special War Transportation Number*, 21 Sept. 1920, pp. 14–20.

'The organisation of war transportation', *Railway Gazette: Special War Transportation Number*, 21 Sept. 1920, p. 1.

Paish, G., *The British Railway Position* (London, 1902).

'The Palestine campaign', *Railway Gazette: Special War Transportation Number*, 21 Sept. 1920, pp. 119–28.

Peschaud, M., *Politique et fonctionnement des transports par chemin de fer pendant la guerre* (Paris, 1926).

Pritchard, H. L. (ed.) *History of the Corps of Royal Engineers* (11 vols., Chatham, 1952).

'Railways and roads on the western front', *Railway Gazette: Special War Transportation Number*, 21 Sept. 1920, pp. 21–9.

'Railways and the Salonica campaign', *Railway Gazette: Special War Transportation Number*, 21 Sept. 1920, pp. 110–18.

Reichsarchiv, *Der Weltkrieg, 1914 Bis 1918: Das Deutsche Feldeisenbahnwesen* (Berlin, 1928).

Richards, F., *Old Soldiers Never Die* (London, 1933).

Robertson, J. K., 'Richborough military transportation depot', *Minutes of the Proceedings of the Institution of Civil Engineers*, ccx (1920), 156–207.

Robertson, W. R., *From Private to Field-Marshal* (London, 1921).

— *Soldiers and Statesmen, 1914–1918* (2 vols., London, 1926).

Royal Commission on London traffic (8 vols., London, 1905–6).

Seely, J. E. B., *Adventure* (London, 1930).

Shakespear, J., *A Record of the 17th and 32nd Service Battalions Northumberland Fusiliers, N.E.R. Pioneers, 1914–1919*, ed. H. Shenton Cole (Newcastle upon Tyne, 1926).

Simpson, L. S., 'Railway operating in France', *Journal of the Institution of Locomotive Engineers*, xii (1922), 697–728.

Smith, F. E., *Contemporary Personalities* (London, 1924).

Stanford, F. O., 'The War Department cross-Channel train ferry', *Minutes of the Proceedings of the Institution of Civil Engineers*, ccx (1920), 208–38.

Stewart, H. A., *From Mons to Loos: Being the Diary of a Supply Officer* (Edinburgh, 1916).

Stewart, T., 'With the Labour Corps in France', *Royal United Services Institution Journal*, lxxiv (1929), 567–71.

Taylor, F. W., *The Principles of Scientific Management* (New York, 1911).

Taylor, M. G., 'Land transportation in the late war', *Royal United Services Institution Journal*, lxvi (1921), 699–722.

Watson, C. M., *History of the Corps of Royal Engineers* (11 vols., Chatham, 1914), iii.

Whitty, A., *A Quartermaster at the Front: the Diary of Lieutenant-Colonel Allen Whitty, Worcestershire Regiment, 1914–1919*, ed. E. Astill (Eastbourne, 2011).

Wilgus, W. J., *Transporting the A.E.F. in Western Europe, 1917–1919* (New York, 1931).

— 'Review of transportation on the western front, 1914–1918. Compiled by Colonel A. M. Henniker', *American Historical Review*, xliv (1939), 386–8.

The Work of the Royal Engineers in the European War, 1914–19. The Organization and Expansion of the Corps, 1914–18 (Uckfield, 2006).

Yelverton, B. J. D. and J. C. Carlile, 'The cross-Channel service', in

Folkestone during the War: a Record of the Town's Life and Work, ed. J. C. Carlile (Folkestone, 1920), pp. 186–98.

Young, H. A., 'Practical economy in the army', *Royal United Services Institution Journal*, l (1906), 1281–5.

Parliamentary records

An Act for regulating railways (Parl. Papers 1840 [97], xcvii).

Army. Report of the advisory board, London School of Economics, on the eighth course at the London School of Economics, 6th October, 1913, to 25th March, 1914, for the training of officers for the higher appointments on the administrative staff of the army and for the charge of departmental services (Parl. Papers 1914 [Cd. 7442], lii).

Army. Report of the advisory board, London School of Economics, on the fifth course at the London School of Economics, October, 1910, to March, 1911, for the training of officers for the higher appointments on the administrative staff of the army and for the charge of departmental services (Parl. Papers 1911 [Cd. 5597], xlvii).

Army. Report of the advisory board, LSE, on the first course at the LSE, January to July, 1907, for the training of officers for the higher appointments on the administrative staff of the army and for the charge of developmental services (Parl. Papers 1907 [Cd. 3696], xlix).

Army. Report of the advisory board, London School of Economics, on the fourth course at the London School of Economics, October, 1909, to March, 1910, for the training of officers for the higher appointments on the administrative staff of the army and for the charge of departmental services (Parl. Papers 1910 [Cd. 5213], ix).

Army. Report of the advisory board, London School of Economics, on the second course at the London School of Economics, October, 1907, to March, 1908, for the training of officers for the higher appointments on the administrative staff of the army and for the charge of departmental services (Parl. Papers 1908 [Cd. 4015], xi).

Army. Report of the advisory board, London School of Economics, on the seventh course at the London School of Economics, 3rd October, 1912, to 19th March, 1913, for the training of officers for the higher appointments on the administrative staff of the army and for the charge of departmental services (Parl. Papers 1913 [Cd. 6693], xlii).

Army. Report of the advisory board, London School of Economics, on the sixth course at the London School of Economics, 5th October, 1911, to 27th March, 1912, for the training of officers for the higher appointments on the administrative staff of the army and for the charge of departmental services (Parl. Papers 1912 [Cd. 6285], li).

Earnings and hours enquiry. Report of an enquiry by the board of trade into the earnings and hours of labour of workpeople of the United Kingdom. VII. Railway service in 1907 (Parl. Papers 1912 [Cd. 6053], cviii).

Report of the advisory board, London School of Economics, on the third course at the London School of Economics, October, 1908, to March, 1909, for the training of officers for the higher appointments on the administrative staff of the army and for the charge of departmental services (Parl. Papers 1909 [Cd. 4610], x).

Report of the commissioners appointed to consider the defences of the United Kingdom (Parl. Papers 1860 [2682], xxiii).

Report of the commissioners appointed to inquire into the system of purchase and sale of commissions in the army (Parl. Papers 1857 [2267], xviii).

Report of the committee appointed to enquire into war office organization (Parl. Papers 1901 [Cd. 580], xl).

War office (reconstitution) committee. Report of the war office (reconstitution) committee. (Part II) (Parl. Papers 1904 [Cd. 1968], viii).

Secondary sources

Adams, R. J. Q., *Arms and the Wizard: Lloyd George and the Ministry of Munitions, 1915–1916* (London, 1978).

Addyman, J. F., 'G. S. Kaye Butterworth, M. C.', *The North Eastern Express*, xxxvii (1998), 64.

Armstrong, J., 'The role of coastal shipping in UK transport: an estimate of comparative traffic movements in 1910', *Journal of Transport History*, viii (1987), 164–78.

Badsey, S., 'Sir John French and command of the BEF', in *Stemming the Tide: Officers and Leadership in the British Expeditionary Force 1914*, ed. S. Jones (Solihull, 2013), pp. 27–50.

Bagwell, P. S. and P. J. Lyth, *Transport in Britain: from Canal Lock to Gridlock* (London, 2006).

Bailey, J. B. A., *The First World War and the Birth of the Modern Style of Warfare* (Camberley, 1996).

Barnes, E. G., *The Midland Main Line, 1875–1922* (London, 1969).

Barnett, C., *The Audit of War: the Illusion and Reality of Britain as a Great Nation* (London, 1986).

Beckett, I., 'Frocks and brasshats', in *The First World War and British Military History*, ed. B. Bond (Oxford, 1991), pp. 89–112.

— 'Going to war: Southampton and military embarkation', in *Southampton: Gateway to the British Empire*, ed. M. Taylor (London, 2007), pp. 133–46.

Beckett, I., T. Bowman and M. Connelly, *The British Army and the First World War* (Cambridge, 2016).

Bell, R., *Twenty-Five Years of the North Eastern Railway, 1898–1922* (London, 1951).

Best, H., *'The Mystery Port', Richborough* (Blackpool, 1929).

Bidwell, S. and D. Graham, *Coalitions, Politicians and Generals: Some Aspects of Command in Two World Wars* (London, 1993).

— *Fire-Power: British Army Weapons and Theories of War, 1904–1945* (Barnsley, 2004).

Boff, J., 'Combined arms during the hundred days campaign, August–November 1918', *War in History*, xvii (2010), 459–78.

— 'Logistics during the hundred days campaign, 1918: British Third Army', *Journal of the Society for Army Historical Research*, lxxxix (2011), 306–21.

— *Winning and Losing on the Western Front: the British Third Army and the Defeat of Germany in 1918* (Cambridge, 2012).

— *Haig's Enemy: Crown Prince Rupprecht and Germany's War on the Western Front* (Oxford, 2018).

Bond, B., *The Victorian Army and the Staff College, 1854–1914* (London, 1972).

— *British Military Policy between the Two World Wars* (Oxford, 1980).

Bond, B. and N. Cave (ed.), *Haig: a Reappraisal 80 Years On* (Barnsley, 2009).

Bourne, J. M., *Britain and the Great War, 1914–1918* (London, 1989).

Broadberry, S. and M. Harrison, 'The economics of World War I: an overview', in *The Economics of World War I*, ed. S. Broadberry and M. Harrison (Cambridge, 2005), pp. 3–40.

Brown, G., *Sabotage: a Study in Industrial Conflict* (Nottingham, 1977).

Brown, I. M., *British Logistics on the Western Front, 1914–1919* (London, 1998).

—— 'Feeding victory: the logistic imperative behind the hundred days', in *1918: Defining Victory*, ed. P. Dennis and J. Grey (Canberra, 1999), pp. 130–47.

—— 'Growing pains: supplying the British Expeditionary Force, 1914–1915', in *Battles Near and Far: a Century of Operational Deployment*, ed. P. Dennis and J. Grey (Canberra, 2004), pp. 33–47.

Bruton, E. and G. Gooday, 'Listening in combat – surveillance technologies beyond the visual in the First World War', *History and Technology*, xxxii (2016), 213–26 .

Buchanan, R. A., 'The diaspora of British engineering', *Technology and Culture*, xxvii (1986), 501–24.

Bucholz, A., *Moltke, Schlieffen, and Prussian War Planning* (New York, 1991).

Bud-Frierman, L., 'Information acumen', in *Information Acumen: the Understanding and Use of Knowledge in Modern Business*, ed. L. Bud-Frierman (London, 1994), pp. 7–25.

Campbell-Kelly, M., 'The Railway Clearing House and Victorian data processing', in *Information Acumen: the Understanding and Use of Knowledge in Modern Business*, ed. L. Bud-Frierman (London, 1994), pp. 51–74.

Carter, E. F., *Railways in Wartime* (London, 1964).

Carver, M., *The National Army Museum Book of the Turkish Front, 1914–1918: the Campaigns at Gallipoli, in Mesopotamia and in Palestine* (London, 2004).

Cassar, G. H., *Lloyd George at War, 1916–1918* (London, 2011).

Cassis, Y., 'Big business in Britain and France, 1890–1990', in *Management and Business in Britain and France: the Age of the Corporate Economy*, ed. Y. Cassis, F. Crouzet and T. R. Gourvish (Oxford, 1995), pp. 214–26.

Chaloner, W. H., *The Social and Economic Development of Crewe, 1780–1923* (Manchester, 1950).

Chandler, A. D., 'The railroads: pioneers in modern corporate management', *Business History Review*, xxxix (1965), 16–40.

—— *The Visible Hand: the Managerial Revolution in American Business* (Cambridge, Mass., 1977).

Churchill, W. S., *The World Crisis* (6 vols., London, 1923).

Clark, C. M., *The Sleepwalkers: How Europe Went to War in 1914* (London, 2012).

Cline, P. K., 'Eric Geddes and the "experiment" with businessmen in government, 1915–22', in *Essays in Anti-Labour History*, ed. K. D. Brown (London, 1974), pp. 74–104.

Cobb, S., *Preparing for Blockade 1885–1914: Naval Contingency for Economic Warfare* (Farnham, 2013).

Coleman, D. C. and C. Macleod, 'Attitudes to new techniques: British businessmen, 1800–1950', *Economic History Review*, xxxix (1986), 588–611.

Coleman, T., *The Railway Navvies* (London, 1981).

Crafts, N., T. Leunig and A. Mulatu, 'Were British railway companies well managed in the early twentieth century?', *Economic History Review*, lxi (2008), 842–66.

Crawford, E., 'Internationalism in science as a casualty of the First World War: relations between German and allied scientists as reflected in nominations for the Nobel prizes in physics and chemistry', *Social Science Information*, xxvii (1988), 163–201.

Crow, D., *A Man of Push and Go: the Life of George Macaulay Booth* (London, 1965).

Damus, S., *Who was Who in Argentine Railways, 1860–1960* (Ottawa, ON, 2008).

Darroch, G. R. S., *Deeds of a Great Railway: a Record of the Enterprise and Achievements of the London and North-Western Railway Company during the Great War* (London, 1920).

Davenport-Hines, R. P. T., 'Girouard, Sir Édouard Percy Cranwill', in *Dictionary of Business Biography: a Biographical Dictionary of Business Leaders Active in Britain in the Period 1860–1980*, ed. D. J. Jeremy (5 vols., London, 1984), ii. 570–4.

Davies, W. J. K., *Light Railways of the First World War: a History of Tactical Rail Communications on the British Fronts, 1914–18* (Newton Abbot, 1967).

De Groot, G. J., *Douglas Haig, 1861–1928* (London, 1988).

Dehne, P. A., *On the Far Western Front: Britain's First World War in South America* (Manchester, 2009).

Doyle, P., *Disputed Earth: Geology and Trench Warfare on the Western Front 1914–18* (London, 2017).

Drummond, D. K., *Crewe: Railway Town, Company, and People, 1840–1914* (Aldershot, 1995).

Duff Cooper, A., *Haig* (2 vols., London, 1935).

Dutton, D. J., 'The Calais conference of December 1915', *Hist. Jour.* , xxi (1978), 143–56.

Edgerton, D., 'The prophet militant and industrial: the peculiarities of Correlli Barnett', *Twentieth Century British History*, ii (1991), 360–79.

— *Warfare State: Britain, 1920–1970* (Cambridge, 2006).

— *The Rise and Fall of the British Nation: a Twentieth-Century History* (London, 2018).

Edwards, R., *Instruments of Control, Measures of Output: Contending Approaches to the Practice of 'Scientific' Management on Britain's Railways in the Early Twentieth Century* (Southampton, 2000).

Emmerson, C., *1913: the World Before the Great War* (London, 2013).

Ernest Fayle, C., 'Carrying-power in war', *Royal United Services Institution Journal*, lxix (1924), 527–41.

Everett, A., *Visionary Pragmatist: Sir Vincent Raven: North Eastern Railway Locomotive Engineer* (Stroud, 2006).

Fell, A. S. and J. Meyer, 'Introduction: untold legacies of the First World War in Britain', *War & Society*, xxxiv (2015), 85–9.

Fergusson, T. G., *British Military Intelligence, 1870–1914: the Development of a Modern Intelligence Organization* (Frederick, MD, 1984).

Fong, G., 'The movement of German divisions to the western front, winter, 1917–1918', *War in History*, vii (2000), 225–35.

Fox, A., *Learning to Fight: Military Innovation and Change in the British Army, 1914–1918* (Cambridge, 2017).

Fraser, P., *Lord Esher: a Political Biography* (London, 1973).

French, D., 'The military background to the "shell crisis" of May 1915', *Journal of Strategic Studies*, ii (1979), 192–205.

— *The Strategy of the Lloyd George Coalition, 1916–1918* (Oxford, 1995).

Funnell, W., 'National efficiency, military accounting and the business of war', *Critical Perspectives on Accounting*, xvii (2006), 719–51.

Gatrell, P., *Russia's First World War: a Social and Economic History* (Harlow, 2005).

Gittins, S., *The Great Western Railway in the First World War* (Stroud, 2010).

Gourvish, T. R. , *Mark Huish and the London & North Western Railway: a Study of Management* (Leicester, 1972).

— 'A British business elite: the chief executive managers of the railway industry, 1850–1922', *Business History Review*, xlvii (1973), 289–316.

—'The rise of the professions', in *Later Victorian Britain, 1867–1900*, ed. T. R. Gourvish and A. O'Day (Basingstoke, 1988), pp. 13–35.

Grant, P., 'Edward Ward, Halford Mackinder and the army administration course at the London School of Economics, 1907–1914', in *A Military Transformed? Adaptation and Innovation in the British Military, 1792–1945*, ed. M. LoCicero, R. Mahoney and S. Mitchell (Solihull, 2014), pp. 97–109.

Greenhalgh, E., *Victory through Coalition: Britain and France during the First World War* (Cambridge, 2005).

— *Foch in Command: the Forging of a First World War General* (Cambridge, 2011).

Gregory, A., 'Railway stations: gateways and termini', in *Capital Cities at War: Paris, London, Berlin 1914–1919*, ed. J. Winter and J. L. Robert (2 vols., Cambridge, 2007), ii. 23–56.

Grieves, K., 'Haig and the government, 1916–1918', in *Haig: a Reappraisal 80 Years On*, ed. B. Bond and N. Cave (Barnsley, 2009), pp. 107–27.

— 'Improvising the British war effort: Eric Geddes and Lloyd George, 1915–18', *War & Society*, vii (1989), 40–55.

— *Sir Eric Geddes: Business and Government in War and Peace* (Manchester, 1989).

— 'Sir Eric Geddes, Lloyd George and the transport problem, 1918–21', *Journal of Transport History*, xiii (1992), 23–42.

— 'The transportation mission to GHQ, 1916', in *'Look to Your Front!' Studies in the First World War by the British Commission for Military History*, ed. B. Bond et al. (Staplehurst, 1999), pp. 63–78.

Griffin, N. J., 'Scientific management in the direction of Britain's military labour establishment during World War I', *Military Affairs*, xlii (1978), 197–201.

Griffith, P., *Battle Tactics on the Western Front: the British Army's Art of Attack, 1916–18* (New Haven, Conn., 1994).

Grigg, J., *Lloyd George: From Peace to War, 1912–1916* (London, 1985).

— *Lloyd George: War Leader, 1916–1918* (London, 2003).

Guinn, P., *British Strategy and Politics, 1914 to 1918* (Oxford, 1965).

Hall, B. N., 'The "life-blood" of command? The British army, communications and the telephone, 1877–1914', *War & Society*, xxvii (2008), 43–65.

— 'Technological adaptation in a global conflict: the British army and communications beyond the western front, 1914–1918', *Journal of Military History*, lxxviii (2014), 37–71.

— *Communications and British Operations on the Western Front, 1914–1918* (Cambridge, 2017).

Hamilton, J. A. B., *Britain's Railways in World War I* (London, 1947).

Hamilton Ellis, C., *The Midland Railway* (London, 1953).

— *British Railway History: an Outline from the Accession of William IV to the Nationalisation of Railways, 1877–1947* (2 vols., London, 1959), ii.

Harris, J. P., *Amiens to the Armistice: the BEF in the Hundred Days' Campaign, 8 August–11 November 1918* (London, 1998).

— *Douglas Haig and the First World War* (Cambridge, 2008).

Harris, P. and S. Marble, 'The 'step-by-step' approach: British military thought and operational method on the western front, 1915–1917', *War in History*, xv (2008), 17–42.

Hart, P., *Fire and Movement: the British Expeditionary Force and the Campaign of 1914* (Oxford, 2015).

Hartcup, G., *The War of Invention: Scientific Developments, 1914–1918* (London, 1988).

Herrera, G. L., 'Inventing the railroad and rifle revolution: information, military innovation and the rise of Germany', *Journal of Strategic Studies*, xxvii (2004), 243–71.

Heywood, A., 'Russia's foreign supply policy in World War I: imports of railway equipment', *Jour. European Econ. Hist.*, xxxii (2003), 77–108.

— *Engineer of Revolutionary Russia: Iurii V. Lomonosov (1876–1952) and the Railways* (Farnham, 2011).

— 'Spark of revolution? Railway disorganisation, freight traffic and Tsarist Russia's war effort, July 1914–March 1917', *Europe-Asia Studies*, lxv (2013), 753–72.

Higgins, J., *Great War Railwaymen: Britain's Railway Company Workers at War 1914–1918* (London, 2014).

Higgs, E., *The Information State in England: the Central Collection of Information on Citizens since 1500* (Basingstoke, 2004).

Hogg, O. F. G., *The Royal Arsenal: Its Background, Origin and Subsequent History* (2 vols., Oxford, 1963), ii.

Hooper, C., *Railways of the Great War* (London, 2014).

Hussey, J., 'The Flanders battleground and the weather in 1917', in *Passchendaele in Perspective: the Third Battle of Ypres*, ed. P. H. Liddle (London, 1997), pp. 140–58.

Irving, R. J., *The North Eastern Railway Company, 1870–1914: an Economic History* (Leicester, 1976).

— 'Gibb, Sir George Stegmann (1850–1925)', in *Dictionary of Business Biography: a Biographical Dictionary of Business Leaders Active in Britain in the Period 1860–1980*, ed. D. J. Jeremy (5 vols., London, 1984), ii. 543–5.

Irving, R. J. and R. P. T. Davenport-Hines, 'Geddes, Sir Eric Campbell (1875–1937)', in *Dictionary of Business Biography: a Biographical Dictionary of Business Leaders Active in Britain in the Period 1860–1980*, ed. D. J. Jeremy (5 vols., London, 1984), ii. 507–16.

James, L., *Imperial Warrior: the Life and Times of Field-Marshal Viscount Allenby 1861–1936* (London, 1993).

Jeffery, K., *Field Marshal Sir Henry Wilson: a Political Soldier* (Oxford, 2006).

— *1916: a Global History* (London, 2015).

Johnson, R., *The Great War and the Middle East: a Strategic Study* (Oxford, 2016).

Jones, H., 'The Great War: how 1914–18 changed the relationship between war and civilians', *RUSI Journal*, xlix (2014), 84–91.

Jones, S., '"To make war as we must, and not as we should like": the British army and the problem of the western front, 1915', in *Courage Without Glory: the British Army on the Western Front 1915*, ed. S. Jones (Solihull, 2015), pp. 31–55.

Kaye Kerr, M., 'Waghorn, Brigadier-General Sir William Danvers', in *Biographical Dictionary of Civil Engineers in Great Britain and Ireland: 1890–1920*, ed. M. M. Chrimes et al. (London, 2014), p. 626.

Kempshall, C., *British, French and American Relations on the Western Front, 1914–1918* (London, 2018).

Kennedy, P., *Engineers of Victory: the Problem Solvers Who Turned the Tide in the Second World War* (London, 2013).

Kirk-Greene, A. H. M., 'Canada in Africa: Sir Percy Girouard, neglected colonial governor', *African Affairs*, lxxxiii (1984), 207–39.

Knight, R., *Britain against Napoleon: the Organization of Victory, 1793–1815* (London, 2014).

Krajewski, M., *Paper Machines: About Cards and Catalogs, 1548–1929*, trans. P. Krapp (Cambridge, Mass., 2011).

Laffin, J., *British Butchers and Bunglers of World War One* (Gloucester, 1988).

Lawrence, J., 'The Transition to War in 1914', in *Capital Cities at War: Paris, London, Berlin, 1914–1919*, ed. J. Winter and J. Robert (2 vols., Cambridge, 1997), i. 135–63.

Liddell Hart, B. H., *Reputations, Ten Years After* (London, 1928).

— *The Real War: 1914–1918* (London, 1930).

Lindemann, M., 'Civilian contractors under military law', *Parameters: US Army War College Quarterly*, xxxvii (2007), 83–94.

Lloyd, N., *Hundred Days: the End of the Great War* (London, 2013).

— *Passchendaele: a New History* (London, 2017).

Lohr, E. and J. Sanborn, '1917: revolution as demobilization and state collapse', *Slavic Review*, lxxvi (2017), 703–9.

Lubenow, W. C., *The Politics of Government Growth: Early Victorian Attitudes toward State Intervention, 1833–1848* (Newton Abbot, 1971).

MacLeod, R. M., 'The "arsenal" in the Strand: Australian chemists and the British munitions effort 1916–1919', *Annals of Science*, xlvi (1989), 45–67.

— 'The chemists go to war: the mobilization of civilian chemists and the British war effort, 1914–1918', *Annals of Science*, l (1993), 455–81.

— 'The scientists go to war: revisiting precept and practice, 1914–1919', *Journal of War and Culture Studies*, ii (2009), 37–51.

Maggs, C., *A History of the Great Western Railway* (Stroud, 2015).

Marx, K., *Lawson Billinton: a Career Cut Short* (Usk, 2007).

McCrae, M., *Coalition Strategy and the End of the First World War: the Supreme War Council and War Planning, 1917–1918* (Cambridge, 2019).

McEwen, J. M., 'Northcliffe and Lloyd George at war, 1914–1918', *Hist. Jour.*, xxiv (1981), 651–72.

McKenna, F., *The Railway Workers, 1840–1970* (London, 1980).

Merridale, C., *Lenin on the Train* (London, 2017).

Messenger, C., *Call-to-Arms: the British Army, 1914–18* (London, 2005).

— *The Day We Won the War: Turning Point at Amiens, 8 August 1918* (London, 2008).

Miller, R. G., 'The logistics of the British Expeditionary Force: 4 August to 5 September 1914', *Military Affairs*, xliii (1979), 133–38.

Morgan, K. O., 'Lloyd George's premiership: a study in "prime ministerial government"', *Hist. Jour.*, xiii (1970), 130–57.

Mullay, A. J., *For the King's Service: Railway Ships at War* (Easingwold, 2008).

— 'Letter from the Somme: the Railway Executive Committee and the military in World War I', *BackTrack*, xxii (2008), 220–3.

Murray Hunter, T., 'Sir George Bury and the Russian Revolution', *The Canadian Historical Association: Report of the Annual Meeting*, cdi (1965), 58–70.

Neillands, R., *The Old Contemptibles: the British Expeditionary Force, 1914* (London, 2004).

Neilson, K., *Strategy and Supply: the Anglo-Russian Alliance, 1914–17* (London, 1984).

Neilson, K. and R. A. Prete (ed.), *Coalition Warfare: an Uneasy Accord* (Waterloo, ON, 1983).

Nelson, D., 'Scientific management, systematic management, and labor, 1880–1915', *Business History Review*, xlviii(1974), 479–500.

'Obituary: William Cawkwell, 1807–1897', *Minutes of the Proceedings of the Institution of Civil Engineers*, cxxix (1897), 398–400.

Palmer, A., *The Gardeners of Salonika* (London, 1965).

Parris, H., *Government and the Railways in Nineteenth-Century Britain* (London, 1965).

Pattenden, N., 'Armageddon? – No just practising', *South Western Circular*, xii (2001).

Phillips, C., 'Early experiments in civil–military cooperation: the South-Eastern and Chatham Railway and the port of Boulogne, 1914–15', *War & Society*, xxxiv (2015), 90–104.

— 'The changing nature of supply: transportation in the BEF during the battle of the Somme', in *At All Costs: the British Army on the Western Front 1916*, ed. S. Jones (Warwick, 2018), pp. 117–38.

Philpott, W., 'The strategic ideas of Sir John French', *Journal of Strategic Studies*, xii (1989), 458–78.

— *Anglo-French Relations and Strategy on the Western Front, 1914–18* (Basingstoke, 1996).

— 'The general staff and the paradoxes of continental war', in *The British General Staff: Reform and Innovation, c.1890–1939*, ed. D. French and B. Holden Reid (London, 2002), pp. 95–111.

— *Bloody Victory: the Sacrifice on the Somme* (London, 2009).

— 'Haig and Britain's European allies', in *Haig: a Reappraisal 80 Years On*, ed. B. Bond and N. Cave (Barnsley, 2009), pp. 128–44.

— *Attrition: Fighting the First World War* (London, 2014).

Pöhlmann, M., 'Return to the Somme 1918', in *Scorched Earth: the Germans on the Somme 1914–1918*, ed. I. Renz, G. Krumeich and G. Hirschfeld (Barnsley, 2009), pp. 179–201.

Pollock, J., *Kitchener* (London, 2002).

Pratt, E. A., *The Rise of Rail Power in War and Conquest, 1833–1914* (London, 1916).

— *British Railways and the Great War; Organization, Efforts, Difficulties and Achievements* (2 vols., London, 1921).

— *War Record of the London and North-Western Railway* (London, 1922).

Prior, R. and T. Wilson, *Command on the Western Front: the Military Career of Sir Henry Rawlinson, 1914–18* (Oxford, 1991).

— *Passchendaele: the Untold Story* (New Haven, Conn., 1996).

Proctor, T. M., *Civilians in a World at War, 1914–1918* (New York, 2010).

Reid, W., *Douglas Haig: Architect of Victory* (Edinburgh, 2006).

Richter, D. C., *Chemical Soldiers: British Gas Warfare in World War I* (London, 1994).

Rogan, E., *The Fall of the Ottomans: the Great War in the Middle East, 1914–1920* (London, 2015).

Rolt, L. T. C., *Red for Danger: a History of Railway Accidents and Railway Safety* (4th edn., Newton Abbot, 1982).

Roy, K., 'From defeat to victory: logistics of the campaign in Mesopotamia, 1914–1918', *First World War Studies*, i (2010), 35–55.

Salter, J. A., *Allied Shipping Control: an Experiment in International Administration* (Oxford, 1921).

Saul, S. B., 'The American impact on British industry 1895–1914', *Business History*, ii (1960), 19–38.

Searle, G. R., *The Quest for National Efficiency: a Study in British Politics and Political Thought, 1899–1914* (Oxford, 1971).

Sheffield, G., *Forgotten Victory. The First World War: Myths and Realities* (London, 2001).

— '"Not the same as friendship": the British empire and coalition warfare in the era of the First World War', in *Entangling Alliances: Coalition*

Warfare in the Twentieth Century, ed. P. Dennis and J. Grey (Canberra, 2005), pp. 38–52.

— *The Chief: Douglas Haig and the British Army* (London, 2011).

Showalter, D. E., 'Soldiers and steam: railways and the military in Prussia, 1832 to 1848', *Historian*, xxxiv (1972), 242–59.

Simkins, P., *Kitchener's Army: the Raising of Britain's New Armies, 1914–1916* (Manchester, 1988).

Simpson, A., *Directing Operations: British Corps Command on the Western Front, 1914–18* (Stroud, 2006).

Sinclair, G. B., *The Staff Corps: the History of the Engineer and Logistic Staff Corps RE* (Chatham, 2001).

Singleton, J., 'Britain's military use of horses 1914–1918', *Past & Present*, cxxxix (1993), 178–203.

Sloan, G., 'Haldane's Mackindergarten: a radical experiment in British military education?', *War in History*, xix (2012), 322–52.

Smith, R., *Jamaican Volunteers in the First World War: Race, Masculinity and the Development of National Consciousness* (Manchester, 2004).

Smithers, M., *The Royal Arsenal Railways: the Rise and Fall of a Military Railway Network* (Barnsley, 2016).

Spencer, J., '"The big brain in the army": Sir William Robertson as quartermaster-general', in *Stemming the Tide: Officers and Leadership in the British Expeditionary Force 1914*, ed. S. Jones (Solihull, 2013), pp. 89–107.

Spiers, E. M., *Haldane: an Army Reformer* (Edinburgh, 1980).

— *Engines for Empire: the Victorian Army and its Use of Railways* (Manchester, 2015).

Stancombe, M., 'The Staff Corps: a civilian resource for the military', *ICE Proceedings*, clvii (2004), 22–6.

Starling, J. and I. Lee, *No Labour, No Battle: Military Labour during the First World War* (Stroud, 2009).

Stevenson, D., *Armaments and the Coming of War: Europe, 1904–1914* (Oxford, 1996).

— 'War by timetable? The railway race before 1914', *Past & Present*, clxii (1999), 163–94.

— *1914–1918: the History of the First World War* (London, 2004).

— '1918 revisited', *Journal of Strategic Studies*, xxviii (2005), 107–39.

— *With Our Backs to the Wall: Victory and Defeat in 1918* (London, 2011).

— *1917: War, Peace, and Revolution* (Oxford, 2017).

Stewart Wallace, W., *The Macmillan Dictionary of Canadian Biography* (3rd edn., Toronto, ON, 1963).

Strachan, H., *The First World War: To Arms* (Oxford, 2001).

— 'The British army, its general staff and the continental commitment, 1904–1914', in *The British General Staff: Reform and Innovation, c.1890–1939*, ed. D. French and B. Holden Reid (London, 2002), pp. 75–94.

Suttie, A., *Rewriting the First World War: Lloyd George, Politics and Strategy, 1914–1918* (Basingstoke, 2005).

Taylor, A. J. P., *The First World War: an Illustrated History* (London, 1963).

— *War by Time-Table: How the First World War Began* (London, 1969).

Taylor, G. W., *The Railway Contractors: the Story of John W. Stewart, His Enterprises and Associates* (Victoria, BC, 1988).

Terraine, J., *Douglas Haig: the Educated Soldier* (London, 1963).

— *The Smoke and the Fire: Myths and Anti-Myths of War, 1861–1945* (London, 1980).

Thompson, J., *The Lifeblood of War: Logistics in Armed Conflict* (Oxford, 1991).

Thompson, J., 'Printed statistics and the public sphere: numeracy, electoral politics, and the visual culture of numbers, 1880–1914', in *Statistics and the Public Sphere: Numbers and the People in Modern Britain, c. 1800–2000*, ed. T. Crook and G. O'Hara (Abingdon, 2011), pp. 121–43.

Thompson, R., 'Mud, blood and wood: BEF operational combat and logistico-engineering during the battle of Third Ypres, 1917', in *Fields of Battle: Terrain in Military History*, ed. P. Doyle and M. R. Bennett (Dordrecht, 2002), pp. 237–55.

— '"Delivering the goods". Operation *Landovery Castle*: a logistical and administrative analysis of Canadian Corps preparations for the battle of Amiens, 8–11 August, 1918', in *Changing War: the British Army, the Hundred Days Campaign and the Birth of the Royal Air Force, 1918*, ed. G. Sheffield and P. Gray (London, 2013), pp. 37–54.

Todman, D. and G. Sheffield, 'Command and control in the British army on the western front', in *Command and Control on the Western Front: the British Army's Experience, 1914–18*, ed. G. Sheffield and D. Todman (Staplehurst, 2004), pp. 1–11.

Tombs, R. and E. Chabal (ed.), *Britain and France in Two World Wars: Truth, Myth and Memory* (London, 2013).

Tomlinson, W. W., *The North Eastern Railway. Its Rise and Development* (Newcastle upon Tyne, 1915).

Torrey, G. E., 'Romania in the First World War: the years of engagement, 1916–1918', *International History Review*, xiv (1992), 462–79.

Towle, P., 'The Russo-Japanese War and the defence of India', *Military Affairs*, xliv (1980), 111–17.

Townsend, C. E. C., *All Rank and No File: a History of the Engineer and Railway Staff Corps RE, 1865–1965* (London, 1969).

Travers, T., 'The hidden army: structural problems in the British officer corps, 1900–1918', *Jour. Contemp. Hist.*, xvii (1982), 523–44.

— 'A particular style of command: Haig and GHQ, 1916–18', *Journal of Strategic Studies*, x (1987), 363–76.

— 'The evolution of British strategy and tactics on the western front in 1918: GHQ, manpower, and technology', *Jour. Military Hist.*, liv (1990), 173–200.

— *The Killing Ground: the British Army, the Western Front and the Emergence of Modern Warfare, 1900–1918* (London, 1990).

— 'Could the tanks of 1918 have been war-winners for the British Expeditionary Force?', *Jour. Contemp. Hist.*, xxvii (1992), 389–406.

— *How the War Was Won: Command and Technology in the British Army on the Western Front, 1917–1918* (London, 1992).

Uglow, J., *In These Times: Living in Britain Through Napoleon's Wars, 1793–1815* (London, 2015).

Ugolini, L., *Civvies: Middle-Class Men on the English Home Front, 1914–18* (Manchester, 2017).

Ulrichsen, K. C., *The Logistics and Politics of the British Campaigns in the Middle East, 1914–22* (Basingstoke, 2011).

— *The First World War in the Middle East* (London, 2014).

Vallance, H. A., *The Highland Railway: the History of the Railways of the Scottish Highlands* (5 vols., Newton Abbot, 1969), ii.

Van Creveld, M., *Supplying War: Logistics from Wallenstein to Patton* (Cambridge, 1977).

— *Technology and War: from 2000 B.C. to the Present* (New York, 1991).

— 'World War I and the revolution in logistics', in *Great War, Total War: Combat and Mobilisation on the Western Front, 1914–18*, ed. R. Chickering and S. Förster (Cambridge, 2005), pp. 57–72.

Van Evera, S., 'The cult of the offensive and the origins of the First World War', *International Security*, ix (1984), 58–107.

Watson, A., *Ring of Steel: Germany and Austria-Hungary at War, 1914–1918* (London, 2015).

Westwood, J. N., *Railways at War* (London, 1980).

Wiener, M. J., *English Culture and the Decline of the Industrial Spirit, 1850–1980* (2nd edn., Cambridge, 2004).

Williams, C. A., *Police Control Systems in Britain, 1775–1975: from Parish Constable to National Computer* (Manchester, 2014).

Williamson, Jr., S. R., *The Politics of Grand Strategy: Britain and France Prepare for War, 1904–1914* (Cambridge, Mass., 1969).

Wilson, B., *Empire of the Deep: the Rise and Fall of the British Navy* (London, 2014).

Winter, D., *Haig's Command: a Reassessment* (London, 1991).

Winton, G. R., 'The British army, mechanization and a new transport system, 1900–14', *Journal of the Society for Army Historical Research*, lxxviii (2000), 197–212.

Wrigley, C., 'The Ministry of Munitions: an innovatory department', in *War and the State: the Transformation of British Government, 1914–1919*, ed. K. Burk (London, 1982), pp. 32–56.

Yates, J., 'Evolving information use in firms, 1850–1920: ideology and information techniques and technologies', in *Information Acumen: the Understanding and Use of Knowledge in Modern Business*, ed. L. Bud-Frierman (London, 1994), pp. 26–50.

Young, M., *Army Service Corps, 1902–1918* (London, 2000).

Zabecki, D. T., *The German 1918 Offensives: a Case Study in the Operational Level of War* (Abingdon, 2006).

Unpublished theses

Brown, I. M., 'The evolution of the British army's logistical and administrative infrastructure and its influence on GHQ's operational and strategic decision-making on the western front, 1914–1918' (University of London PhD thesis, 1996).

Pelizza, S., 'Geopolitics, education, and empire: the political life of Sir Halford Mackinder, 1895–1925' (University of Leeds PhD thesis, 2013).

Stoneman, M. R., 'Wilhelm Groener, officering, and the Schlieffen plan' (Georgetown University PhD thesis, 2006).

Strangleman, T., 'Railway and grade: the historical construction of contemporary identities' (Durham University PhD thesis, 1998).

Turner, D. A., 'Managing the "royal road": the London & South Western Railway 1870–1911' (University of York PhD thesis, 2013).

Websites

'1922 who's who in engineering: Name H', *Grace's Guide to British Industrial History* <https://www.gracesguide.co.uk/1922_Who's_Who_In_Engineering:_Name_H>.

'Alfred Howe Collinson', *Grace's Guide to British Industrial History* <https://www.gracesguide.co.uk/Alfred_Howe_Collinson>.

Baker, C., 'The RE Railway construction companies', *The Long, Long Trail: the British Army of 1914–1918 – For Family Historians* <http://www.1914-1918.net/re_rlwy_cos.htm>.

Bourne, J., 'William Danvers Waghorn', *Lions Led by Donkeys* <http://www.birmingham.ac.uk/research/activity/warstudies/research/projects/lionsdonkeys/t.aspx>.

Carpenter, G. W., 'Fowler, Sir Henry (1870–1938)', *ODNB* <https://doi.org/10.1093/ref:odnb/37427>.

Fair, C., 'The Folkestone harbour station canteen and the visitors' books', *Step Short: Remembering the Soldiers of the Great War* <http://www.kentfallen.com/PDF%20REPORTS/FOLKESTONE%20HARBOUR%20STATION.pdf>.

Flint, J., 'Girouard, Sir (Édouard) Percy Cranwill (1867–1932)', *ODNB* <https://doi.org/10.1093/ref:odnb/33415>.

'Inflation calculator', *Bank of England* <http://www.bankofengland.co.uk/education/Pages/resources/inflationtools/calculator/default.aspx>.

'Philip Arthur Manley Nash', *Grace's Guide to British Industrial History* <http://www.gracesguide.co.uk/Philip_Arthur_Manley_Nash>.

Philpott, W., 'Beyond the "learning curve": the British army's military transformation in the First World War', *RUSI* <https://rusi.org/commentary/beyond-learning-curve-british-armys-military-transformation-first-world-war>.

Scott, P. and J. T. Walker, 'Demonstrating distinction at "the lowest edge of the black-coated class": the family expenditures of Edwardian railway clerks', *Centre for International Business History* <http://www.henley.ac.uk/files/pdf/research/papers-publications/IBH-2014-04%20Scott%20and%20Walker.pdf>.

Shaw-Taylor, L. and X. You, 'The development of the railway network in Britain 1825–1911', *The Online Historical Atlas of Transport, Urbanization and Economic Development in England and Wales c.1680–1911* <https://www.campop.geog.cam.ac.uk/research/projects/transport/onlineatlas/railways.pdf>.

Turner, D., 'Unlocking the early railway manager – a project to follow', *Turnip Rail* <http://turniprail.blogspot.com/2011/04/unlocking-early-railway-manager-project.html>.

'Winter 1916–17', *IWM Voices of the First World War* <http://www.iwm.org.uk/history/podcasts/voices-of-the-first-world-war/podcast-25-winter-1916-17>.

Index

 New Historical
PERSPECTIVES

Recent and forthcoming titles

The Family Firm: Monarchy, Mass Media and the British Public, 1932–53
(2019)
Edward Owens

*Cinemas and Cinema-Going in the United Kingdom: Decades of Decline,
1945–65* (2020)
Sam Manning

*Civilian Specialists at War: Britain's Transport Experts and the First World
War* (2020)
Christopher Phillips

Individuals and Institutions in Medieval Scholasticism (2020)
edited by Antonia Fitzpatrick and John Sabapathy

*Unite, Proletarian Brothers! Radicalism and Revolution in the Spanish Second
Republic* (2020)
Matthew Kerry

Masculinity and Danger on the Eighteenth-Century Grand Tour (2020)
Sarah Goldsmith

*The Memory and Meaning of Coalfields: Deindustrialization and Scottish
Nationhood* (2021)
Ewan Gibbs

Lightning Source UK Ltd.
Milton Keynes UK
UKHW052037090420
361596UK00001B/2